Lecture Notes in Mathematics 2091

Editors-in-Chief:
J.-M. Morel, Cachan
B. Teissier, Paris

Advisory Board:
Camillo De Lellis (Zurich)
Mario Di Bernardo (Bristol)
Alessio Figalli (Pisa/Austin)
Davar Khoshnevisan (Salt Lake City)
Ioannis Kontoyiannis (Athens)
Gabor Lugosi (Barcelona)
Mark Podolskij (Heidelberg)
Sylvia Serfaty (Paris and NY)
Catharina Stroppel (Bonn)
Anna Wienhard (Heidelberg)

For further volumes:
http://www.springer.com/series/304

Miguel A. Sainz • Joaquim Armengol
Remei Calm • Pau Herrero • Lambert Jorba
Josep Vehi

Modal Interval Analysis

New Tools for Numerical Information

Miguel A. Sainz
Informática Matemática
Aplicada y Estadística
Escola Politecnica Superior
University of Girona
Girona, Spain

Remei Calm
Informática Matemática
Aplicada y Estadística
Escola Politecnica Superior
University of Girona
Girona, Spain

Lambert Jorba
Matemática Económica
Financiera y Actuarial
Facultad de Economia y Empresa
Universitat de Barcelona
Barcelona, Spain

Joaquim Armengol
Enginyeria Elèctrica
Electrònica i Automàtica
Escola Politecnica Superior
University of Girona
Girona, Spain

Pau Herrero
Imperial College London
London, United Kingdom

Josep Vehi
Enginyeria Elèctrica
Electrònica i Automàtica
Escola Politecnica Superior
University of Girona
Girona, Spain

ISBN 978-3-319-01720-4   ISBN 978-3-319-01721-1 (eBook)
DOI 10.1007/978-3-319-01721-1
Springer Cham Heidelberg New York Dordrecht London

Lecture Notes in Mathematics ISSN print edition: 0075-8434
                             ISSN electronic edition: 1617-9692

Mathematics Subject Classification (2010): 65G40, 65G50, 03B10, 26A24, 65F10

© Springer International Publishing Switzerland 2014
This work is subject to copyright. All rights are reserved by the Publisher, whether the whole or part of the material is concerned, specifically the rights of translation, reprinting, reuse of illustrations, recitation, broadcasting, reproduction on microfilms or in any other physical way, and transmission or information storage and retrieval, electronic adaptation, computer software, or by similar or dissimilar methodology now known or hereafter developed. Exempted from this legal reservation are brief excerpts in connection with reviews or scholarly analysis or material supplied specifically for the purpose of being entered and executed on a computer system, for exclusive use by the purchaser of the work. Duplication of this publication or parts thereof is permitted only under the provisions of the Copyright Law of the Publisher's location, in its current version, and permission for use must always be obtained from Springer. Permissions for use may be obtained through RightsLink at the Copyright Clearance Center. Violations are liable to prosecution under the respective Copyright Law.
The use of general descriptive names, registered names, trademarks, service marks, etc. in this publication does not imply, even in the absence of a specific statement, that such names are exempt from the relevant protective laws and regulations and therefore free for general use.
While the advice and information in this book are believed to be true and accurate at the date of publication, neither the authors nor the editors nor the publisher can accept any legal responsibility for any errors or omissions that may be made. The publisher makes no warranty, express or implied, with respect to the material contained herein.

Printed on acid-free paper

Springer is part of Springer Science+Business Media (www.springer.com)

Preface

The basic idea of Interval Mathematics is that ordinary set-theoretical intervals $I(\mathbb{R})$ provide a consistent support for numerical computing. The set-theoretical form of Interval Analysis has produced a large amount of work since its initiation in the late 1950s, [1, 60, 61]. There are some later papers devoted to the structural analysis of the method or its completion [13, 14, 49, 66, 72, 87, 96, 97], but with no fundamental departure from its initial set-theoretical foundations.

This book presents a new interval theory, the Modal Interval Analysis (MIA), as a structural, algebraic, and logical completion of the classical intervals. The starting point of MIA is quite simple: to define a modal interval attaching a quantifier to a classical interval, and to introduce the basic relation of inclusion between modal intervals by means of the inclusion between the sets of predicates they accept. So a modal interval consists in a classical interval, which defines its domain, and a quantifier, which defines its modality. This modal approach introduces interval extensions of the real continuous functions, gives equivalences between logical formulas and interval inclusions, and provides the semantic theorems that justify these equivalences and guidelines to get these inclusions.

The significant change of perspective in the treatment of information, coming from this new approach, makes Modal Interval Analysis more a new tool for the general practice of Numerical Applied Mathematics than a contribution to the previous Interval Theory. It supposes a complete philosophy of numerical information which is, or can be, its best virtue and produces at each stage of its development not only one body of solutions, but also questions leading to the construction of the next stage of the theory. Modal intervals system is not a breaking-off with the classic intervals, but a algebraic, structural, and logic completion of them that opens a way to new forms of numerical information treatment.

This book summarizes the most relevant results and features of MIA and also provides several application examples that illustrate the use of them in different problems and domains. The book contains the detailed development of the theory, the main concepts and results, together with several examples to clarify their meaning and to balance the mathematical items and proofs. Definitions contain the concepts, theorems provide the main results, and corollaries and lemmas embrace detailed logical developments.

Implementation of arithmetics, computation rules, and some algorithms is in the software MISO developed by the research group MiceLab of the University of Girona (Spain) and available at http://www.cs.utep.edu/interval-comp/intsoft.html

After the introductory Chap. 1, about real, digital numbers and intervals, and limitations of classical interval theory, Chap. 2 gives an account of the fundamental definitions and structures which support the semantically oriented system of modal intervals $I^*(\mathbb{R})$. Basic concepts such as predicates, canonical coordinates, modal inclusion, and equality, duality, interval predicates and co-predicates, rounding, and lattice operators are presented in detail. The set $I^*(\mathbb{R})$ of modal intervals turns out to be a completion of $I(\mathbb{R})$, in a similar way to that in which the complex numbers are a completion of the real numbers. So, a subset of $I^*(\mathbb{R})$, the "proper" modal interval $([a,b]$ with $a \leq b)$ is identifiable with a classical interval $[a,b]$ and all the results of Classical Interval Analysis are also results of Modal Interval Analysis.

In accordance with the sense of the term *analysis* in Mathematics, as a discipline in which the objects of study are, first and foremost, functions, Modal Interval theory can be considered, indeed, as an analysis because it studies a numerical field, the modal intervals, and the functions defined on it. So, Chap. 3 deals with the interval extension of the real continuous functions. The historic reason for the theory of these extensions of continuous real functions is to overcome the limited character of the classical set-theoretical approach. The geometrical semantics of $(n+1)$-dimensional real space, $\mathbb{R}^{n+1}$, is basically defined by the continuous functions f from $\mathbb{R}^n$ to $\mathbb{R}$. The semantic interval functions f^* and f^{**} from $I^*(\mathbb{R}^n)$ to $I^*(\mathbb{R})$, consistently referring to the continuous functions f from $\mathbb{R}^n$ to $\mathbb{R}$, are obtained by translating to modal terms the set-theoretical definition of a simple interval extension of a real continuous function. When the continuous real function is considered as a syntactic tree, it can also be extended to a rational interval function fR from $\mathbb{R}^n$ to $\mathbb{R}$, by using the computing program implicitly defined by the syntax of the expression defining the function. The idea of *interpretability* is given as a definite formulation via the cornerstones which are the *Semantic Theorems*.

Chapter 4 is devoted to characterizing the existence of optimal computations for a semantic function. Both modal extensions f^* and f^{**} are semantically interpretable, but not computable in general. When f^* and f^{**} are computed through the modal rational extension fR, a null, partial, or complete loss of information is generated. The point is to find functions for which the program fR is optimal, that is, for which $fR(X)$ equals $f^*(X)$ and $f^{**}(X)$.

Modal interval arithmetic operators and metric functions are considered in Chap. 5. The modal arithmetic coincides, certainly, with the arithmetic of the Kaucher's Extended Interval Space $\mathbb{IR}$ [49] with an important difference: in $I^*(\mathbb{R})$ interval results provided by the arithmetic have a logical meaning related to the points of the operand intervals domains, thanks to the semantic and interpretability theorems. Thus, unlike Extended Interval Space, which is a formal and algebraic completion of the Classic Intervals Space $I(\mathbb{R})$, Gardenyes' Modal Intervals are also a semantic completion of $I(\mathbb{R})$.

Chapter 6 contains procedures for solving interval linear equations and systems. The Jacobi method is adapted to interval systems together with convergence and non-convergence conditions. An important point is to provide a logical meaning to the solution using the semantic theorems.

The definition of the semantic extension of a real continuous function does not provide any indication about how to compute it. Some conditions under which f^* can be computed through the syntactic extension are given in Chap. 3, but in the most general case it is obtained by means of an algorithm developed in Chap. 7, referred as f^*-*algorithm*. First, some considerations about twins (intervals of intervals) are given to provide a background for this f^*-algorithm.

The matter of the necessary rounding is introduced in Chap. 1 and dealt with in the following chapters. Nevertheless, a shortcoming of the modal theory is managing the rounding of an interval when it appears both as it is and dualized in the same computation, for example in the solution of a linear system. To overcome this difficulty, in Chap. 8 a new object based on modal intervals is introduced: marks. Definitions, relations, the extension of a continuous function to a function of marks, operators of marks, and the corresponding semantic results are given in detail together with examples, not only to illustrate the different concepts and results, but also to show that marks can be used in a very practical way to aware about ill computations which can appear in the use of algorithms with real numbers.

Chapter 9 closes the loop opened in Chap. 2 dealing with intervals and modal intervals of marks, following a parallel development to the one started in Chap. 2 for modal intervals of real numbers $I^*(\mathbb{R})$. Predicates, relations, lattice operators, semantic and syntactic functional extensions to intervals of marks, and the semantic theorem, together with the arithmetic operators, are outlined throughout the chapter.

Finally, Chap. 10 is devoted to showing some applications of modal intervals. Specifically, they are used to deal with three problems: minimax, characterization of solution sets of quantified constraint satisfaction problems, and statement of problems in control engineering or process control from a semantic point of view. Algorithms and procedures about these topics are presented, together with examples to illustrate the procedures.

The beginnings of MIA can be situated in the SIGLA project, developed at the University of Barcelona in the late 1970s. In the 1980s and 1990s it was further continued by the SIGLA/X group (University of Barcelona and Polytechnical University of Catalonia in Spain), some of whose results can be found in [20–27, 86, 90]. The kernel of its main applications has been developed inside the MiceLab of the University of Girona (Spain) from the 1990s to the present.

We, the authors, are indebted to many people who have played significant roles in the development of Modal Interval Theory. First of all, to Dr. E. Gardenyes, founder of the Modal Interval Analysis since his first works in the 1980s until early 2000s together with a set of coworkers, Dr. H. Mielgo, Dr. A. Trepat, Dr. J.M. Janer, Dra. R. Estela, and some of us. With this book we want to render him tribute. Along the hardcore of the theory, we have wanted to preserve, in some way, his conceptualist style and notation, except for some adaptations to the standards. Also, we wish to thank several colleagues for their valuable comments and criticisms,

Dr. V. Kreinovich, Dr. A. Neumaier, Dr. L. Jaulin, Dr. A. Goldsztein, Dr. S. Ratschan and, in a very special way, we thank Dr. S.P. Shary and Dr. E. Walter, for the patient reading of the manuscript. Nevertheless, any error, omission, or obscurity are entirely our responsibility.

Girona, Spain	Miguel A. Sainz
Girona, Spain	Joaquim Armengol
Girona, Spain	Remei Calm
London, UK	Pau Herrero
Barcelona, Spain	Lambert Jorba
Girona, Spain	Josep Vehi

Notations

In order to make clear enough the main mathematical subjects put to work along the text, we have used the following typefaces and notations:

- Lowercase and italic x for a *real number*.
- Lowercase and italic f for a *real function of one variable*.
- Lowercase, bold and italic $\boldsymbol{x}$ for a *vector* with real components

$$\boldsymbol{x} = (x_1, x_2, \ldots, x_m).$$

- Lowercase, bold and italic $\boldsymbol{f}$ for a *vectorial function* with functional components

$$\boldsymbol{f} = (f_1, f_2, \ldots, f_m).$$

- Uppercase, italic A and apostrophe for a *classical interval* with real bounds

$$A' = [a_1, a_2]' \quad \text{or} \quad A' = [\underline{a}, \overline{a}]'.$$

- Uppercase and italic A for a *modal interval* with real bounds

$$A = [a_1, a_2] \quad \text{or} \quad A = [\underline{a}, \overline{a}].$$

- Uppercase, bold and italic for an *modal interval vector* with modal interval components

$$\boldsymbol{A} = (A_1, A_2, \ldots, A_m).$$

- Uppercase and bold for a *real matrix* with real elements

$$\mathbf{A} = (a_{ij}).$$

- Also uppercase, bold and italic for an *interval matrix* with interval elements

$$\boldsymbol{A} = (\boldsymbol{A}_{ij}).$$

The context prevents any lack of distinction between interval vector and interval matrix, which is often irrelevant because a vector in a finite-dimension vectorial space can be identified with a row or column matrix.
- wid(X) for the *width*, mid(X) for the *midpoint*, mig(X) for the *mignitude* and abs(X) for the *absolute value* of an interval X
- $|X|$ for the interval *absolute value function*.
- dist(X, Y) for the *Hausdorff distance* between two intervals X and Y.
- Q(x, X) for the *modal quantifier*.
- Lowercase mathfrak $\mathfrak{m}$ for a *mark* with real attributes

$$\mathfrak{m} = \langle c, t, g, n, b \rangle.$$

- Uppercase mathfrak and apostrophe $\mathfrak{A}'$ for a *set-theoretical interval of marks* with marks bounds

$$\mathfrak{A}' = [\underline{\mathfrak{a}}, \overline{\mathfrak{a}}]'.$$

- Uppercase mathfrak $\mathfrak{A}$ for a *modal interval of marks* with marks bounds

$$\mathfrak{A} = [\underline{\mathfrak{a}}, \overline{\mathfrak{a}}].$$

- $\mathbb{A}$ for a *twin* with interval bounds

$$\mathbb{A} = |[\underline{A}, \overline{A}]|.$$

- $\mathbb{R}$ for the *set of real numbers*.
- $I(\mathbb{R})$ for the *set of classical intervals*, as subsets of $\mathbb{R}$.
- $I^*(\mathbb{R})$ for the *set of modal intervals*.
- $\mathbb{M}(t, n, b)$ for the *set of marks* with tolerance t, number of digits n and scale basis b.
- $I(\mathbb{M}(t, n))$ for the *set of proper intervals of marks*, abridged to $I(\mathbb{M})$ when the type of the marks is arranged in advance.
- $I^*(\mathbb{M}(t, n))$ for the *set of modal intervals of marks*, abridged to $I^*(\mathbb{M})$ when the type of the marks is arranged in advance.
- $I(I^*(\mathbb{R}))$ for the *set of proper twins*.
- $I^*(I^*(\mathbb{R}))$ for the *set of twins*.

Contents

1 **Intervals** 1
   1.1   Introduction: The Classical Interval System 1
         1.1.1   Real Numbers and Numerical Computation 7
         1.1.2   Essential Reason for Computation with Intervals 7
         1.1.3   Why Intervals? 8
         1.1.4   Specificity of the System of Intervals 9
   1.2   Limitations of System $I(\mathbb{R})$ 10
         1.2.1   Equations $A + X = [0, 0]$ and $A * X = [1, 1]$ 11
         1.2.2   Solution of the Equation $A + X = B$ 11
         1.2.3   Interpretation of the Relations on Intervals 12
         1.2.4   Standards, Specifications and Quantifiers 13
         1.2.5   Semantic Vacuity of the Solution of $A + X = B$ 13
         1.2.6   Logical Insufficiency of the Digital Outer Rounding 14
         1.2.7   Arithmetical Limitation of Digital Outer Rounding 15
         1.2.8   Semantic Drawbacks 15

2 **Modal Intervals** 17
   2.1   Introduction 17
   2.2   Construction of the Modal Intervals 18
         2.2.1   Concepts, Definitions and Notations 18
         2.2.2   Set of Predicates Accepted by $A \in I^*(\mathbb{R})$ 20
         2.2.3   Canonical Coordinates of the Modal Intervals 21
         2.2.4   Modal Inclusion and Equality 23
         2.2.5   Duality and Co-predicates on $I^*(\mathbb{R})$ 26
         2.2.6   Rounding on $(I^*(\mathbb{R}), I^*(\mathbb{D}))$ 27
         2.2.7   Operations in the Lattice $(I^*(\mathbb{R}), \subseteq, \leq)$ 28
         2.2.8   Interval Predicates and Co-predicates 31
         2.2.9   The k-Dimensional Case 33
   2.3   Concluding Remarks 35

3	**Modal Interval Extensions**...	39	
	3.1 Introduction..	39	
	3.1.1 Poor Computational Extension	41	
	3.1.2 Modal Interval Extension	42	
	3.2 Semantic Functions	43	
	3.2.1 Properties of the Semantic Extensions...................	45	
	3.2.2 Characterization of JM-Commutativity	48	
	3.3 Semantic Theorems	53	
	3.4 Syntactic Functions	59	
	3.4.1 Syntactic Extensions	59	
	3.4.2 Modal Syntactic Operators	62	
	3.4.3 Modal Syntactic Computations with Rounding	70	
	3.5 Concluding Remarks	71	
4	**Interpretability and Optimality**	73	
	4.1 Introduction..	73	
	4.2 Interpretability and Optimality	73	
	4.2.1 Uni-incidence and Multi-incidence	74	
	4.2.2 Interpretability in the Uni-incidence Case	75	
	4.2.3 Optimality in the Uni-incidence Case	78	
	4.2.4 Tree-Optimality	82	
	4.2.5 Interpretability in the Multi-incidence Case	83	
	4.2.6 Optimality in the Multi-incidence Case	93	
	4.3 Conditional Optimality	98	
	4.3.1 Conditional Tree-Optimality	104	
	4.4 m-Dimensional Computations	106	
	4.5 Additional Examples	115	
5	**Interval Arithmetic**	121	
	5.1 Introduction..	121	
	5.2 One-Variable Function	122	
	5.3 Arithmetic Operators...	122	
	5.3.1 Properties of the Arithmetic Operations	125	
	5.3.2 Inner-Rounding and Computations	128	
	5.3.3 Sub-distributivity of the Operations $*$ and $+$	128	
	5.3.4 Metric Functions...	129	
	5.4 Interval Arithmetic for the C++ Environment	131	
	5.4.1 About the Library..	132	
	5.4.2 Available Functions and Operators	133	
6	**Equations and Systems**...	143	
	6.1 Introduction..	143	
	6.2 Linear Equation	143	
	6.3 Formal Solutions to a Linear System..............................	144	
	6.3.1 Solving a Linear System	146	
	6.3.2 Algorithm with Rounding	147	

		6.3.3	Sufficient Conditions for Convergence	150
		6.3.4	Sufficient Condition for Non-convergence	152
		6.3.5	Solution in the Case of Non-convergence	152
	6.4	Logical Meaning of the Solution		155
		6.4.1	Semantics in the General Case	156
	6.5	System Solution Sets		157

7 Twins and f^* Algorithm ... 159
7.1 Introduction ... 159
7.2 Twins ... 160
7.2.1 Twins Associated to the Relations $\leq$ and $\subseteq$ ... 160
7.2.2 Proper Twins ... 166
7.2.3 The Set of Twins ... 169
7.3 Symmetries ... 170
7.4 Semantic Extension of a Function to $I^*(I^*(\mathbb{R}))$ ... 172
7.5 The f^* Algorithm ... 172
7.5.1 Approximate *-Semantic Extension ... 173
7.5.2 Basic Algorithm ... 177
7.5.3 Improvements ... 179
7.5.4 Termination, Soundness and Completeness ... 183

8 Marks ... 185
8.1 Introduction ... 185
8.2 Marks ... 186
8.2.1 Real Line, Digital Line and Interval Analysis ... 186
8.2.2 From the Set-Theoretical Interval System to Modal Intervals ... 187
8.2.3 Deficiencies in the System $I^*(\mathbb{R})$ ... 187
8.2.4 The System of Marks ... 189
8.3 Marks and Associated Intervals ... 190
8.3.1 Mark on a Digital Scale ... 191
8.3.2 Imprecision and Validity of a Mark ... 193
8.3.3 Associated Intervals to a Mark ... 193
8.4 Relations in the Set of the Marks ... 195
8.4.1 Equality Relations ... 195
8.4.2 Inequality Relations ... 196
8.4.3 Strict Inequality Relations ... 197
8.5 Mark Operators ... 197
8.5.1 Mark Operators over $I^*(\mathbb{R})$ ... 198
8.5.2 Mark Operators over $\mathbb{M}(t,n)$ ... 199
8.6 Max and Min Operators ... 201
8.6.1 Maximum ... 201
8.6.2 Minimum ... 203
8.7 Arithmetic Operators ... 204
8.7.1 Product Operator ... 204
8.7.2 Quotient Operator ... 205

	8.7.3	Sum of Operands Having the Same Sign	207
	8.7.4	Sum with Operators of Different Signs (Subtraction)	208
8.8	Semantic Interpretations		219
8.9	Functions of Marks		221
8.10	Remarks About Granularity		227

9 Intervals of Marks ... 229
   9.1 Introduction ... 229
   9.2 Intervals of Marks ... 230
   9.3 Relations in the Set of Intervals of Marks ... 234
      9.3.1 Material Relations ... 234
      9.3.2 Weak Relations ... 235
      9.3.3 Interval Lattices ... 238
      9.3.4 Interval Predicates and Co-predicates ... 240
      9.3.5 k-Dimensional Intervals of Marks ... 241
   9.4 Interval Extensions of Functions of Marks ... 242
      9.4.1 Semantic Functions ... 244
      9.4.2 Properties of the $*$- and $**$-Semantics Functions ... 246
      9.4.3 Semantic Theorems ... 251
   9.5 Syntactic Extensions ... 254
      9.5.1 Arithmetic Operations for Intervals of Marks ... 259

10 Some Related Problems ... 265
   10.1 Introduction ... 265
   10.2 Minimax ... 268
      10.2.1 Solution of Unconstrained Problems ... 269
      10.2.2 Minimax Algorithm ... 272
      10.2.3 Solution of Constrained Problems ... 276
   10.3 Solution Sets ... 279
      10.3.1 Pavings ... 281
      10.3.2 Interval Estimations ... 290
   10.4 A Semantic View of Control ... 298
      10.4.1 Measurements and Uncertainty ... 301
      10.4.2 An Application to Temperature Control ... 302
   10.5 Concluding Remarks ... 305

References ... 307

Index ... 313

List of Figures

Fig. 2.1	Modal intervals	18
Fig. 2.2	(Inf, Sup)-diagram	22
Fig. 2.3	Inclusion diagram	25
Fig. 2.4	Less than or equal to relations	25
Fig. 2.5	Inclusion and less than or equal to relations	26
Fig. 2.6	Meet, Join, Min and Max	30
Fig. 2.7	Modal intervals as meet–join	31
Fig. 3.1	Poor extension diagram	42
Fig. 3.2	Saddle-point	49
Fig. 3.3	Function $f(x_1, x_2) = x_1^2 + x_2^2$ in $\boldsymbol{X}' = ([-1, 1]', [0, 2]')$	51
Fig. 3.4	Function $f(x_1, x_2) = x_1^2 + x_2^2$ in $\boldsymbol{X}' = ([-1, 1]', [-1, 1]')$	52
Fig. 3.5	Gas containers	58
Fig. 3.6	Saddle point for a two-variables operator	63
Fig. 3.7	Saddle points for addition	63
Fig. 3.8	Saddle points for the product function	64
Fig. 4.1	Electrical circuits	96
Fig. 4.2	Two tanks system	108
Fig. 4.3	Lens	116
Fig. 4.4	Circuit	118
Fig. 7.1	$\leq$-proper twin	161
Fig. 7.2	$\leq$-proper twins	162
Fig. 7.3	$\leq$-infimum and supremum, $\subseteq$-infimum and supremum	163
Fig. 7.4	$\subseteq$-proper twin	163
Fig. 7.5	$\subseteq$-proper twins	165
Fig. 7.6	$\leq$-infimum and supremum, $\subseteq$-infimum and supremum	165
Fig. 7.7	Proper twin	166
Fig. 7.8	$\leq$-twin, $\subseteq$-twin and $\sqsubseteq$-twin relations between proper twin	167
Fig. 7.9	Meet and join with proper twins	168

Fig. 7.10	Partition, *Strips* and *Cells*	174
Fig. 8.1	Physical system	187
Fig. 9.1	Physical system	262
Fig. 10.1	Feasibility region and partitions	277
Fig. 10.2	Intervals (F_{exp}, V_{exp})	284
Fig. 10.3	Ξ paving	285
Fig. 10.4	Graphical output for Example 10.3.2	286
Fig. 10.5	Graphical output for Example 10.3.3	287
Fig. 10.6	Simulation with the obtained parameters	293

Chapter 1
Intervals

1.1 Introduction: The Classical Interval System

Classical, or set-theoretical intervals [1, 60–62] are a conceptual tool of computation with a sufficiently mature theoretical background to make the development of its techniques of application a major center of interest [33, 44].

Interval mathematics identifies a classical interval $[a, b]$ with the set of numerical values x that lie between a and b, and operates with intervals instead of numbers.

$$[a, b] = \{x \in \mathbb{R} \mid a \leq x \leq b\}.$$

In the system of intervals

$$I(\mathbb{R}) = \{[a, b] \mid a, b \in \mathbb{R}, a \leq b\}$$

the real arithmetic operators are introduced: if ω is an arithmetical operation for real numbers, $\omega \in \{+, -, *, /\}$, the corresponding operation for intervals, denoted by

$$[a, b]\omega[c, d],$$

is defined by

$$[a, b]\omega[c, d] = \{x\omega y \mid x \in [a, b], y \in [c, d]\},$$

that is, the set of values of all possible ω-operations between a first operand x of $[a, b]$ and a second operand y of $[c, d]$ (it can be observed that interval division is not defined when $c \leq 0 \leq d$) [1, 8, 61]. This definition leads to the following operation rules

$$[a, b] + [c, d] = [a + b, c + d]$$
$$[a, b] - [c, d] = [a - d, b - c]$$
$$[a, b] * [c, d] = [\min(ac, ad, bc, bd), \max(ac, ad, bc, bd)]$$

and

$$1/[c,d] = [1/d, 1/c]$$
$$[a,b]/[c,d] = [a,b] * (1/[c,d]).$$

when $0 \notin [c,d]$.

In the system of intervals with digital bounds

$$I(\mathbb{D}) = \{[a,b] \mid a,b \in \mathbb{D}, a \leq b\},$$

if $A, B \in I(\mathbb{D})$, any general interval operation $A \,\omega\, B$ must be defined by an outer rounding of the set of values $\{x \omega y \mid x \in A, y \in B\}$ in order to guarantee the implication

$$(x \in A, y \in B) \Rightarrow x \omega y \in A \,\omega\, B.$$

Example 1.1.1 Let us consider a physical system consisting of a tank of volume $v\,l$, with one input and one output, which contains saline solution. Designating by $x(t)g$ and $y(t)g/l$ the mass and concentration of salt in the tank, a constant flow of $q\,l/s$ of saline solution, with an amount of $u(t)g/s$ of salt, is entering into the tank, and the same outflow $q\,l/s$ of saline solution with a concentration of $y(t)g/l$ is leaving the tank. If the initial mass of salt in the tank is $x(0)g$, the problem is to know the evolution of the concentration of salt in the outflow, along the time of simulation.

Taking into account the mass balance of salt, the discrete mathematical model after Euler discretization for this physical system is

$$\begin{aligned} x(t+\Delta_t) &= (1-\Delta_t q/v)x(t) + \Delta_t u(t) \\ y(t+\Delta_t) &= x(t+\Delta_t)/v, \end{aligned} \quad (1.1)$$

where Δ_t is the discretization time-step, which provides the variation of the output concentration of salt along the time. Thus, for

$$v = 11\,l$$
$$q = 0.7\,l/s$$
$$u(t) = \begin{cases} 0.7\,g/s & \text{for } t < 20 \\ 0.0\,g/s & \text{for } t \geq 20 \end{cases}$$
$$x(0) = 7\,g$$
$$\Delta_t = 1\,s,$$

the obtained results of the simulation are summarized in Table 1.1, which shows the evolution of the mass and concentration of salt in the tank. Obviously the same

1.1 Introduction: The Classical Interval System

Table 1.1 Simulations

t	x(t)	y(t)
1	7.5091	0.6826
2	7.9534	0.7230
...	...	...
20	10.7372	0.9761
21	10.7706	0.9791
...	...	...
70	0.0136	0.0012
71	0.0119	0.0011
...	...	...

results can be obtained when any equivalent formulation (i.e. *syntactic tree*) for the output function

$$y(t + \Delta_t) = x(t + \Delta_t)/v,$$
$$y(t + \Delta_t) = ((1 - \Delta_t q/v)x(t) + \Delta_t u(t))/v, \quad (1.2)$$
$$y(t + \Delta_t) = ((1/v - \Delta_t q/v^2)x(t) + \Delta_t u(t)/v$$

is used.

Introducing interval uncertainties in the physical system, let $X(t)$ and $Y(t)$ be intervals of variation for the mass and concentration of salt, and $X(0)$ be for the initial state. Substituting the real variables of the model equations (1.1) in their corresponding variation intervals, the model becomes the interval model

$$X(t + \Delta_t) = (1 - \Delta_t Q/V) * X(t) + \Delta_t U(t)/V$$
$$Y(t + \Delta_t) = (X(t + \Delta_t)/V.$$

Let us consider the interval formulation of the different syntactic trees (1.2) for the output function $y(t)$

$$Y1(t + \Delta_t) = (X(t + \Delta_t)/V$$
$$Y2(t + \Delta_t) = ((1 - \Delta_t Q/V) * X(t) + \Delta_t U(t))/V$$
$$Y3(t + \Delta_t) = ((1/V - \Delta_t Q/V^2) * X(t) + \Delta_t U(t)/V,$$

together with a fourth different formulation

$$Y4(t + \Delta_t) = ((1/V - \Delta_t Q/(\text{Dual}(V)^2) * X(t) + \Delta_t U(t)/V$$

where the second occurrence of v, has been replaced by the dual interval of V (dual of an interval $[a, b]$ will be defined as the "interval" $[b, a]$). It is expected that these equations will give intervals of variation of the concentration along the time. Thus, running a simulation for

Table 1.2 Interval simulations

t	$X(t)$	$Y1(t)$	$Y2(t)$	$Y3(t)$	$Y4(t)$
1	[6.9638, 8.0488]	[0.6055, 0.7666]	[0.6055, 0.7666]	[0.5969, 0.7728]	[0.6133, 0.7598]

$$V = [10.5, 11.5] \, l$$
$$Q = [0.6, 0.8] \, l/s$$
$$U(t) = \begin{cases} [0.6, 0.8] \, g/s & \text{for } t < 20 \\ [0.0, 0.0] \, g/s & \text{for } t \geq 20 \end{cases}$$
$$X(0) = [6.8, 7.2] \, g$$
$$Y(0) = [3.5, 4.5] \, g$$
$$\Delta_t = 1 \, s,$$

the results, just for the first iteration, are in Table 1.2, with different intervals in line with the different syntactic trees of the output function, contrary to it could be expected. Why? The equality between $Y1$ and $Y2$ indicates that the interval computations can be associative. The different result for $Y3$ could indicate the non-distributivity of the quotient of intervals. The result for $Y4$ is also different and, curiously, it is the true interval of variation for the output (its bounds can be separately computed as a problem of maxima an minima with the bounds of the intervals for $X(0)$, Q and V).

This example illustrates in some way the behavior of intervals, different from the real numbers, and the mistake of making a hasty and naive substitution of real numbers by their intervals of variation.

An interval extension of a continuous function from $\mathbb{R}^k$ to $\mathbb{R}$, $z = f(x_1, \ldots, x_k)$, is the *united extension* R_f of f, defined as the range of the f-values on X and an important objective of classic interval computations is to estimate this range. Since this range can be hard to compute, an interval syntactic extension fR is defined by replacing the real operands and operators of the real function by the homonymous operands and operators defined on the system $(I(\mathbb{R}), I(\mathbb{D}))$. The crucial relation between both extensions is

$$R_f(X_1, \ldots, X_k) \subseteq fR(X_1, \ldots, X_k),$$

This inclusion can represent an important loss of information, but there exist different methods, such as centered forms and Taylor series [1, 61, 71], that allow to obtain tighter inclusions of fR.

A critical basic fact is that the interval syntactic extension fR satisfies only one kind of interval predicate compatible with the outer rounding

$$(\forall x_1 \in X_1) \cdots (\forall x_k \in X_k) \, (\exists z \in \text{Out}(fR(X_1, \ldots, X_k))) \, z = f(x_1, \ldots, x_k).$$

1.1 Introduction: The Classical Interval System

It would be very interesting to get interval syntactic extensions satisfying more general predicates, with existential and universal quantifiers combined following some kind of rules.

Example 1.1.2 Consider an electrical circuit with a voltage source e of 11 V and a rheostat which provides a resistance r of 10 Ω. In accordance with the Ohm's law, the current i is

$$i = \frac{e}{r} = \frac{11}{10} = 1.1 \text{ A}.$$

Let us consider variations of these quantities: the voltage source e is between 10 and 12 V and the resistance r can take any value between 7 and 40 Ω. Representing these variations in an interval way, $e \in E = [10, 12]$ and $r \in R = [7, 40]$ and converting this quotient in its interval counterpart, the current i will be inside the interval

$$I = \frac{E}{R} = \frac{[10, 12]}{[7, 40]} \subseteq [0.25, 1.72]. \tag{1.3}$$

Would it be possible to know the value of $r \in [7, 40]$ that provides a current of e.g. 1.7 A when e is 10 V? The answer is that this value does not exist, because

$$r = \frac{10}{1.7} = 5.88.. \notin [7, 40].$$

In fact, the interval result in Equation (1.3) has a unique and precise meaning (i. e., a *semantics*) that is: for any value of e between 10 and 12 V and any value of r between 7 and 40 Ω, the current will take a value between 0.25 and 1.72 A. In a formal way using a formula of the first order logic,

$$(\forall e \in [10, 12]) \, (\forall r \in [7, 40]) \, (\exists i \in [0.25, 1.72]) \, i = \frac{e}{r}.$$

Nevertheless, there exists a value of $r \in [7, 40]$ providing a current $i = 1$ A when $e = 10$ V, because

$$r = \frac{10}{1} = 10 \in [7, 40].$$

So, a new problem can be stated: to find an interval R such that for every $r \in R$, there exist values of $i \in [0.25, 1.72]$ and $e \in [10, 12]$ such that $i = \frac{e}{r}$, i.e.

$$(\forall r \in R) \, (\exists i \in [0.25, 1.72])(\exists e \in [10, 12]) \, i = \frac{e}{r}.$$

The semantic of classic intervals does not allow to find a direct solution inside $I(\mathbb{R})$, but the modal interval computation

$$R = \frac{[12, 10]}{[1.72, 0.25]} \subseteq [48, 5.82]$$

solves this problem, using again dual intervals. The semantical rigidity of $I(\mathbb{R})$ will be broken with modal intervals to obtain other different semantics, depending on the kind of the involved intervals.

This example indicates that it can exist a strong relationship between the results of an interval computation and some type of logical formulas, relating intervals with the values they contain.

A series of works appeared at the beginning of interval analysis, showing various attempts to overcome the limitations which, significantly, move this system away from the structural and operative regularity of the classical numerical system $\mathbb{R}$ and which always present implicit and effective barriers to its application [27, 49]. In this book, intervals, considered in their specialized function of elementary parts of computation, are framed inside Modal Interval Analysis (MIA) by formally subordinating them to some semantic functions. The model framework of the modal intervals has sufficient conceptual richness to pose and gradually solve central questions about the processing of numerical data which remain out of reach of the traditional model of intervals.

Probably the innovation which best characterizes the system of modal intervals is the identification of an interval with a record of values either autonomous or regulating. However, even though the modal theory reveals significant conceptual and computational resources hidden under the apparent simplicity of the classical intervals, this theory confirms also the logical solidity of a part of the limitations involved in the traditional system, showing that they come from the internal logic of the information provided by numerical measurements. The example

$$[1, 3] * ([1, 1] + [-1, -1]) = [0, 0] \subseteq [1, 3] * [1, 1] + [1, 3] * [-1, -1] = [-2, 2]$$

illustrates one of these limitations which must be accepted: the weakening of the ordinary distributive law to a sub-distributive law for the product of intervals, whether classical or modal.

Apart from the conceptual and operational suitability of the modal theory for facing problems which require a treatment by intervals, there are specific difficulties due to moving away from the traditional model of classical intervals to the model of modal intervals. In synthesis, these difficulties are:

- The set-theoretical intervals, in spite of their conceptual and operational limitations, cohere extremely well with ordinary numerical intuition.
- To make use of modal intervals, it is necessary to add a good number of basic concepts to the ordinary baggage of numerical mathematics.

The analysis by modal intervals introduces new conceptual and operational possibilities, but reveals also significant limitations regarding the possibilities of numerical data processing, if one wants to push the intervals up to the level of an analysis which is logically coherent with the numerical models.

1.1 Introduction: The Classical Interval System

1.1.1 Real Numbers and Numerical Computation

The system $\mathbb{R}$ of the numbers, known as the *real numbers*, became, after its theoretical consolidation at the end of the nineteenth century, the abstract and computational horizon of the mathematics applied to the experimental systems ruled by measurements. One can identify this system either with the Euclidean real line provided with a system of coordinates, the point of view of synthetic geometry, or the set of Cauchy series of encased intervals with widths decreasing towards zero, if one chooses the constructive perspective of mathematical analysis.

In a more pragmatic approach, $\mathbb{R}$ can almost be identified with the system of the numerical values obtained with all the precision that is relevant for each application. The fact that numerical values are exact only to a limited extent, that is, that they bear a certain amount of numerical error, is very well known by everyone dealing with numerical information. No matter its origin and application, numerical results other that those carried by small integer numbers, are, inescapably, only estimated or approximated results of some measurement or computation. This fact leads to an obvious conclusion: single numbers are unable to usefully represent numerical information.

However, although a more or less abstract use of the system $\mathbb{R}$ was established, giving its familiar sense to the practice of numerical mathematics, this pragmatic use provides reliable reasons because it is based in the consistent theoretical construction of the system $\mathbb{R}$ of the numbers called "real". But one cannot forget that this construction of $\mathbb{R}$ is always an essentially infinite construction: the exact effective computation of a particular value of a real value would inevitably presuppose, except for trivial cases, a process of successive approximations with an infinite number of iterations. For this reason the real values, in general, only allow of being indicated either by a theoretical definition or by some approximate value obtained by a finite process.

This is the motive for the need of analytically coherent but computationally finite tools of the numerical calculus supported by the two basic concepts of exact value and sufficient approximation. It is the mathematical development of these two notions on the double system $(\mathbb{R}, \mathbb{D})$ of the real numbers $\mathbb{R}$ and the *digital numbers* $\mathbb{D}$, which makes possible a mutually significant reference between effective computations and mathematical models.

1.1.2 Essential Reason for Computation with Intervals

Numerical applied mathematics is grounded on the use of the numerical values $d \in \mathbb{D}$ actually obtainable from some digital process which can be related, after some measuring/computing technical protocol, to the real line's numerical values $x \in \mathbb{R}$ which are fitted to some geometrical model on the real line $\mathbb{R}$. In addition to some criterion of proximity to the exact value x, d will satisfy one of the two relations of side delimitation $d \leq x$ or $d \geq x$. The relation of equality $d = x$ will be always deprived of sense in the analytic/computational context of the system $(\mathbb{R}, \mathbb{D})$, with the values d and x pertaining to completely different sources of experiment.

This is the framing of the fundamental problem, faced by any numerical computation, of keeping a single digital value for each ideal value aimed at in $\mathbb{R}$. The basic idea of any interval computation is to keep, for each level of the computing process and both its data and the final result, a pair of possible approximations (d_1, d_2) satisfying the conditions $d_1 \leq x \leq d_2$. The pair (d_1, d_2) replaces the singular approximate value d which would keep a classical numerical computation.

For those parts of the numerical information are able to be treated systematically, it will be necessary to identify the pairs (d_1, d_2) with some mathematical object endowing them with a precise analytical significance and providing the possibility of their playing the role of elementary objects for each level of computation. This identification will turn out to be the base for the different versions of interval analysis.

This computational context is fully recognized and put to work by the approach of classical interval analysis to numerical mathematics, when it decides to keep systematically the two nearest procedurally discernible digital bounds, a lower bound $d_1 \in \mathbb{D}$ and an upper bound $d_2 \in \mathbb{D}$, to represent any real value $x \in \mathbb{R}$, conceptually compatible with a definite measurement or actual computation and consistent with the geometrical model guiding the interpretation of the referenced measurement or computational operation.

Example 1.1.3 It is hardly acceptable to guess from the number 15 which of $15 \pm 0.1 = [14.9, 15.1]$ or $15 \pm 5 = [10, 20]$ is meant. Neither the indication of the step in the scale of measurement or numeration is enough, because of the widening effect that comes out from theoretical operations themselves. The only way to indicate the spread of a numerical result is by pointing to a lower and an upper limit of its possible values, maybe through a direct interval notation such as $[14.9, 15.1]$ or $[10, 20]$ or maybe through a more indirect one such as 15 ± 0.1 or 15 ± 5.

To operate or to perform some mathematics with numerical information is the departing point of interval mathematics from the usual way of handling numeric information through numbers. The main decision in the analysis by classical set-theoretical intervals consists in identifying the pairs $(d_1, d_2) \in \mathbb{D}^2$ with the real line subsets $\mathbb{R}$

$$[d_1, d_2] = \{x \in \mathbb{R} \mid d_1 \leq x \leq d_2\}$$

and intervals become the actual elementary items of numerical information.

1.1.3 Why Intervals?

The system $\mathbb{R}$ only provides logical support for models about real processes dealing with measures, supposedly objective, of quantities, supposedly continuous. The practical process of computation/measurement is only able to define an interval

1.1 Introduction: The Classical Interval System

which provides the optimal operational identification with the computed/measured value. It has been proved [54] that, in a similar way to the central limit theorem used in statistics, for any error which is a sum of a large number of independent small components, the set of its values is close (in the Hausdorff distance sense) to an interval. Moreover, computational feasible problems involving a family of intervals, describing the uncertainty of the inputs, become intractable by increasing the family with any non-interval set [68].

Intervals make possible:

1. bounding truncation errors, for example $\pi = 3.14159\ldots \in [3.14, 3.15]$,
2. the automatic control of operational and rounding errors in a digital computation,
3. representing errors in physical quantities coming from experimental measurements, for example the acceleration due to gravitational $g = 9.807 \pm 0.027\, \text{m/s}^2 \in [9.780, 9.834]$, and
4. introducing variables representing uncertainty or variation, for example the temperature of a room varying between 18 and 25 °C within 24 h can be represented by the interval $T = [18, 25]$.

In many cases, the probability distribution of the measurement uncertainty is known, but there are some other situations when only the lower and upper bounds of such uncertainty are known, and this is when intervals are useful. The use of the set of intervals $I(\mathbb{R})$ will allow handling the physical systems as numerical models which take into account the unavoidable uncertainty associated with any computation/measurement process and holds a computational representation for uncertainties and errors.

In the construction of $I(\mathbb{R})$ some essential properties of the real numbers are lost, for example distributivity, or gained, for example the inclusion relation. So the numerical structure $I(\mathbb{R})$ is not only a simple completion of $\mathbb{R}$ [89].

From the identification of the pair of digital numbers (d_1, d_2) with a set-theoretical interval $[d_1, d_2]$, the set $I(\mathbb{D})$ of intervals with bounds in $\mathbb{D}$ becomes the operative background for computations which take into account the automatic control of the uncertainties inherent to the computed/measured numerical values.

The set $I(\mathbb{R})$ does not allow an algorithmic use, but it provides an analytical formal model for interval relations and operations. The set $I(\mathbb{D})$ does not assure exact relations or computations, but allows approximating $I(\mathbb{R})$ and controlling the round-off errors and truncations of the algorithms defined in $I(\mathbb{R})$ but performed in $I(\mathbb{D})$.

1.1.4 Specificity of the System of Intervals

The system $I(\mathbb{R})$, with its equality and inclusion relations and its algebraic operators [1,8,61], has a structure much more complex than $\mathbb{R}$ and, thereafter, both the models and the algorithms defined on $\mathbb{R}$ will be too poor, systematically, to determine a corresponding model or algorithm on $I(\mathbb{R})$. That is shown by the following structural peculiarities of $I(\mathbb{R})$:

1. The difference, even with arithmetic ideally exact on $\mathbb{R}$, between the set of values of a continuous real function

$$\{f(x_1,\ldots,x_n) \mid x_1 \in X_1,\ldots,x_n \in X_n\}$$

   and the value of its interval syntactic computation $fR(X_1,\ldots,X_n)$ is often too big, far from that which one could expect for a reasonable approximation, when some of the parameters $(x_1,\ldots,x_n)$ appear in a multiple way in the expression of the function f, i.e., when some components of the argument $x = (x_1,\ldots,x_n)$ are multi-incident on the computation program formalized by the syntactic tree of the function f. To see this, it is enough to compare, for $x \in X = [1,2]$, the set of values of the continuous real function $f(x) = x - x$, which is $\{x - x \mid x \in [1,2]\} = [0,0]$, with the result of the syntactic operation on $I(\mathbb{R})$, $fR(X) = [1,2] - [1,2] = [-1,1]$. This phenomenon is called the "amplification of dependence".

2. The distributivity of multiplication over a sum becomes regionalized, or reduced to a sub-distributive law:

$$A * (B + C) \subseteq (A * B) + (A * C).$$

3. The breaking points of the algorithms, where the alternative

$$(a \leq x \mid\mid a \geq x)$$

   on $\mathbb{R}$ is binary, become breaking points with four branches on $I(\mathbb{R})$, according to the alternative

$$(A \subseteq X \mid\mid A \supseteq X \mid\mid A \leq X \mid\mid A \geq X),$$

   where the system of relations $(I(\mathbb{R}), \leq, \geq)$ is the partial order complementary to the partial order $(I(\mathbb{R}), \subseteq, \supseteq)$ defined by inclusion in the system of subsets real line $I(\mathbb{R})$.

These three features of the structure $I(\mathbb{R})$ have as a consequence that neither $I(\mathbb{R})$ nor any space admitting a subsystem isomorphic with $I(\mathbb{R})$, i.e., the space obtained from $I(\mathbb{R})$ preserving the ability to state and solve the problems that $I(\mathbb{R})$ makes it possible to state and solve correctly, will be able to fill the frame of implementation for the algorithms designed on $I(\mathbb{R})$.

1.2 Limitations of System $I(\mathbb{R})$

Interval mathematics uses, through the entire process of a computation, all the range of possible values that correspond to every item of numerical information, and there are a lot of successful interval computing methods, but there exists some very important difficulties in the interval approach.

1.2.1 Equations $A + X = [0, 0]$ and $A * X = [1, 1]$

Maybe the most fundamental difficulty comes from the fact that interval subtraction is not the inverse operation of interval addition, as the following example shows:

$$[1, 2] - [1, 2] = [-1, 1].$$

Even more: if $[a, b]$ is an interval with $a \neq b$, there exists no interval $[x, y]$ such that

$$[a, b] + [x, y] = [0, 0],$$

because the addition rule would imply

$$[a, b] + [x, y] = [a + x, b + y] = [0, 0]$$

and then

$$[x, y] = [-a, -b],$$

with $a < b$ implying $-a > -b$, this is to say $[-a, -b]$ would be no interval at all.

A quite analogous reasoning leads to the non-existence of any interval $[x, y]$ such that $[a, b] * [x, y] = [1, 1]$ if $a \neq b$.

So, the basic equation $A + X = [0, 0]$ have no solution in $I(\mathbb{R})$ because subtraction is not the opposite operation to the addition in $I(\mathbb{R})$. The same for the equation $A * X = [1, 1]$, because division is not the opposite to multiplication. These anomalies can be solved by adding certain elements to the system $I(\mathbb{R})$, the dual intervals (for example the new element $[2, 1] = \text{Dual}([1, 2])$), which extend to the new system the preliminary structure $(I(\mathbb{R}), +, -, *, /)$ by means of compatibility criteria. The completed structure, named $(\mathbb{IR}, +, -, *, /)$, is a group for the operation of addition, as well as for the multiplication of intervals not containing zero [48, 49].

1.2.2 Solution of the Equation $A + X = B$

The solution $X = [x_1, x_2]$ of the equation $A + X = B$ satisfies $a_1 + x_1 = b_1$ and $a_2 + x_2 = b_2$. This solution exists on $I(\mathbb{R})$ only under the condition $\text{wid}(A) \leq \text{wid}(B)$, where wid is the *width* of the interval $A = [a_1, a_2]$, defined by

$$\text{wid}(A) = a_2 - a_1$$

But even when the equation $A + X = B$ has a solution because the condition wid(A) ≤ wid(B) holds, this solution cannot be obtained by any interval syntactic computation $X = fR(A, B)$ on $I(\mathbb{R})$. This comes from the fact that if there were such a computation, for $A \subseteq A_1$, $A \neq A_1$, the solution of the equation $A_1 + X = B$ would also be obtained from this same syntactic computation $X_1 = fR(A_1, B)$. But, since fR is inclusive, X_1 satisfies the relations $X_1 \supseteq X$, wid(X_1) ≥ wid(X), incompatible with wid(X) = wid(B) − wid(A), wid(X_1) = wid(B) − wid(A_1) and wid(A_1) > wid(A).

Example 1.2.1 The equation $[2, 5] + [x_1, x_2] = [3, 7]$ has the solution $[x_1, x_2] = [1, 2]$, which can not be obtained by any operation within the system of classical set-theoretical intervals, because $[x_1, x_2] = [3, 7] − [2, 5] = [−2, 5]$.

In short, the existence of a solution for the equation $A + X = B$ on $I(\mathbb{R})$ does not imply that this solution has a syntactic computation in $I(\mathbb{R})$.

1.2.3 Interpretation of the Relations on Intervals

Let us suppose that it is necessary to extend a line at a known distance between 100 and 120 m, and that one initially has a reel of cable measuring between 60 and 70 m length. The solution $X = [40, 50]$ of the equation $[60, 70] + X = [100, 120]$ would provide limits acceptable for the necessary additional length if one can accept any overall length ranging between 100 and 120 m (for example if one needs to link two rims of a deep canyon).

By changing the scene one supposes that the goal is now to cover a distance between 100 and 120 m, the precise value being unknown when the problem has been just posed (the real distance inside the interval $[100, 120]$ could not be accessible to measurement because of a wood or fog preventing one from seeing the point where the other extreme of the cable should be connected). For this new situation one should thus get an additional reel of cable allowing the unfolding of a certain length in the interval $[30, 60]$, different from the preceding solution $X = [40, 50]$.

If one revises the statement of these two situations, on each one of them, the intervals $[100, 120]$ and X are associated to different processes, due to the selection from the operational values which they delimit.

This example illustrates the problems related to the fact that the relations on intervals are computationally interpreted. To specify the logical mode of this interpretation, and doing it in a way that will be accessible to the analysis, one introduces an essential tool for the analysis by modal intervals: quantifiers: universal, $\forall$, and existential, $\exists$. The formula $(\forall x \in X) \, P(x)$ means "for any value x pertaining to the set-interval X for which the property $P(x)$ is true", and the formula $(\exists x \in X) \, P(x)$ means "there exists at least a value x pertaining to the set-interval X for which the property $P(x)$ is true".

1.2 Limitations of System $I(\mathbb{R})$

With this formal tool, the validity of the solutions in the two suggested addition scenarios is expressed by the propositions

$$(\forall a \in [60, 70]) \, (\forall x \in [40, 50]) \, (\exists t \in [100, 120]) \, a + x = t$$

and

$$(\forall a \in [60, 70]) \, (\forall t \in [100, 120]) \, (\exists x \in [30, 60]) \, a + x = t$$

One thus realizes that for problems arising in terms of $I(\mathbb{R})$ one cannot translate, without more examination, intuitive addition problems into addition equations.

1.2.4 Standards, Specifications and Quantifiers

We must associate to the intervals the universal and the existential quantifiers to formalize the expression of compatibility between the standard imposed on one of its measurable characteristics of an unspecified product and the specification which a process of production can guarantee.

For concreteness, let us suppose that the standard for a batch of bars of length l is defined by the expression $(\forall l \in [99.6, 100.4]) Acceptable(l)$, meaning that the lengths in this interval will not cause trouble in their use. Thus the bars produced by a factory will be accepted if it can guarantee a specification, ensuring for the length of each unit the validity, e.g., of the expression $(\exists l \in [99.8, 100.2]) EffectiveLength(l)$. Under these circumstances, one can show the validity of the statement

$$(\exists l \in [99.8, 100.2]) \, (EffectiveLength(l), Acceptable(l)),$$

translating the compatibility of the standard for the product with the quality of the effective manufacture process aiming to satisfy it.

1.2.5 Semantic Vacuity of the Solution of $A + X = B$

We will now propose an addition problem which, in spite of the lack of a solution for the corresponding interval equation in the interesting cases, will be significant and will have, in the case supposed, a good practical solution.

Let us imagine that A, B and X are intervals. A contains the spontaneous monetary offer of a country, B the tolerable monetary offer for this country, and X the monetary intervention of its government. If one reflects this situation by the countable equation $A + X = B$, this problem will have a solution in system $I(\mathbb{R})$ only under the condition $wid(A) \leq wid(B)$ which would make it possible for the government to spend money freely, within the limits imposed by the interval X. The interpretation of the equation $A + X = B$ would be formalized in these circumstances by the proposition

$$(\forall a \in A)\,(\forall x \in X)\,(\exists b \in B)\; a + x = b.$$

But we can suppose that a certain government might face less flexible circumstances resulting from the condition wid$(A) >$ wid(B). One can prove that such a government, with this simple economy, could arithmetically solve this particular problem with the solution X of the equation $X = B-A$ in $I(\mathbb{R})$. But, is it acceptable that it is necessary to change the form of the equations to address the variability of the conditions? The combinatorial explosion threatened by such a process is not acceptable. However the interpretation of the solution to this formulation of the problem would be formalized by the expression

$$(\forall a \in A)\,(\exists b \in B)\,(\exists x \in X)\; a + x = b,$$

or

$$(\forall a \in A)\,(\forall b \in B)\,(\exists x \in X)\; a + x = b,$$

if the government were obliged to ensure one of the values limited by B, unknown *ante facto*.

1.2.6 Logical Insufficiency of the Digital Outer Rounding

The only digital rounding allowed in the context $(I(\mathbb{R}), I(\mathbb{D}))$ is the outer rounding, Out$[a, b] = [\text{Left}(a), \text{Right}(b)]$, because it guarantees the implication $x \in [a, b] \Rightarrow x \in [\text{Left}(a), \text{Right}(b)]$. This theoretical decision leads nevertheless to another dead-end of the interval analysis on $(I(\mathbb{R}), I(\mathbb{D}))$.

As the computational interpretation of the intervals formally results in the use of two quantifiers $(\forall x \in [a, b])$ and $(\exists x \in [a, b])$, if the interval value $[a, b]$ is relevant because it validates, for example the statement

$$(\exists x \in [a, b])\; x = (a + b)/2,$$

this property will preserve its validity in the statement

$$(\exists x \in [\text{Left}(a), \text{Right}(b)])\; x = (a + b)/2,$$

obtained from the former one by outer-rounding the interval.

1.2 Limitations of System $I(\mathbb{R})$

One must however pose another question: what does occur if the pertinent property for the interval $[a, b]$ were correctly proposed by the statement $(\forall x \in [a,b])$ $x \geq a$? Then the validity of this statement wouldn't absolutely guarantee the validity of

$$(\forall x \in [\text{Left}(a), \text{Right}(b)])\ x \geq a$$

obtained by outer-rounding of $[a, b]$. But it is obvious that the statement

$$(\forall x \in [\text{Right}(a), \text{Left}(b)])\ x \geq a,$$

obtained from an inner rounding $\text{Inn}[a, b] := [\text{Right}(a), \text{Left}(b)]$, would preserve the validity of the original statement on $[a, b]$, if the condition $\text{Right}(a) \leq \text{Left}(b)$ guarantees the existence of the inner rounding of $[a, b]$.

The problem for Inn comes from the fact that this operation does not exist unconditionally on $I(\mathbb{D})$, as the case $\text{Inn}[x, x]$ shows when $x \in \mathbb{D}$.

1.2.7 Arithmetical Limitation of Digital Outer Rounding

The intervals of $I(\mathbb{R})$ rounded outside are not always enough to compute other results rounded outside. For example an outer rounded solution of the equation $A+X = B$ could be obtained from the transformed equation $\text{Inn}(A)+Y = \text{Out}(B)$, but not from $\text{Out}(A) + Z = \text{Out}(B)$.

This is clear, because from to the relations $a_1 + x_1 = b_1, a_2 + x_2 = b_2$, one obtains $y_1 = \text{Left}(b_1) - \text{Right}(a_1)$ and $y_2 = \text{Right}(b_2) - \text{Left}(a_2)$, which makes $[y_1, y_2]$ an outer rounding of $[x_1, x_2]$, i.e., $[y_1, y_2] \supseteq [x_1, x_2]$, because the subtraction increases with the first operand and decreases with the second. A completely parallel reasoning shows that one could not affirm the same thing for $[z_1, z_2]$, with $z_1 = \text{Left}(b_1) - \text{Left}(a_1)$ and $z_2 = \text{Right}(b_2) - \text{Right}(a_2)$.

1.2.8 Semantic Drawbacks

The fact of not having the additional information provided by the association of quantifiers to interval bounds, can be better illustrated by the following example of four interval statements referring to the relation $a + x = b$ on the real line and keeping the requirements $a \in [1, 2], b \in [3, 7]$:

$$(\forall a \in [1, 2])\ (\forall x \in [2, 5])\ (\exists b \in [3, 7])\ a + x = b \qquad (1.4)$$

$$(\forall a \in [1, 2])\ (\forall b \in [3, 7])\ (\exists x \in [1, 6])\ a + x = b \qquad (1.5)$$

$$(\forall x \in [1, 6])\ (\exists a \in [1, 2])\ (\exists b \in [3, 7])\ a + x = b \qquad (1.6)$$

$$(\forall b \in [3, 7])\ (\exists a \in [1, 2])\ (\exists x \in [2, 5])\ a + x = b. \qquad (1.7)$$

Statements (1.4) and (1.7) hold for the well known solution $X = [2, 5]$ of the classical set-theoretical interval equation $[1, 2] + X = [3, 7]$. Statements (1.5) and (1.6), in spite of making complete sense, are out of reach of this classical interval equation of the same form as the "real" one $a + x = b$. If the addition was a digital operation, the practical interval equation could be $[1, 2] + X \subseteq [2.9, 7.1]$ with the only possible rounding for set-theoretical digital intervals: the classical "outer rounding". In this case the statement (1.4) would become

$$(\forall a \in [1, 2]) \, (\forall x \in [2, 5]) \, (\exists b \in [2.9, 7.1]) \, a + x = b, \qquad (1.8)$$

but statement (1.7) would cease to be valid, since the outer rounding of $[3, 7]$ would be incompatible with the ∀-quantifier. Moreover, we should not forget that, even in the very simple case of the statement (1.4), the solution X of the corresponding interval equation $[1, 2] + X \subseteq [3, 7]$ could not be obtained by any operation within the system of classical set-theoretical intervals.

Chapter 2
Modal Intervals

2.1 Introduction

The semantical lack of the classical system of intervals cannot be resolved by remaining bound to the idea that identifies each interval $[a, b]$ with the set of numerical values x for which the condition $a \leq x \leq b$ holds. The way out must be found, therefore, through a restatement of the problem.

The first hint of a solution comes from the pragmatics of the standard interval mathematics itself: an interval $[a, b]$ is most frequently looked upon as the representative of some unknown value (or values) x that satisfies the condition $x \in [a, b]$, rather than being looked on as representing all of these values. This perspective is specially obvious in processes of successive approximations, when the approximated numerical value belongs to every interval of the approximating succession.

So, if for example we consider the interval $[1, 2]$ as representing an "unknown" value $x \in [1, 2]$, is there anything against denoting $[2, 1]$ the "interval" representing a "selectable" value $y \in [1, 2]$? It is obvious that if the addition rule for standard intervals is kept

$$[1, 2] + [2, 1] = [3, 3]$$

moreover for every $x \in [1, 2]$ (for example $x = 1.8$) a $y \in [1, 2]$ (in this example $y = 1.2$) can be selected such that $x + y = 3$.

Let us introduce some terminology:

"Proper interval" means any "interval" $[a, b]$ satisfying the condition $a \leq b$. In this case it is equivalent to assert that x is representable by $[a, b]$, to say that x belongs to the set $\{x \mid a \leq x \leq b\}$. A proper interval is interpreted as a "tolerance interval".

An "improper interval" is any "interval" $[a, b]$ satisfying the condition $a \geq b$; for the moment let us simply take it as a control able to provide any numerical

Fig. 2.1 Modal intervals

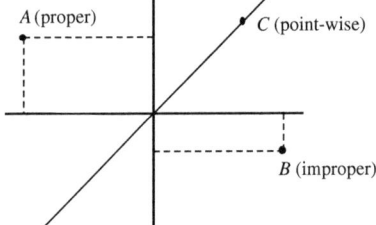

value (control value) x between b and a. An improper interval is interpreted as a "control interval".

"Point-wise interval" or "degenerate interval" is any "interval" $[a, b]$ satisfying the condition $a = b$, just representing the single numerical value $x = a = b$.

Now, to visualize these different intervals, it is useful to represent them in a system of coordinates, as depicted in Fig. 2.1.

This allows of understanding geometrically the completed intervals, with the half plane of proper or tolerance intervals (the standard intervals), with the half plane of the improper or control intervals, and with their common border of point-wise intervals.

2.2 Construction of the Modal Intervals

To obtain the system of objects able to provide a coherent use of the intervals, with a philosophy of application to the double system of real and digital numbers $(\mathbb{R}, \mathbb{D})$, we underline a fundamental fact: a datum in the form of an interval X is interesting owing to it indicates, by limiting it, an ideal value x validating a property $P(x)$, which must be relevant for the reasons which will have guided to obtain the interval X delimiting an environment of x.

But if x cannot have a digital designation because, contrary to X, it is not the result of a measurement and/or a computation, the only ways to formalize the reference of the interval X to a value $x \in \mathbb{R}$ and to the property P, will be one of the two expressions $(\exists x \in X)\, P(x)$ and $(\forall x \in X)\, P(x)$, where the variable x plays only the role of an abstract index. Then the classical intervals X become ambiguous, since they do not contain any attribute to determine univocally which one of the two possible quantified expressions must be selected as a consistent reference to x and the property P.

2.2.1 Concepts, Definitions and Notations

So far and from now on $\mathbb{R}$, $I(\mathbb{R})$ and $I^*(\mathbb{R})$ represent the set of the real numbers, the set of classical intervals, and the set of modal intervals together with the following notations:

2.2 Construction of the Modal Intervals

- $[a, b]' = \{x \in \mathbb{R} \mid a \leq x \leq b\}$ is a classical or set-theoretical interval; the notation $[a, b]'$ will be used to distinguish $[a, b]' \in I(\mathbb{R})$ from a modal interval $[a, b] \in I^*(\mathbb{R})$.
- $\text{Pred}(\mathbb{R}) = \{P \mid P : \mathbb{R} \to \{0, 1\}\}$ is the set of classical predicates, with values 0 or 1, on $\mathbb{R}$;
- $\text{Set}(P) = \{x \in \mathbb{R} \mid P(x) = 1\}$ is the domain or characteristic set of the predicate P;
- $\text{Pred}(x) := \{P \in \text{Pred}(\mathbb{R}) \mid P(x) = 1\}$ is the set of the real predicates defined on the set $\mathbb{R}$, which are validated for the point $x \in \mathbb{R}$.

To relate intervals with the sets $\text{Pred}(x)$ corresponding to their points, we define:

Definition 2.2.1 (Modal interval) A modal interval A is an element of the Cartesian product $(I(\mathbb{R}), \{\forall, \exists\})$

$$A = (A', Q_A),$$

where Q_A is one of the classical quantifiers $\forall$ or $\exists$.

Definition 2.2.2 (Modal coordinates of a modal interval) For $A = (A', Q_A) \in I^*(\mathbb{R})$:

$$Domain : \text{Set}(A', Q_A) = A'$$
$$Modality : \text{Mod}(A', Q_A) = Q_A.$$

Definition 2.2.3 (Set of modal intervals) The set of modal intervals is denoted by $I^*(\mathbb{R})$ and defined as

$$I^*(\mathbb{R}) = \{(A', Q_A) \mid A' \in I(\mathbb{R}), Q_A \in \{\exists, \forall\}\}. \tag{2.1}$$

This fundamental definition leads naturally to outline the following subsets:

Definition 2.2.4 (Sets of existential, universal and point-wise intervals)

$$I_e(\mathbb{R}) = \{(A', \exists) \mid A' \in I(\mathbb{R})\};$$
$$I_u(\mathbb{R}) = \{(A', \forall) \mid A' \in I(\mathbb{R})\};$$
$$I_p(\mathbb{R}) = \{([a, a]', \exists) \mid a \in \mathbb{R}\} \text{ or } I_p(\mathbb{R}) = \{([a, a]', \forall) \mid a \in \mathbb{R}\}.$$

Obviously, $I_p(\mathbb{R}) \subseteq I_e(\mathbb{R})$, $I_p(\mathbb{R}) \subseteq I_u(\mathbb{R})$ and $I^*(\mathbb{R}) = I_e(\mathbb{R}) \cup I_u(\mathbb{R})$.

The main instrument of Modal Interval Analysis is the modal quantifier Q delimited by modal intervals, associating to every real predicate a unique hereditary interval predicate on the modal intervals $I^*(\mathbb{R})$ by means of the following rule defining the modal quantifier Q in function of the classical ones $\exists$ and $\forall$.

Definition 2.2.5 (Modal quantifier Q) For a variable x on $\mathbb{R}$ and $A \in I^*(\mathbb{R})$,

$$Q(x, A)P(x) = (Q_A x \in A')P(x).$$

For instance,

$$Q(x, ([-3, 1]', \exists)) \, x \geq 0 = (\exists x \in [-3, 1]') \, x \geq 0$$

and

$$Q(x, ([1, 2]', \forall)) \, x \geq 0 = (\forall x \in [1, 2]') \, x \geq 0.$$

Remark 2.2.1 If $A = [a, a]$ is a point-wise interval,

$$Q(x, A)P(x) = (\forall x \in A')P(x) = (\exists x \in A')P(x) = P(a).$$

and, consequently, both definitions of A decide the acceptance or rejection of the predicate P by the value of $P(a)$.

2.2.2 Set of Predicates Accepted by $A \in I^*(\mathbb{R})$

The main idea underlying the definition of the expression $Q(x, A)P(x)$ is to play the role of a formalized test. Their results 1 or 0, depending on the domain and the modality of A, identify a modal interval A as an acceptor of predicates making true or false the real predicate $P(x)$.

Definition 2.2.6 (Set of real predicates validated—or accepted—by a modal interval) Given a modal interval $A = (A', Q_A)$,

$$\text{Pred}(A) = \{P \in \text{Pred}(\mathbb{R}) \mid Q(x, A)P(x)\}.$$

is the set of predicates accepted by A.

This definition allows identifying each modal interval A with the set $\text{Pred}(A)$ of the real predicates P which are true according to its attributes $\text{Set}(A)$ and $\text{Mod}(A)$.

The two following lemmas show the close geometrical and logical relation between a modal interval and the real predicates which it accepts.

Lemma 2.2.1 (Predicate of modal intervals) *If* $\bigcup_{x \in A'}$ *is the "union operator" of a family of sets of index x ranging on the interval A', and* $\bigcap_{x \in A'}$ *is the corresponding "intersection operator",*

2.2 Construction of the Modal Intervals

$$\text{Pred}(A', \exists) = \{P \in \text{Pred}(\mathbb{R}) \mid (\exists x \in A') P(x)\} = \bigcup_{x \in A'} \text{Pred}(x)$$

$$\text{Pred}(A', \forall) = \{P \in \text{Pred}(\mathbb{R}) \mid (\forall x \in A') P(x)\} = \bigcap_{x \in A'} \text{Pred}(x)$$

Proof

$$P \in \text{Pred}(A', \exists) \Leftrightarrow (\exists x \in A') P(x)$$
$$\Leftrightarrow (\exists x \in A')(P \in \{P \in \text{Pred}(\mathbb{R}) \mid P(x) = 1\})$$
$$\Leftrightarrow P \in \bigcup_{x \in A'} \text{Pred}(x)$$

and

$$P \in \text{Pred}(A', \forall) \Leftrightarrow (\forall x \in A') P(x)$$
$$\Leftrightarrow (\forall x \in A')(P \in \{P \in \text{Pred}(\mathbb{R}) \mid P(x) = 1\})$$
$$\Leftrightarrow P \in \bigcap_{x \in A'} \text{Pred}(x). \qquad \blacksquare$$

Lemma 2.2.2 (Set-theoretical meaning of the acceptance of predicates)

If $(\text{Mod}(A) = \exists \text{ then } P \in \text{Pred}(A)) \Leftrightarrow \text{Set}(A) \cap \text{Set}(P) \neq \emptyset$
If $(\text{Mod}(A) = \forall \text{ then } P \in \text{Pred}(A)) \Leftrightarrow \text{Set}(A) \subseteq \text{Set}(P)$

Proof If $\text{Mod}(A) = \exists$,

$$P \in \text{Pred}(A) \Leftrightarrow (\exists x \in A') P(x) \Leftrightarrow \text{Set}(A) \cap \text{Set}(P) \neq \emptyset$$

If $\text{Mod}(A) = \forall$,

$$P \in \text{Pred}(A) \Leftrightarrow (\forall x \in A') P(x) \Leftrightarrow \text{Set}(A) \subseteq \text{Set}(P). \qquad \blacksquare$$

In other words, the modal intervals of form $(A', \exists)$ will verify only the predicates which are validated by unspecified points of the set A'. On the other hand, an interval $(A', \forall)$ will verify only the predicates $P(x)$ which are validated by all the points of the set A'.

2.2.3 Canonical Coordinates of the Modal Intervals

Let us define the following canonical notation for the modal intervals:

Fig. 2.2 (Inf, Sup)-diagram

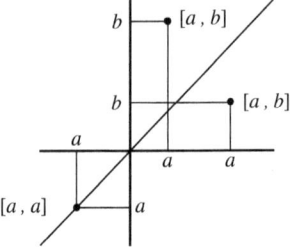

Definition 2.2.7 (Canonical coordinates)

$$[a,b] = \begin{cases} ([a,b]', \exists) & \text{if } a \leq b \\ ([b,a]', \forall) & \text{if } a \geq b \end{cases}$$

Moreover, let us make the following definitions:

Definition 2.2.8 (Infimum and supremum)

$$\text{Inf}([a,b]) = a$$
$$\text{Sup}([a,b]) = b.$$

They are the infimum and the supremum of an interval of $I^*(\mathbb{R})$, i.e., its first and second coordinates.

Definition 2.2.9 (Minimum and maximum)

$$\min([a,b]) = \min(a,b)$$
$$\max([a,b]) = \max(a,b).$$

With these notations one can indicate the previous sets of modal intervals as

$$I^*(\mathbb{R}) = \{[a,b] \mid a, b \in \mathbb{R}\};$$
$$I_e(\mathbb{R}) = \{[a,b] \in I^*(\mathbb{R}) \mid a \leq b\};$$
$$I_u(\mathbb{R}) = \{[a,b] \in I^*(\mathbb{R}) \mid a \geq b\};$$
$$I_p(\mathbb{R}) = \{[a,b] \in I^*(\mathbb{R}) \mid a = b\}.$$

Remembering that $[a,b] \in I_e(\mathbb{R})$ is called a "proper interval", an interval $[a,b] \in I_u(\mathbb{R})$ is called an "improper interval", and an interval $[a,b] \in I_p(\mathbb{R})$ is called a "point-wise interval" or "degenerated interval", graphical representations in a *(Inf,Sup)-diagram* are in Fig. 2.2, where the diagonal contains the point-wise intervals, proper intervals are represented by points in the left half-plane, and improper intervals are in the right half-plane.

2.2.4 Modal Inclusion and Equality

The identification of a modal interval with the set of its accepted predicates, $A \leftrightarrow \text{Pred}(A)$, allows the definition of modal inclusion and modal equality in terms of the inclusion and equality of the sets of predicates.

Definition 2.2.10 (Modal inclusion and modal equality) If $A, B \in I^*(\mathbb{R})$,

$$A \subseteq B : \text{Pred}(A) \subseteq \text{Pred}(B)$$
$$A = B : (A \subseteq B, A \supseteq B) \Leftrightarrow \text{Pred}(A) = \text{Pred}(B)$$

The set-theoretical projection of modal inclusion is next established:

Lemma 2.2.3 (Modal components analysis of modal inclusion) For $A, B \in I^*(\mathbb{R})$,

$$A \subseteq B \Leftrightarrow \begin{cases} 1) \; A' \subseteq B' \text{ if } \text{Mod}(A) = \text{Mod}(B) = \exists \\ 2) \; A' \supseteq B' \text{ if } \text{Mod}(A) = \text{Mod}(B) = \forall \\ 3) \; A' \cap B' \neq \emptyset \text{ if } (\text{Mod}(A) = \forall, \text{Mod}(B) = \exists) \\ 4) \; A' = B' = [a, a]' \text{ if } (\text{Mod}(A) = \exists, \text{Mod}(B) = \forall). \end{cases}$$

Proof By Definition 2.2.10, $A \subseteq B \Leftrightarrow \text{Pred}(A) \subseteq \text{Pred}(B)$. So

1) $\Rightarrow: (A \subseteq B; A, B \in I_e(\mathbb{R}))$
  $\Leftrightarrow (\forall P \in \text{Pred}(A)) ((\exists x \in A') \, P(x) \Rightarrow (\exists x \in B') \, P(x))$
  // Particularizing $P(x) : x = a$ for every $a \in A'$.
  $\Rightarrow (\forall a \in A') ((\exists x \in A') \, x = a \Rightarrow (\exists x \in B') \, x = a) \Leftrightarrow A' \subseteq B'$
1) $\Leftarrow: A' \subseteq B' \Rightarrow (\forall P \in \text{Pred}(A)) ((\exists x \in A') \, P(x) \Rightarrow (\exists x \in B') \, P(x))$
  $\Leftrightarrow (A', \exists) \subseteq (B', \exists)$.
2) $\Rightarrow: (A \subseteq B; A, B \in I_u(\mathbb{R}))$
  $\Leftrightarrow (\forall P \in \text{Pred}(A)) ((\forall x \in A') \, P(x) \Rightarrow (\forall x \in B') \, P(x))$
  // Particularizing $P(x) : x \in A'$.
  $\Rightarrow ((\forall x \in A') \, x \in A' \Rightarrow \forall (x \in B') \, x \in A') \Leftrightarrow A' \supseteq B'$.
2) $\Leftarrow: A' \supseteq B' \Rightarrow (\forall P \in \text{Pred}(A)) ((\forall x \in A') \, P(x) \Rightarrow (\forall x \in B') \, P(x))$
  $\Leftrightarrow (A', \forall) \subseteq (B', \forall)$.
3) $\Rightarrow: (A \subseteq B; A \in I_u(\mathbb{R}), B \in I_e(\mathbb{R}))$
  $\Leftrightarrow (\forall P \in \text{Pred}(A)) ((\forall x \in A') \, P(x) \Rightarrow (\exists x \in B') \, P(x))$
  // Particularizing $P(x) : x \in A'$.
  $\Rightarrow ((\forall x \in A') \, x \in A' \Rightarrow \exists (x \in B') \, x \in A') \Leftrightarrow A' \cap B' \neq \emptyset$.
3) $\Leftarrow: A' \subseteq B' \Rightarrow (\forall P \in \text{Pred}(A)) ((\forall x \in A') \, P(x) \Rightarrow (\exists x \in B') \, P(x))$
  $\Leftrightarrow (A', \forall) \subseteq (B', \exists)$.
4) $\Rightarrow: (A \subseteq B; A \in I_e(\mathbb{R}), B \in I_u(\mathbb{R}))$
  $\Leftrightarrow (\forall P \in \text{Pred}(A)) ((\exists x \in A') \, P(x) \Rightarrow (\forall x \in B') \, P(x))$
  // $P(x) : x = a$ for every $a \in A'$.
  $\Rightarrow (\forall a \in A') ((\exists x \in A') \, x = a \Rightarrow (\forall x \in B') \, x = a)$

// $(\forall a \in A')\, (\exists x \in A')\, x = a$ is true.
$\Leftrightarrow (\forall a \in A')\, (\forall x \in B')\, x = a$
$\Leftrightarrow (\forall a \in A')\, B' = [a, a]'$
$\Leftrightarrow A' = B' = [a, a]'$.

4) $\Leftarrow: A' = B' = [a, a]' \Rightarrow (A', \exists) \subseteq (B', \forall)$. ∎

Inclusion and equality can be reduced to relations between the infimum and supremum of the involved intervals.

Lemma 2.2.4 (Programming of inclusion and equality)

$$[a_1, a_2] \subseteq [b_1, b_2] \Leftrightarrow (a_1 \geq b_1,\, a_2 \leq b_2);$$
$$[a_1, a_2] = [b_1, b_2] \Leftrightarrow (a_1 = b_1,\, a_2 = b_2).$$

Proof The possible cases of inclusion are classified after the modalities of the intervals $A = [a_1, a_2]$ and $B = [b_1, b_2]$:

1. $(\text{Mod}(A) = \text{Mod}(B) = \exists) \Leftrightarrow (a_1 \leq a_2,\, b_1 \leq b_2)$:
   Since $A \subseteq B \Leftrightarrow [a_1, a_2]' \subseteq [b_1, b_2]'$.
2. $(\text{Mod}(A) = \text{Mod}(B) = \forall) \Leftrightarrow (a_1 \geq a_2,\, b_1 \geq b_2)$:
   Since $A \subseteq B \Leftrightarrow [a_2, a_1]' \supseteq [b_2, b_1]' \Leftrightarrow b_2 \geq a_1,\, b_1 \leq a_2$
   $\Leftrightarrow b_2 \geq a_1 \geq a_2,\, b_2 \leq b_1 \leq a_2$.
3. $(\text{Mod}(A) = \forall,\, \text{Mod}(B) = \exists) \Leftrightarrow (a_1 \geq a_2,\, b_1 \leq b_2)$:
   $A \subseteq B \Leftrightarrow [a_2, a_1]' \cap [b_1, b_2]' \neq \emptyset$
   $\Leftrightarrow \neg(a_2 > b_2 \text{ or } a_1 < b_1) \Leftrightarrow (a_2 \leq b_2,\, a_1 \geq b_1)$.
4. $(\text{Mod}(A) = \exists,\, \text{Mod}(B) = \forall) \Leftrightarrow (a_1 \leq a_2,\, b_1 \geq b_2)$:
   $A \subseteq B \Leftrightarrow A' = B' = [a, a]' \Leftrightarrow (a_1 = a_2 = b_1 = b_2 = a_2 = b_2 = a)$
   $\Leftrightarrow (a_1 \leq a_2,\, b_1 \geq b_2,\, a_2 \leq b_2,\, a_1 \geq b_1)$. ∎

Figure 2.3 illustrates this result

Example 2.2.1 Since $[7, 0] \subseteq [-3, 1]$ because $(7 \geq -3,\, 0 \leq 1)$, it follows that

$$(x \geq 0) \in \text{Pred}([7, 0]) \Leftrightarrow Q(x, [7, 0])\, x \geq 0 \Leftrightarrow (\forall x \in [0, 7]')\, x \geq 0$$

and

$$(x \geq 0) \in \text{Pred}([-3, 1]) \Leftrightarrow Q(x, [-3, 1])\, x \geq 0 \Leftrightarrow (\exists x \in [-3, 1]')\, x \geq 0.$$

Remark 2.2.2 These programming results for the relations "=" and "⊆" on $I^*(\mathbb{R})$ their formal identity to the corresponding ones in $I(\mathbb{R})$.

Specifically, the formal identity which this lemma shows for the existential or proper intervals $[a, b]$, where $a \leq b$, and for the classical intervals $[a, b]'$, which meet also the condition $a \leq b$, establishes a formal correspondence between the systems $I(\mathbb{R})$ and $I_e(\mathbb{R})$ that one must take care not to see as an identity, since the

2.2 Construction of the Modal Intervals

Fig. 2.3 Inclusion diagram

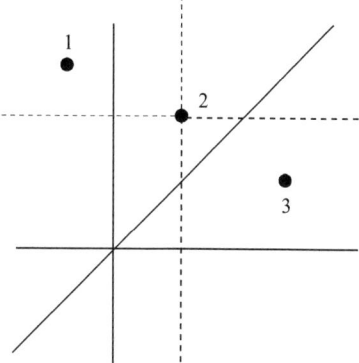

Fig. 2.4 Less than or equal to relations

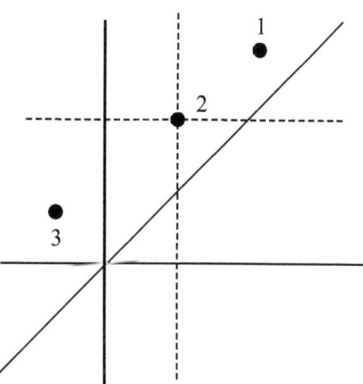

elements of $I_e(\mathbb{R})$ have a domain on $\mathbb{R}$ but also a modality, the existential one, unlike the elements of $I(\mathbb{R})$ whose definition only considers the domain.

The relation of inequality is formally generated by the closed complement of modal inclusion.

Definition 2.2.11 (Less than or equal to for modal intervals) For $A = [a_1, a_2]$ and $B = [b_1, b_2]$,

$$A \leq B : (a_1 \leq b_1, a_2 \leq b_2).$$

In the (Inf, Sup)-diagram, a representation for the "less than or equal to" relation is in Fig. 2.4.

Lemma 2.2.5 (Order structure of the system $(I^*(\mathbb{R})), \subseteq, \leq))$ *Both $\leq$ and $\subseteq$ are partial order relations on $I^*(\mathbb{R})$ and between two intervals A and B there always exists at least one of the following situations*

$$A \subseteq B \text{ or } A \supseteq B \text{ or } A \leq B \text{ or } A \geq B.$$

Fig. 2.5 Inclusion and less than or equal to relations

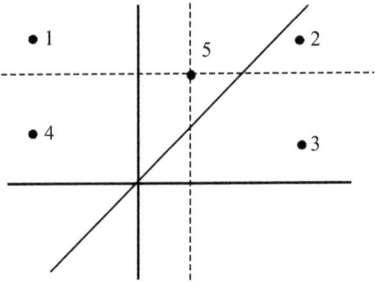

Proof Obviously they are the only possible situations of any interval in relation with interval A in Fig. 2.5. ∎

2.2.5 Duality and Co-predicates on $I^*(\mathbb{R})$

Modal intervals may also be associated with the sets of real predicates they reject, establishing a dual semantics for $I^*(\mathbb{R})$:

Definition 2.2.12 (Set of real "co-predicates" rejected by a modal interval) Given a modal interval A by its modal coordinates (A', Q_A),

$$\text{Copred}(A) = \{P \in \text{Pred}(\mathbb{R}) \mid \neg Q(x, A) P(x)\}.$$

When $P \in \text{Copred}(A)$, P is *rejected* by A, i.e., P doesn't belong to the set of predicates accepted by A. Obviously

$$\text{Copred}(A) = \text{Pred}(\mathbb{R}) - \text{Pred}(A),$$

Example 2.2.2 $(x \geq 0) \in \text{Copred}([7, 1])$ is equivalent to

$$\neg((x \geq 0) \in \text{Pred}([7, 1])) \Leftrightarrow \neg Q(x, [7, 1]) \, x \geq 0$$
$$\Leftrightarrow \neg(\forall x \in [1, 7]') \, x \geq 0$$
$$\Leftrightarrow (\exists x \in [1, 7]') \, x < 0,$$

a final proposition whose logical value is, like the initial ones, utterly false.

This complementary semantics is related to the symmetry of $I^*(\mathbb{R})$ established by the "duality" operator:

Definition 2.2.13 ($\subseteq$-Duality on $I^*(\mathbb{R})$).

$$\text{Dual}([a, b]) = [b, a].$$

2.2 Construction of the Modal Intervals

Lemma 2.2.6 (⊆-monotonicity of the Dual and Copred operators) *For $A, B \in I^*(\mathbb{R})$,*

$$A \subseteq B \Leftrightarrow \text{Dual}(A) \supseteq \text{Dual}(B) \Leftrightarrow \text{Copred}(A) \supseteq \text{Copred}(B).$$

Proof The first equivalence for $A \subseteq B$ arises from the definitions of the Dual operator and of the programming theorem for the modal inclusion. The second one results from the definition of modal inclusion and from the Pred($\mathbb{R}$)-complementarity of Pred(A) and Copred(A). ∎

Lemma 2.2.7 (¬-commutation with the modal quantifier) *For $A \in I^*(\mathbb{R})$,*

$$\neg Q(x, A)\, P(x) \Leftrightarrow Q(x, \text{Dual}(A))\, \neg P(x);$$
$$P \in \text{Copred}(A) \Leftrightarrow (\neg P) \in \text{Pred}(\text{Dual}(A)).$$

Proof. From the commutation rule of the "not"-operator with the classical quantifiers. ∎

2.2.6 Rounding on $(I^*(\mathbb{R}), I^*(\mathbb{D}))$

For any interval $[a, b] \in I^*(\mathbb{R})$ an inner and outer rounding on a digital line $I^*(\mathbb{D})$ must be an interval of $I^*(\mathbb{D})$ contained and containing, respectively, $[a, b]$.

Definition 2.2.14 (Modal rounding)

$$\text{Inn}([a, b]) = [\text{Right}(a), \text{Left}(b)] \in \{[x, y] \mid x, y \in \mathbb{D}, [x, y] \subseteq [a, b]\},$$
$$\text{Out}([a, b]) = [\text{Left}(a), \text{Right}(b)] \in \{[x, y] \mid x, y \in \mathbb{D}, [x, y] \supseteq [a, b]\},$$

where Left(a) is the greatest element of $\mathbb{D}$ less than or equal to a, and Right(a) is the least element of $\mathbb{D}$ greater than or equal to a.

Theorem 2.2.1 (Transmission of the information associated with modal truncations)

$$\text{Pred}(\text{Inn}(A)) \subseteq \text{Pred}(A) \subseteq \text{Pred}(\text{Out}(A))$$
$$\text{Copred}(\text{Inn}(A)) \supseteq \text{Copred}(A) \supseteq \text{Copred}(\text{Out}(A)).$$

Proof By the previous definition,

$$\text{Inn}([a, b]) \subseteq [a, b] \subseteq \text{Out}([a, b]),$$

which is equivalent to the corresponding relations of inclusion among predicates and co-predicates. ∎

After the identification of the modal intervals with the sets of real predicates they validate, this theorem states that if Inn(A) and Out(A) are the rounded results supplied by some computing algorithm and/or some observation about a modal interval theoretical exact A, these relations mean that only the predicates of Pred(Inn(A)) and the co-predicates of Copred(Out(A)) are a posteriori decidable for the interval A. That is, the predicates which are true for Inn(A) are also true for A, and the predicates which are false for Out(A) are also false for A. Similarly only the predicates of Pred(Out(A)) and the co-predicates of Copred(Inn(A)) are a priori decidable for the interval A. That is, the predicates which are true for A are also true for Out(A), and the predicates which are false for A are also false for Inn(A).

In short, the assertive information is transmitted by the first chain of inclusions equivalent to the implications

$$Q(x, \text{Inn}(A)) P(x) \Rightarrow Q(x, A) P(x) \Rightarrow Q(x, \text{Out}(A)) P(x)$$

and the negative information is transmitted by the second chain equivalent to the system of implications

$$\neg Q(x, \text{Out}(A)) P(x) \Rightarrow \neg Q(x, A) P(x) \Rightarrow \neg Q(x, \text{Inn}(A)) P(x)$$

Having a dual operator avoids a double implementation for rounding, since inner rounding can be reduced to outer rounding.

Theorem 2.2.2 (Unnecessary implementation of the inner rounding)

$$\text{Inn}(A) = \text{Dual}(\text{Out}(\text{Dual}(A))).$$

Proof If $A = [a, b]$,

$$\text{Inn}(A) = [\text{Right}(a), \text{Left}(b)] = \text{Dual}[\text{Left}(b), \text{Right}(a)]$$
$$= \text{Dual}(\text{Out}[b, a]) = \text{Dual}(\text{Out}(\text{Dual}(A))). \quad \blacksquare$$

2.2.7 Operations in the Lattice ($I^*(\mathbb{R}), \subseteq, \leq$)

We will introduce the infimum and supremum operations for the lattices defined by the partial order relations $\subseteq$ and $\leq$ on $I^*(\mathbb{R})$.

Lemma 2.2.8 (The lattice ($I^*(\mathbb{R}), \subseteq$)) *The structure ($I^*(\mathbb{R}), \subseteq$) is a lattice, isomorphic to (($\mathbb{R}, \mathbb{R}), (\geq, \leq)$).*

Proof From Lemma 2.2.4 there exist a supremum and infimum

$$\text{Inf}([a_1, a_2], [b_1, b_2]) = [\max(a_1, b_1), \min(a_2, b_2)]$$
$$\text{Sup}([a_1, a_2], [b_1, b_2]) = [\min(a_1, b_1), \max(a_2, b_2)]. \quad \blacksquare$$

2.2 Construction of the Modal Intervals

The lattice operations "Meet" and "Join" on $I^*(\mathbb{R})$, i.e., the Inf() and Sup() of the modal inclusion order on $I^*(\mathbb{R})$, are defined on bounded families of modal intervals as follows.

Definition 2.2.15 ("Meet" and "Join" operators on $(I^*(\mathbb{R}), \subseteq)$) For a bounded family $A(I) = \{A(i) \in I^*(\mathbb{R}) \mid i \in I\}$ of modal intervals (I is the index's domain),

$$\bigwedge_{i \in I} A(i) = A \in I^*(\mathbb{R}) \text{ is such that } (\forall i \in I)\, X \subseteq A(i) \Leftrightarrow X \subseteq A,$$

$$\bigvee_{i \in I} A(i) = B \in I^*(\mathbb{R}) \text{ is such that } (\forall i \in I)\, X \supseteq A(i) \Leftrightarrow X \supseteq B,$$

annotated $(A \wedge B)$ and $(A \vee B)$ for the corresponding two-operand case.

The structure $(I^*(\mathbb{R}), \leq)$ is a lattice with the lattice operations "Min" and "Max" defined by

Definition 2.2.16 ("Min" and "Max" operators on $(I^*(\mathbb{R}), \leq)$) For a bounded family $A(I) = \{A(i) \in I^*(\mathbb{R}) \mid i \in I\}$,

$$\operatorname*{Min}_{i \in I} A(i) = A \in I^*(\mathbb{R}) \text{ is such that } (\forall i \in I)\, X \leq A(i) \Leftrightarrow X \leq A;$$

$$\operatorname*{Max}_{i \in I} A(i) = B \in I^*(\mathbb{R}) \text{ is such that } (\forall i \in I)\, X \geq A(i) \Leftrightarrow X \geq B.$$

These operators can be easily obtained by means of operations on the bounds of the intervals.

Lemma 2.2.9 (Lattice operator programming) *For a bounded family of $A(I) \in I^*(\mathbb{R})$, if $A(i) = [a_1(i), a_2(i)]$:*

$$\bigwedge_{i \in I} A(i) = [\max_{i \in I} a_1(i), \min_{i \in I} a_2(i)]$$

$$\bigvee_{i \in I} A(i) = [\min_{i \in I} a_1(i), \max_{i \in I} a_2(i)]$$

$$\operatorname*{Min}_{i \in I} A(i) = [\min_{i \in I} a_1(i), \min_{i \in I} a_2(i)]$$

$$\operatorname*{Max}_{i \in I} A(i) = [\max_{i \in I} a_1(i), \max_{i \in I} a_2(i)].$$

Proof From Definitions 2.2.15, 2.2.11, and 2.2.16, and from Lemma 2.2.4. ∎

In the (Inf, Sup)-diagram, a representation for Meet, Join, Min and Max is in Fig. 2.6.

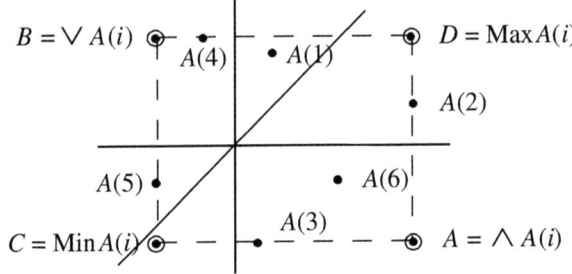

Fig. 2.6 Meet, Join, Min and Max

Lemma 2.2.10 (Duality of lattice operators) *For a bounded family of $A(I) \in I^*(\mathbb{R})$, if $A(i) = [a_1(i), a_2(i)]$:*

$$\mathrm{Dual}(\bigvee_{i \in I} A(i)) = \bigwedge_{i \in I} \mathrm{Dual}(A(i))$$

$$\mathrm{Dual}(\bigwedge_{i \in I} A(i)) = \bigvee_{i \in I} \mathrm{Dual}(A(i))$$

Proof From Definition 2.2.13 and Lemma 2.2.4. ∎

Definition 2.2.17 ("Meet–join" operator on $I^*(\mathbb{R})$) For $A \in I^*(\mathbb{R})$

$$\Omega_{(a,A)} = \begin{cases} \bigwedge_{a \in A'} & \text{if } A \text{ is improper} \\ \bigvee_{a \in A'} & \text{if } A \text{ is proper} \end{cases}$$

Lemma 2.2.11 (Modal intervals as meet–join)

$$\text{For } A \in I_u(\mathbb{R}) \text{ is } A = \bigwedge_{a \in A'} [a, a].$$

$$\text{For } A \in I_e(\mathbb{R}) \text{ is } A = \bigvee_{a \in A'} [a, a].$$

$$\text{For } A \in I^*(\mathbb{R}) \text{ is } A = \Omega_{(a,A)} [a, a].$$

Proof From Lemma 2.2.9. ∎

Figure 2.7 illustrates this result.

It would be lovely if $\mathrm{Pred}(A \vee B)$ were equal to $\mathrm{Pred}(A) \cup \mathrm{Pred}(B)$ and $\mathrm{Pred}(A \wedge B)$ equal to $\mathrm{Pred}(A) \cap \mathrm{Pred}(B)$. This is far from the truth, as the following example shows.

Example 2.2.3 Considering, for example these relations among the predicate sets $\mathrm{Pred}([1, 2])$, $\mathrm{Pred}([3, 4])$ and $\mathrm{Pred}([1, 2] \wedge [3, 4]) = \mathrm{Pred}([3, 2])$, the predicate $x \in \{1.5, 3.5\}$ belongs to $\mathrm{Pred}([1, 2])$ and to $\mathrm{Pred}([3, 4])$ and therefore to the

2.2 Construction of the Modal Intervals

Fig. 2.7 Modal intervals as meet–join

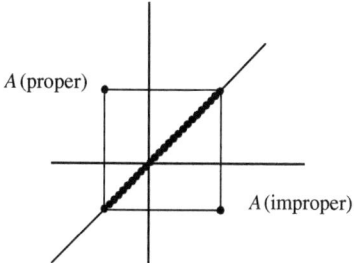

intersection of these two sets of predicates Pred([1, 2]) ∩ Pred([3, 4]), but not to their meet Pred([3, 2]). Similarly, the predicate $x = 2.5$ belongs to Pred([1, 4]) which is equal to Pred([1, 2]) ∨ Pred([3, 4]), but it belongs neither to Pred([1, 2]) nor to Pred([3, 4]), and consequently does not belong to Pred([1, 2]) ∪ Pred([3, 4]).

In view of this example, about the definition of modal interval inclusion and the related lattice operations meet and join, the following result holds:

Lemma 2.2.12 (Non triviality of the sublattice $(\text{Pred}(X), \subseteq)$**)**

1. $\text{Pred}(A \wedge B) \subseteq \text{Pred}(A) \cap \text{Pred}(B)$;
2. $\text{Pred}(A \vee B) \supseteq \text{Pred}(A) \cup \text{Pred}(B)$;
3. $\text{Copred}(A \wedge B) \supseteq \text{Copred}(A) \cup \text{Copred}(B)$;
4. $\text{Copred}(A \vee B) \subseteq \text{Copred}(A) \cap \text{Copred}(B)$.

Proof

1. and 2. Because $A \wedge B \subseteq A$, $A \wedge B \subseteq B$, $A \vee B \supseteq A$ and $A \vee B \supseteq B$. The equalities are not true as the previous example shows.
3. Pred($\mathbb{R}$)-complementary of 1.
4. Pred($\mathbb{R}$)-complementary of 2.

2.2.8 Interval Predicates and Co-predicates

We define, in what follows, a few particular classes of sets of predicates which bear a stronger structural relationship with modal intervals.

Definition 2.2.18 (Set of interval predicates)

$$\text{Pred}^*(\mathbb{R}) = \{x \in X' \mid X \in I^*(\mathbb{R})\}.$$

Definition 2.2.19 (Set of interval co-predicates)

$$\text{Copred}^*(\mathbb{R}) = \{x \notin X' \mid X \in I^*(\mathbb{R})\}.$$

Definition 2.2.20 (Set of interval predicates validated—or accepted—by A)

$$\text{Pred}^*(A) = \{(x \in X') \in \text{Pred}^*(\mathbb{R}) \mid Q(x, A)\ x \in X'\}.$$

Definition 2.2.21 (Set of interval co-predicates rejected by A)

$$\text{Copred}^*(A) = \{(x \notin X') \in \text{Copred}^*(\mathbb{R}) \mid \neg Q(x, A)\ x \notin X'\}.$$

Definition 2.2.22 ("Proper" and "improper" operators)

$$\text{Prop}(X) = \text{Prop}(X') = (X', \exists);$$
$$\text{Impr}(X) = \text{Impr}(X') = (X', \forall).$$

The operators meet and join bring about the equivalences contained in the following lemma.

Lemma 2.2.13 (Predicates and modality) *For $A \in I^*(\mathbb{R})$:*

1. $(x \in X') \in \text{Pred}^*(A) \Leftrightarrow \text{Impr}(X') \subseteq A$;
2. $(x \notin X') \in \text{Copred}^*(A) \Leftrightarrow \text{Prop}(X') \supseteq A$.

Proof

1. A proper: $(\exists x \in A')\ x \in X' \Leftrightarrow A' \cap X' \neq \emptyset \Leftrightarrow (A', E) \supseteq (X', \forall)$.
   A improper: $(\forall x \in A')\ x \in X' \Leftrightarrow A' \subseteq X' \Leftrightarrow (A', \forall) \supseteq (X', \forall)$.
2. $(x \notin X') \in \text{Copred}^*(A) \Leftrightarrow (x \in X') \in \text{Pred}^*(\text{Dual}(A))$
   $\Leftrightarrow \text{Impr}(X') \subseteq \text{Dual}(A) \Leftrightarrow \text{Prop}(X') \supseteq A$.

Remark 2.2.3 This latter result allows embedding the set $\text{Pred}^*(\mathbb{R})$ onto $I_u(\mathbb{R})$ by means of the correspondence $(x \in X') \leftrightarrow \text{Impr}(X')$ and $\text{Copred}^*(\mathbb{R})$ onto $I_e(\mathbb{R})$ by means of $(\neg(x \in X')) \leftrightarrow \text{Prop}(X')$, i.e., it is possible to identify the canonical predicates and co-predicates $(x \in X')$ and $(x \notin X')$ with the elements $\text{Impr}(X)$ of $I_u(\mathbb{R})$ and $\text{Prop}(X')$ of $I_e(\mathbb{R})$, and to identify also the set-theoretical expressions on the left of these equivalences by the analytical and programmable relations on the right.

The results of Lemma 2.2.12, because of their failure to provide the expected equalities, obstructs any straight path from the predicate-theoretical semantics of modal intervals to the semantics of their inclusion-lattice. Its ultimate meaning is the fact, common to all non-cheating real-life information processing, that interval processing of digital numerical information necessarily implies a certain degree of information loss.

For a better understanding of this particularity, we shall consider the case for the more restricted sets $\text{Pred}^*(X)$. In this case, the equalities missing in Lemma 2.2.12 for $A \wedge B$ and $A \vee B$, are shown to hold in some cases for interval predicates and co-predicates.

2.2 Construction of the Modal Intervals

Lemma 2.2.14 (Relations in the sublattice $(\{\text{Pred}^*(X) \mid X \in I^*(\mathbb{R})\}, \subseteq)$)

1. $\text{Pred}^*(A \wedge B) = \text{Pred}^*(A) \cap \text{Pred}^*(B)$;
2. $\text{Pred}^*(A \vee B) \supseteq \text{Pred}^*(A) \cup \text{Pred}^*(B)$;
3. $\text{Copred}^*(A \wedge B) \supseteq \text{Copred}^*(A) \cup \text{Copred}^*(B)$;
4. $\text{Copred}^*(A \vee B) = \text{Copred}^*(A) \cap \text{Copred}^*(B)$.

Proof

1. $(x \in X') \in \text{Pred}^*(A \wedge B) \Leftrightarrow \text{Impr}(X') \subseteq (A \wedge B)$

    $\Leftrightarrow (\text{Impr}(X') \subseteq A, \ \text{Impr}(X') \subseteq B)$

    $\Leftrightarrow (x \in X') \in (\text{Pred}^*(A) \cap \text{Pred}^*(B))$.

2. From Lemma 2.2.12

    $(x \in X') \in (\text{Pred}(A) \cup \text{Pred}(B)) \Rightarrow (x \in X') \in \text{Pred}(A \vee B)$,

    and asterisks can be added to $\text{Pred}(A)$, $\text{Pred}(B)$ and $\text{Pred}(A \vee B)$, after the form of $(x \in X')$. Moreover $\text{Pred}^*(A \vee B)$ can be larger than $\text{Pred}^*(A) \cup \text{Pred}^*(B)$, as Example 2.2.3 about $(x \in [2.5, 2.5]) \in \text{Pred}^*([1, 2] \vee [3, 4]) = \text{Pred}^*([1, 4])$ shows.

3. It is the dual of 2. (see Lemma 2.2.12). Also, $\text{Copred}^*(A \wedge B)$ can be larger than $\text{Copred}^*(A) \cup \text{Copred}^*(B)$, as comes out from Example 2.2.3 about $(x \notin 2.5) \in \text{Copred}^*([2, 1] \wedge [4, 3]) = \text{Copred}^*([4, 1])$.

4. It is the dual statement of 1:

$(x \notin X') \in \text{Copred}^*(A \vee B)$

$\Leftrightarrow (x \in X') \in \text{Pred}^*(\text{Dual}(A \vee B))$

$\Leftrightarrow (x \in X') \in \text{Pred}^*(\text{Dual}(A) \wedge \text{Dual}(B))$

$\Leftrightarrow (x \in X') \in (\text{Pred}^*(\text{Dual}(A)) \cap \text{Pred}^*(\text{Dual}(B)))$

$\Leftrightarrow (x \notin X') \in (\text{Copred}^*(A) \cap \text{Copred}^*(B))$. ∎

2.2.9 The k-Dimensional Case

To obtain the theoretical instruments which allow a logical formulation of the interval extension of a function $f : \mathbb{R}^k \to \mathbb{R}$, it is necessary to give some preliminary definitions which will make it possible to avoid the use of the set-theoretical extension.

We will use the symbol $I^*(\mathbb{R}^k)$ for the set of k-dimensional modal intervals.

Definition 2.2.23 (Set of k-dimensional modal intervals)

$$I^*(\mathbb{R}^k) = \{([a_1, b_1], \ldots, [a_k, b_k]) \mid [a_1, b_1] \in I^*(\mathbb{R}), \ldots, [a_k, b_k] \in I^*(\mathbb{R})\}.$$

Previous definitions and relationships in $I^*(\mathbb{R})$ are generalized in a natural way.

Definition 2.2.24 (k-dimensional inclusion and equality) For $A = (A_1, \ldots, A_k) \in I^*(\mathbb{R}^k)$, $B = (B_1, \ldots, B_k) \in I^*(\mathbb{R}^k)$,

$$A \subseteq B \Leftrightarrow (A_1 \subseteq B_1, \ldots, A_k \subseteq B_k)$$
$$A = B \Leftrightarrow (A_1 = B_1, \ldots, A_k = B_k).$$

Definition 2.2.25 (Proper and Improper operators) For $X = (X_1, \ldots, X_k) \in I^*(\mathbb{R}^k)$, $X' = (X'_1, \ldots, X'_k) \in I(\mathbb{R}^k)$:

$$\text{Prop}(X) = \text{Prop}(X') = ((X'_1, \exists), \ldots, (X'_k, \exists))$$
$$\text{Impr}(X) = \text{Impr}(X') = ((X'_1, \forall), \ldots, (X'_k, \forall)).$$

Definition 2.2.26 (Proper and improper sub-vectors of a modal interval vector) To single out the sub-vectors of proper and improper components of a modal vector $A \in I^*(\mathbb{R}^k)$, we will use the notational convention $A = (A_p, A_i) \in (I_e(\mathbb{R}^{k_p}), I_u(\mathbb{R}^{k_i}))$, where $k_p + k_i = k$, and the original indices are supposed maintained.

Remark 2.2.4 The definition of the vectors A_p and A_i would actually imply the rigorous definition of vectors with void components with their corresponding operations. It should be noted that this notation does not imply any permutation of the components of A, the vector A is not modified and, consequently, each component preserves its original index on A.

Definition 2.2.27 (Join and meet for families of indexed intervals) If $X' = (X'_1, \ldots, X'_k) \in I(\mathbb{R}^k)$, $x = (x_1, \ldots, x_k) \in \mathbb{R}^k$ and $F(x) = [F_1(x), F_2(x)] \in I^*(\mathbb{R})$,

$$\bigwedge_{x \in X'} F(x) = \bigwedge_{x_1 \in X'_1} \ldots \bigwedge_{x_k \in X'_k} F(x) = [\max_{x \in X'} F_1(x), \min_{x \in X'} F_2(x)]$$

$$\bigvee_{x \in X'} F(x) = \bigvee_{x_1 \in X'_1} \ldots \bigvee_{x_k \in X'_k} F(x) = [\min_{x \in X'} F_1(x), \max_{x \in X'} F_2(x)]$$

(the order of the component operators is irrelevant in both cases).

Definition 2.2.28 (Sets of k-dimensional interval predicates) For the vectors $A = (A_1, \ldots, A_k) \in I^*(\mathbb{R}^k)$, $X' = (X'_1, \ldots, X'_k) \in I(\mathbb{R}^k)$ and $x = (x_1, \ldots, x_k) \in \mathbb{R}^k$

$(x \in X') = (x_1 \in X'_1, \ldots, x_k \in X'_k);$

$\text{Pred}^*(A) = \{(x \in X') \mid (x_1 \in X'_1) \in \text{Pred}^*(A_1), \ldots, (x_k \in X'_k) \in \text{Pred}^*(A_k)\}.$

Remark 2.2.5 The condition defining the set Pred*(A) is equivalent to

$$Q(x_1, A_1) \ldots Q(x_k, A_k) \, (x_1 \in X'_1, \ldots, x_k \in X'_k),$$

where the order of the modal quantifiers does not matter, because of the x_i-arguments' independence among the predicates $x_i \in X'_i$.

Definition 2.2.29 (Sets of k-dimensional interval co-predicates) For $A = (A_1, \ldots, A_k) \in I^*(\mathbb{R}^k)$, $X' = (X'_1, \ldots, X'_k) \in I(\mathbb{R}^k)$ and $x = (x_1, \ldots, x_k) \in \mathbb{R}^k$

$(x \notin X') = \neg(x \in X') = (x_1 \notin X'_1 \text{ or} \ldots \text{or } x_k \notin X'_k);$

Copred*(A) = $\{(x \notin X') \mid (x_1 \notin X'_1) \in \text{Copred}^*(A_1), \ldots, (x_k \notin X'_k) \in \text{Copred}^*(A_k)\}.$

Lemma 2.2.15 (Interval representation of Pred*(A) and Copred*(A)) With the hypotheses of the previous definitions,

$$(x \in X') \in \text{Pred}^*(A) \Leftrightarrow \text{Impr}(X') \subseteq A$$
$$(x \notin X') \in \text{Copred}^*(A) \Leftrightarrow \text{Prop}(X') \supseteq A.$$

Proof See Lemma 2.2.13 and the previous definitions. ∎

2.3 Concluding Remarks

Modal Interval Analysis takes as its grounding principle that a real value in an applied context is actually not only worthy, but determined, by the set of properties, of predicates, which it validates. Therefore, Modal Interval Analysis extends the real numbers to intervals starting from the identification of real numbers with the set of predicates they validate.

The defining relation for modal intervals is

$$X = (X', Q_X) \leftrightarrow \text{Pred}(X),$$

where $\text{Pred}(X) = \{P \mid Q(x, X) P(x)\}$, $X' \in I(\mathbb{R})$, $Q_X \in \{\exists, \forall\}$ and Q is the newly introduced logical constant, the modal quantifier. The canonical notation, directly related to the inclusion completion of the set-theoretical intervals, is

$$[a_1, a_2] = \text{if } a_1 \leq a_2 \text{ then } ([a_1, a_2]', \exists)$$
$$\text{if } a_1 \geq a_2 \text{ then } ([a_2, a_1]', \forall).$$

Less formally, if modal intervals are canonically denoted like [1, 2] or [2, 1], when the lower limit is written to the left, as in [1, 2], this means that the interval is just an "existential interval" that we can also write as $([1,2]', \exists)$ to make explicit the "set component" $[1,2]'$ of the interval $[1,2]$ and its "selection modality" $\exists$. When the upper limit is written to the left, as in $[2,1]$, this indicates that the interval is a "universal interval", which we can denote by $([1,2]', \forall)$ to make explicit its set and modality components.

Analogously to the role of classical intervals in quantifying existential and universal prefixes ($\exists x \in X'$) and ($\forall x \in X'$), modal intervals bound the "modal quantifier" Q to give bounded quantifying prefixes such as $Q(x, [1,2])$ or $Q(x, [2,1])$. The ground definition of these modal quantifying prefixes depends on the modality of the bounding interval: so,

$$Q(x, [1,2]) \, P(x) = (\exists x \in [1,2]') \, P(x),$$

and

$$Q(x, [2,1]) \, P(x) = (\forall x \in [1,2]') \, P(x).$$

Now, after having developed the basic tools, the parallel relation to the inclusion of set-theoretical intervals can be introduced into the system of modal intervals. The inclusion relation of classical intervals, $A' \subseteq B'$, is equivalent to the validity of the implication $(x \in A' \Rightarrow x \in B')$ for any real number x, and is determined by the computable relation $(a_1 \geq b_1, a_2 \leq b_2)$ among their coordinates. The inclusion of modal intervals, $A \subseteq B$, is defined by the validity of the implication

$$Q(x, A) P(x) \Rightarrow Q(x, B) P(x) \text{ (or } \neg Q(x, B) P(x) \Rightarrow \neg Q(x, A) P(x))$$

for any property $P(x)$ on the real numbers, and happens to be determined also by the same relation $(a_1 \geq b_1, a_2 \leq b_2)$ among their canonical coordinates which marks the parent set-theoretical inclusion of set-intervals. Let's insist: if, for $A, B \in I^*(\mathbb{R})$,

$$A \subseteq B \Leftrightarrow \text{Pred}(A) \subseteq \text{Pred}(B),$$

the result $A \subseteq B \Leftrightarrow (a_1 \geq b_1, a_2 \leq b_2)$, makes of the set of modal intervals with its inclusion relation $(I^*(\mathbb{R}), \subseteq)$ the structural completion of set-theoretical intervals $(I(\mathbb{R}), \subseteq)$ guided by the semi-lattice structure of the set of classical intervals.

Outer and inner interval rounding are defined by

$$\text{Inn}([x_1, x_2]) = [\text{Right}(x_1), \text{Left}(x_2)];$$
$$\text{Out}([x_1, x_2]) = [\text{Left}(x_1), \text{Right}(x_2)].$$

These two interval rounding are universally possible within the limits of a given digital scale, and satisfy the property $\text{Inn}(X) \subseteq X \subseteq \text{Out}(X)$. Given the logical meaning of modal inclusion, in case X is the exact—not rounded—value of an interval defined by some analytical procedure, this relationship allows applying to

2.3 Concluding Remarks

the computed values Out(X) and Inn(X) the analytical properties of the possibly unknown value X, and also of applying to X the experimental properties of the computed results Inn(X) and Out(X), since Inn(X) $\subseteq X \subseteq$ Out(X) is equivalent to

$$\text{Pred}(\text{Inn}(X)) \subseteq \text{Pred}(X) \subseteq \text{Pred}(\text{Out}(X))$$

and

$$\text{Copred}(\text{Inn}(X)) \supseteq \text{Copred}(X) \supseteq \text{Copred}(\text{Out}(X)).$$

Finally, the double—analytical and logical—face of the modal inclusion selects both the inner and outer truncation as the normative modes of digital computation. On this basis, the semantics of interval values do not come from some external consideration about any particular interval relation, but they are given with and by the data of each problem or are obtained by mechanical computations and measurements.

Chapter 3
Modal Interval Extensions

3.1 Introduction

The problem discussed in this chapter is that of obtaining a class of interval functions $F : I^*(\mathbb{R}^k) \to I^*(\mathbb{R})$, consistently referring to the continuous functions f from $\mathbb{R}^k$ to $\mathbb{R}$.

In classical interval analysis, an interval extension of a $\mathbb{R}^k$ to $\mathbb{R}$ continuous function $z = f(x_1, \ldots, x_k)$ is the *interval united extension* R_f of f. Given an interval argument $X = (X_1, \ldots, X_k) \in I(\mathbb{R}^k)$, it is defined as the range of f-values on X

$$R_f(X_1, \ldots, X_k) = \{f(x_1, \ldots, x_k) \mid x_1 \in X_1, \ldots, x_k \in X_k\}$$
$$= [\min\{f(x_1, \ldots, x_k) \mid x_1 \in X_1, \ldots, x_k \in X_k\},$$
$$\max\{f(x_1, \ldots, x_k) \mid x_1 \in X_1, \ldots, x_k \in X_k\}],$$

which can be considered as a "semantic extension" of f, since it admits the logical interpretations

$$(\forall x_1 \in X_1) \cdots (\forall x_k \in X_k) (\exists z \in R_f(X_1, \ldots, X_k)) z = f(x_1, \ldots, x_k)$$

and

$$(\forall z \in R_f(X_1, \ldots, X_k)) (\exists x_1 \in X_1) \cdots (\exists x_k \in X_k) z = f(x_1, \ldots, x_k).$$

Since the domain of values of a continuous function is generally not easily computable, an interval syntactic extension $fR(X_1, \ldots, X_k)$ is defined by replacing the real operators of the real functions $f(x_1, \ldots, x_k)$ on $\mathbb{R}$ by the homonymous operators defined on the system $(I(\mathbb{R}), I(\mathbb{D}))$, that is, replacing

1. their numerical arguments $x_1, \ldots, x_k$ by the interval arguments $X_1, \ldots, X_k$, and
2. their real arithmetic operators ω by the corresponding interval operations ω which, in the common case of the truncated computations of any actual arithmetic, must be the outwards directed ω_R because of the inclusion

$$X\omega Y \subseteq X\omega_R Y = \text{Out}(X\omega Y).$$

The crucial relation between both extensions is

$$R_f(X_1, \ldots, X_k) \subseteq fR(X_1, \ldots, X_k),$$

under the condition that the function $fR(X_1, \ldots, X_k)$ is well defined, i.e., that it does not imply division by an unspecified interval containing the value zero. Therefore a syntactic extension $fR(X_1, \ldots, X_k)$ is computable from the bounds of the intervals $X_1, \ldots, X_k$, and usually represents an overestimation of $R_f(X_1, \ldots, X_k)$.

Example 3.1.1 The united extension of the continuous real function

$$f(x) = \frac{x}{1+x}$$

to the interval $[2, 4]$ is $R_f([2, 4]) = [2/3, 4/5]$, the range of f in this interval. The syntactic extension for the same interval is

$$fR([2, 4]) = \frac{[2, 4]}{1 + [2, 4]} = [2/5, 4/3],$$

and, in fact, $R_f([2, 4]) = [2/3, 4/5] \subseteq [2/5, 4/3] = fR([2, 4])$.

Syntactic interval functions have the property, fundamental to the whole field of Interval Analysis, of being "inclusive", that is, for $A_1 \subseteq B_1, \ldots, A_k \subseteq B_k$, the relation

$$fR(A_1, \ldots, A_k) \subseteq fR(B_1, \ldots, B_k)$$

holds.

A basic critical fact is that the interval syntactic extension fR of f satisfies only one kind of interval predicate compatible with outer rounding:

$$(\forall x_1 \in X_1) \cdots (\forall x_k \in X_k) \, (\exists z \in \text{Out}(fR(X_1, \ldots, X_k))) \, z = f(x_1, \ldots, x_k).$$

In the context of modal intervals, it may be expected, as a starting point, that as soon as the $\mathbb{R}$-predicate $P(x)$ results in the modal interval predicate $Q(x, X)P(x)$, the relation $z = f(x_1, \ldots, x_k)$ must become some kind of interval relation $Z = F(X_1, \ldots, X_k)$ guaranteeing some sort of $(k+1)$-dimensional interval predicate of the form

3.1 Introduction

$$Q_1(x_1, X_1) \ldots Q_k(x_k, X_k) Q_z(z, Z) z = f(x_1, \ldots, x_k),$$

where an ordering problem obviously arises since the quantifying prefixes are not generally commutable.

3.1.1 Poor Computational Extension

To find a more general approach, we scale down the problem of digital computation to its bare essentials and start by considering the most elementary sort of computational functions able to get actual information about the ideal connections established by the continuous real functions $f : \mathbb{R}^k \to \mathbb{R}$. To prevent the restrictions of any extension by set of values due to the limited character of system $(I(\mathbb{R}), +, *)$, illustrated in Chap. 1, we will first introduce the definition of a "poor" computational extension of a continuous real function.

Definition 3.1.1 (Poor computational extension) The function $F : \mathbb{R}^k \to I(\mathbb{R})$ is a *poor computational extension* of a continuous real function $f : \mathbb{R}^k \to \mathbb{R}$ if the existence of $F(a)'$ implies that $f(a) \in F(a)'$.

These simplest partial computational functions are defined on a subset of $\mathbb{R}$ and have, wherever defined, the two values $\text{Sup}(F(a)')$ and $\text{Inf}(F(a)')$, upper and lower bounds of the analytically defined value $f(a)$, the value of which the exact determination is, as a general matter of fact, out of reach for digital processing.

The usefulness of this definition is to induce a more general one for extensions of the kind $F : I^*(\mathbb{R}^k) \to I^*(\mathbb{R})$, as it results from the lemma that follows.

Lemma 3.1.1 (Semantic formulation of a poor computational extension) *Let $F : \mathbb{R}^k \to I(\mathbb{R})$ be a poor computational extension of f, and let $f : \mathbb{R}^k \to \mathbb{R}$ be a continuous function. Supposing that $F(a)' \in I(\mathbb{R})$ exists, the condition $f(a) \in F(a)'$ is equivalent to*

$$(\forall X' \in I(\mathbb{R}^k)) ((x \in X') \in \text{Pred}^*([a, a]) \Rightarrow (z \in f(X')) \in \text{Pred}^*(\text{Prop}(F(a)))),$$

where $f(X')$ is the united extension or domain of values of f on X'.

Proof This logical formula is equivalent to

$$(\forall X' \in I(\mathbb{R}^k)) (a \in X' \Rightarrow F(a)' \cap f(X') \neq \emptyset),$$

which is equivalent to $f(a) \in F(a)'$ because:

(1) particularizing X' to $[a, a]'$, becomes

$$(a \in [a, a]' \Rightarrow F(a)' \cap f([a, a]') \neq \emptyset) \Rightarrow f(a) \in F(a)';$$

Fig. 3.1 Poor extension diagram

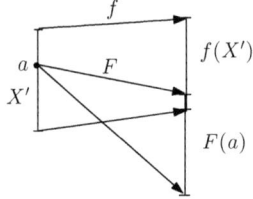

(2) for the reverse implication, if $f(a) \in F(a)'$,

$$(\forall X' \in I(\mathbb{R}^k))\,(a \in X' \Rightarrow f(a) \in f(X') \Rightarrow F(a)' \cap f(X') \neq \emptyset).$$

■

Figure 3.1 illustrates this proof.

Example 3.1.2 For the function $f : \mathbb{R} \to \mathbb{R}$ given by $f(x) = x/3$, a poor computational extension can be $F : \{1\} \to I(\mathbb{R})$ given by $F(1)' = [0.33, 0.34]'$. In this case, $a = 1$ and $f(1) = 1/3$ satisfy $f(1) \in F(1)'$. This relation is equivalent to

$$(\forall X' \in I(\mathbb{R}))(1 \in X' \Rightarrow [0.33, 0.34]' \cap f(X') \neq \emptyset),$$

which can be written in the form

$$(\forall X' \in I(\mathbb{R}))(1 \in X' \Rightarrow (\exists z \in [0.33, 0.34]')\, z \in f(X'))$$

or

$$(\forall X' \in I(\mathbb{R}))((x \in X') \in \text{Pred}^*([1,1]) \Rightarrow (z \in f(X')) \in \text{Pred}^*([0.33, 0.34]')).$$

Now, the equivalent definition for poor computational extensions, made available by this lemma, can be extended to define logically the "modal interval extensions" of continuous functions by formally substituting the element $[a, a]$ by a general modal interval $A \in I^*(\mathbb{R}^k)$, overcoming the rigidities of the theory of functions of the ordinary interval analysis which are induced by the set-theoretical domain-of-values approach.

3.1.2 Modal Interval Extension

In the logical formulation of the poor computational extension of a continuous function, let us replace the argument $[a, a]$ and its modal image $\text{Prop}(F(a)')$ by the more general argument and image, A and $F(A)$.

Definition 3.1.2 (Modal interval extension) If $f : \mathbb{R}^k \to \mathbb{R}$ is a real continuous function, then $F : I^*(\mathbb{R}^k) \to I^*(\mathbb{R})$ is its *modal interval extension*, if, wherever $F(A)$ exists,

$$(\forall X' \in I(\mathbb{R}^k)) \, ((x \in X') \in \mathrm{Pred}^*(A) \Rightarrow (z \in f(X')) \in \mathrm{Pred}^*(F(A))).$$

The logical form of the condition is acceptable. It is thus necessary only to emphasize the properties that make these functions interesting for a computation by intervals. The first indication of the nature of this definition is that it does not give a univocal value for each $F(A)$ which exists, but it gives only a lower limit for the modal inclusion. Indeed, by Lemma 2.2.13, the condition that $F(A)$ is a modal extension of f can be written in the following equivalent form:

$$(\forall X' \in I(\mathbb{R}^k)) \, (\mathrm{Impr}(X') \subseteq A \Rightarrow \mathrm{Impr}(f(X')) \subseteq F(A)).$$

There remains the task of uncovering the properties which characterize analytically and semantically these formally constructed "modal interval extensions".

3.2 Semantic Functions

We will define the two "semantic" interval functions which play a grounding role in the theory because they are in close relation with the modal interval extensions of continuous functions.

Definition 3.2.1 (*-semantic extension) If f is an $\mathbb{R}^k$ to $\mathbb{R}$ continuous function and if $x = (x_p, x_i)$ is the component-splitting corresponding to $X = (X_p, X_i) \in I^*(\mathbb{R}^k)$,

$$f^*(X) = \bigvee_{x_p \in X'_p} \bigwedge_{x_i \in X'_i} [f(x_p, x_i), f(x_p, x_i)]$$

$$= [\min_{x_p \in X'_p} \max_{x_i \in X'_i} f(x_p, x_i), \max_{x_p \in X'_p} \min_{x_i \in X'_i} f(x_p, x_i)].$$

and is called the *-semantic extension of f.

Definition 3.2.2 (-semantic extension)** With the same hypotheses as the previous definition,

$$f^{**}(X) = \bigwedge_{x_i \in X'_i} \bigvee_{x_p \in X'_p} [f(x_p, x_i), f(x_p, x_i)] =$$

$$= [\max_{x_i \in X'_i} \min_{x_p \in X'_p} f(x_p, x_i), \min_{x_i \in X'_i} \max_{x_p \in X'_p} f(x_p, x_i)].$$

and is called the **-semantic extension of f.

Remark 3.2.1 If all the X-components are proper intervals, i.e., $X_i = \emptyset$ allowing for the abuse of language, then

$$f^*(X) = f^{**}(X) = [\min\{f(x_1,\ldots,x_k) \mid x_1 \in X'_1,\ldots,x_k \in X'_k\},$$
$$\max\{f(x_1,\ldots,x_k) \mid x_1 \in X'_1,\ldots,x_k \in X'_k\}],$$

which corresponds to the interval united extension R_f of the classical interval analysis, and $\text{Mod}(f^*(X)) = \exists$.

If all the X-components are improper intervals, i.e., $X_p = \emptyset$ allowing for the abuse of language, then, one has instead

$$f^*(X) = f^{**}(X) = [\max\{f(x_1,\ldots,x_k) \mid x_1 \in X'_1,\ldots,x_k \in X'_k\},$$
$$\min\{f(x_1,\ldots,x_k) \mid x_1 \in X'_1,\ldots,x_k \in X'_k\}],$$

with $\text{Set}(f^*(X)) = R_f$ and $\text{Mod}(f^*(X)) = \forall$.

Example 3.2.1 For the continuous real continuous $f(x_1,x_2) = x_1^2 + x_2^2$, the computation of the *-semantic and the **-semantic functions for $X = ([-1,1],[1,-1])$ yields the following results:

$$f^*([-1,1],[1,-1]) = \bigvee_{x_1 \in [-1,1]'} \bigwedge_{x_2 \in [-1,1]'} [x_1^2 + x_2^2, x_1^2 + x_2^2]$$
$$= \bigvee_{x_1 \in [-1,1]'} [x_1^2 + 1, x_1^2] = [1,1];$$
$$f^{**}([-1,1],[1,-1]) = \bigwedge_{x_2 \in [-1,1]'} \bigvee_{x_1 \in [-1,1]'} [x_1^2 + x_2^2, x_1^2 + x_2^2]$$
$$= \bigwedge_{x_2 \in [-1,1]'} [x_2^2, 1 + x_2^2] = [1,1].$$

For the continuous real continuous function $g(x_1,x_2) = (x_1 + x_2)^2$, the corresponding *-semantic and **-semantic functions for $X = ([-1,1],[1,-1])$ don't have coincident values:

$$g^*([-1,1],[1,-1]) = \bigvee_{x_1 \in [-1,1]'} \bigwedge_{x_2 \in [-1,1]'} [(x_1+x_2)^2, (x_1+x_2)^2]$$
$$= \bigvee_{x_1 \in [-1,1]'} [\text{if } x_1 < 0 \text{ then } (x_1-1)^2 \text{ else } (x_1+1)^2, 0]$$
$$= [1,0];$$
$$g^{**}([-1,1],[1,-1]) = \bigwedge_{x_2 \in [-1,1]'} \bigvee_{x_1 \in [-1,1]'} [(x_1+x_2)^2, (x_1+x_2)^2]$$
$$= \bigwedge_{x_2 \in [-1,1]'} [0, \text{if } x_2 < 0 \text{ then } (x_2-1)^2 \text{ else } (x_2+1)^2]$$
$$= [0,1].$$

3.2 Semantic Functions

In general, both semantic extensions are out of reach of any direct computation except for some very simple continuous real functions, as the previous example may suggest, as unary operator (exp, ln, ...) and the arithmetic operators of which semantic computations, properties and implementation are in Chap. 5.

The case of equality between both extensions characterizes the following important concept.

Definition 3.2.3 (*JM-commutativity*) A continuous real function $f : \mathbb{R}^k \to \mathbb{R}$ is JM-commutable for $A \in I^*(\mathbb{R}^k)$ if $f^*(A) = f^{**}(A)$.

3.2.1 Properties of the Semantic Extensions

The semantic extensions are not independent: there exists a relation of duality.

Theorem 3.2.1 (**Duality of the semantic functions**) *If f is an $\mathbb{R}^k$ to $\mathbb{R}$ continuous function and $X \in I^*(\mathbb{R}^k)$,*

$$\mathrm{Dual}(f^*(X)) = f^{**}(\mathrm{Dual}(X)).$$

Proof From the definitions of f^* and f^{**} and Lemma 2.2.10

$$\mathrm{Dual}(f^*(X)) = \mathrm{Dual}(\bigvee_{x_p \in X'_p} \bigwedge_{x_i \in X'_i} [f(x), f(x)])$$

$$= \bigwedge_{x_p \in X'_p} \bigvee_{x_i \in X'_i} [f(x), f(x)] = f^{**}(\mathrm{Dual}(X)).$$

∎

The following result yields the basic relation of inclusion between the semantic extensions.

Theorem 3.2.2 (**Min–max**) *If f is an $\mathbb{R}^k$ to $\mathbb{R}$ continuous function, and (X'_1, X'_2) is any component splitting of $X' \in I(\mathbb{R}^k)$, then*

$$(\forall (x_1, x_2) \in (X'_1, X'_2)) \max_{x_1 \in X'_1} \min_{x_2 \in X'_2} f(x_1, x_2) \leq \min_{x_2 \in X'_2} \max_{x_1 \in X'_1} f(x_1, x_2)$$

and

$$\max_{x_1 \in X'_1} \min_{x_2 \in X'_2} f(x_1, x_2) \leq f(x_{1m}, x_{2M}) \leq \min_{x_2 \in X'_2} \max_{x_1 \in X'_1} f(x_1, x_2),$$

where x_{1m} is a point on which the function $\min_{x_2 \in X'_2} f(x_1, x_2)$ reaches its maximum and x_{2M} is a point on which the function $\max_{x_1 \in X'_1} f(x_1, x_2)$ reaches its minimum.

Proof The first inequality is true since

$$(\forall x_1 \in X'_1) \min_{x_2 \in X'_2} f(x_1, x_2) \leq f(x_1, x_2)$$

$$(\forall x_2 \in X'_2) \max_{x_1 \in X'_1} f(x_1, x_2) \geq f(x_1, x_2)$$

For the second inequality, defining

$$f_m(x'_1) = \min_{x_2 \in X'_2} f(x'_1, x_2),$$

$$f_M(x'_2) = \max_{x_1 \in X'_1} f(x_1, x'_2)$$

it follows that

$$(\forall x'_1 \in X'_1)(\forall x'_2 \in X'_2)(f_m(x'_1) \leq f(x'_1, x'_2) \leq f_M(x'_2)).$$

■

Remark 3.2.2 Since all the values $f_m(x'_1)$ are less than or equal to all the values of $f_M(x'_2)$, and the functions f_m and f_M are continuous, the sets $F'_m = \{f_m(x'_1) \mid x'_1 \in X'_1\}$ and $F'_M = \{f_M(x'_2) \mid x'_2 \in X'_2\}$ are intervals such that $\text{Sup}(F'_m) \leq \text{Inf}(F'_M)$, as is partially stated by this theorem.

Next, firstly, the inclusion relation between $f^*(X)$ and $f^{**}(X)$ will be shown:

Theorem 3.2.3 (Inclusion of f^* in f^{**}) *If f is an $\mathbb{R}^k$ to $\mathbb{R}$ continuous real function and $X \in I^*(\mathbb{R}^k)$, then*

$$f^*(X) \subseteq f^{**}(X).$$

Proof

$$f^*(X) = \bigvee_{x_p \in X'_p} \bigwedge_{x_i \in X'_i} [f(x_p, x_i), f(x_p, x_i)]$$

$$= [\min_{x_p \in X'_p} \max_{x_i \in X'_i} f(x_p, x_i), \max_{x_p \in X'_p} \min_{x_i \in X'_i} f(x_p, x_i)]$$

$$\subseteq [\max_{x_i \in X'_i} \min_{x_p \in X'_p} f(x_p, x_i), \min_{x_i \in X'_i} \max_{x_p \in X'_p} f(x_p, x_i)]$$

$$= \bigwedge_{x_i \in X'_i} \bigvee_{x_p \in X'_p} [f(x_p, x_i), f(x_p, x_i)]$$

$$= f^{**}(X).$$

■

Secondly, there follows the $\subseteq$-monotonicity of f^* and f^{**}.

3.2 Semantic Functions

Lemma 3.2.1 *For $X \in I^*(\mathbb{R})$ and any $\mathbb{R}$ to $\mathbb{R}$ continuous functions F_1, F_2*

$$F_1(x) \subseteq F_2(x) \Rightarrow \underset{(x,X)}{\Omega} F_1(x) \subseteq \underset{(x,X)}{\Omega} F_2(x).$$

Proof In agreement with the Definition 2.2.17 of the meet–join operator,

a) If X is a proper interval,

$$\underset{(x,X)}{\Omega} F_1(x) = \underset{x \in X'}{\vee} F_1(x)$$

$$= [\underset{x \in X'}{\min} \operatorname{Inf}(F_1(x)), \underset{x \in X'}{\max} \operatorname{Sup}(F_1(x))]$$

$$\subseteq [\underset{x \in X'}{\min} \operatorname{Inf}(F_2(x)), \underset{x \in X'}{\max} \operatorname{Sup}(F_2(x))]$$

$$= \underset{x \in X'}{\vee} F_2(x)$$

$$= \underset{(x,X)}{\Omega} F_2(x).$$

b) A dual proof is valid if X is an improper interval.

Lemma 3.2.2 *For $X_1, X_2 \in I^*(\mathbb{R})$ and $F : \mathbb{R} \to I^*(\mathbb{R})$,*

$$X_1 \subseteq X_2 \Rightarrow \underset{(x,X_1)}{\Omega} F(x) \subseteq \underset{(x,X_2)}{\Omega} F(x).$$

Proof

a) If X_1 is a proper interval,

$$\underset{(x,X_1)}{\Omega} F(x) = \underset{x \in X_1'}{\vee} F(x) = [\underset{x \in X_1'}{\min} \operatorname{Inf}(F(x)), \underset{x \in X_1'}{\max} \operatorname{Sup}(F(x))],$$

a1) if X_2 is a proper interval, then $X_1' \subseteq X_2'$ and therefore

$$[\underset{x \in X_1'}{\min} \operatorname{Inf}(F(x)), \underset{x \in X_1'}{\max} \operatorname{Sup}(F(x))]$$

$$\subseteq [\underset{x \in X_2'}{\min} \operatorname{Inf}(F(x)), \underset{x \in X_2'}{\max} \operatorname{Sup}(F(x))]$$

$$= \underset{x \in X_2'}{\vee} F(x)$$

$$= \underset{(x,X_2)}{\Omega} F(x),$$

a2) if X_2 is an improper interval, then $X_1' = X_2' = \{a\}$ and $\underset{(x,[a,a])}{\Omega}$ reduces to the identity operator.

b) The proof is completed by a similar reasoning if X_1 is an improper interval, and by a two step process through a point $x_0 \in X_1' \cap X_2'$ for the case of X_1 improper and X_2 proper. ∎

Lemma 3.2.3 For $X_1, X_2 \in I^*(\mathbb{R})$ and $F_1, F_2 : \mathbb{R} \to I^*(\mathbb{R})$,

$$(X_1 \subseteq X_2, F_1(x) \subseteq F_2(x)) \Rightarrow \underset{(x,X_1)}{\Omega} F_1(x) \subseteq \underset{(x,X_2)}{\Omega} F_2(x).$$

Proof From Lemmata 3.2.1 and 3.2.2,

$$\underset{(x,X_1)}{\Omega} F_1(x) \subseteq \underset{(x,X_2)}{\Omega} F_1(x) \subseteq \underset{(x,X_2)}{\Omega} F_2(x).$$
∎

Theorem 3.2.4 (Inclusivity of the semantic extensions) *If $X, Y \in I^*(\mathbb{R}^k)$ and f from $\mathbb{R}^k$ to $\mathbb{R}$ is continuous,*

$$X \subseteq Y \Rightarrow (f^*(X) \subseteq f^*(Y), \ f^{**}(X) \subseteq f^{**}(Y)).$$

Proof From the previous lemma,

$$\left. \begin{array}{l} f^*(X) \\ f^{**}(X) \end{array} \right\} = \underset{(x_1,X_1)}{\Omega} \cdots \underset{(x_k,X_k)}{\Omega} [f(x_1,\cdots,x_k), f(x_1,\cdots,x_k)]$$

$$\subseteq \underset{(x_1,Y_1)}{\Omega} \cdots \underset{(x_k,Y_k)}{\Omega} [f(x_1,\cdots,x_k), f(x_1,\cdots,x_k)] = \left\{ \begin{array}{l} f^*(Y) \\ f^{**}(Y) \end{array} \right.$$
∎

3.2.2 Characterization of JM-Commutativity

Next the case $f^*(X) = f^{**}(X)$ will be characterized, when X is not uni-modal. The main role in this characterization is played by the saddle-points of the function f.

Definition 3.2.4 (Saddle-points set) Let $(X_1', X_2') = X'$ be a component splitting of $X' \in I(\mathbb{R}^k)$, and f be a continuous function from $\mathbb{R}^k$ to $\mathbb{R}$. The *set of saddle points* of f in X' is

$$\text{SDP}(f, X_1', X_2') = \{(\pmb{x}_{1m}, \pmb{x}_{2M}) \mid (\forall \pmb{x}_1 \in X_1')(\forall \pmb{x}_2 \in X_2') \, (f(\pmb{x}_{1m}, \pmb{x}_2)$$
$$\leq f(\pmb{x}_{1m}, \pmb{x}_{2M}) \leq f(\pmb{x}_1, \pmb{x}_{2M}))\}.$$

3.2 Semantic Functions

Fig. 3.2 Saddle-point

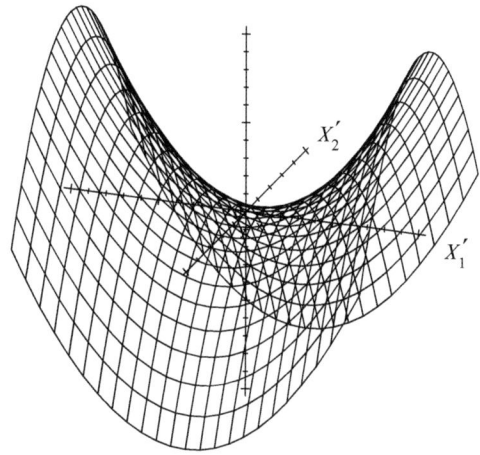

Figure 3.2 illustrates this definition.

Definition 3.2.5 (Saddle value). Let $(X'_1, X'_2) = X'$ be a component splitting of $X' \in I(\mathbb{R}^k)$, and f be a continuous function from $\mathbb{R}^k$ to $\mathbb{R}$. The set of *saddle values* of f in X' is

$$\text{SDV}(f, X'_1, X'_2) = (x_{1m}, x_{2M})$$

if $(\text{SDP}(f, X'_1, X'_2) \neq \emptyset$ and $(x_{1m}, x_{2M}) \in \text{SDP}(f, X'_1, X'_2))$. Otherwise, it is undefined.

A well known property of saddle points is in the next lemma.

Lemma 3.2.4 *If (x_{1m}, x_{2M}) and (x'_{1m}, x'_{2M}) are two saddle points of f in (X'_1, X'_2), then (x_{1m}, x'_{2M}) and (x'_{1m}, x_{2M}) are also saddle points. Moreover,*

$$f(x_{1m}, x_{2M}) = f(x_{1m}, x'_{2M}) = f(x'_{1m}, x_{2M}) = f(x'_{1m}, x'_{2M})$$

Proof For any $x_1 \in X'_1$ and $x_2 \in X'_2$ the inequalities

$$f(x_{1m}, x_2) \leq f(x_{1m}, x_{2M}) \leq f(x_1, x_{2M})$$

$$f(x'_{1m}, x_2) \leq f(x'_{1m}, x'_{2M}) \leq f(x_1, x'_{2M})$$

are true. So, particularizing the first one to x'_{1m} and x'_{2M} and the second one to x_{1m} and x_{2M},

$$f(x'_{1m}, x_{2M}) \leq f(x'_{1m}, x'_{2M}) \leq f(x_{1m}, x'_{2M}) \leq f(x_{1m}, x_{2M}) \leq f(x'_{1m}, x_{2M})$$

which implies the result. Moreover

$$f(x'_{1m}, x_{2M}) = f(x_{1m}, x_{2M}) \leq f(x_1, x_{2M})$$

and

$$f(x'_{1m}, x_2) \leq f(x'_{1m}, x'_{2M}) = f(x'_{1m}, x_{2M})$$

imply

$$f(x'_{1m}, x_2) \leq f(x'_{1m}, x_{2M}) \leq f(x_1, x_{2M}).$$

Therefore (x'_{1m}, x_{2M}) is a saddle point. Similarly for (x_{1m}, x'_{2M}). ∎

In accordance with this result, the set of saddle values of f in (X'_1, X'_2) is either empty or contains a unique point.

Lemma 3.2.5 *In the context of the previous definition, if there exists a saddle point (x_{1m}, x_{2M}) of f in X',*

$$\text{SDV}(f, X'_1, X'_2) = f(x_{1m}, x_{2M})$$
$$= \min_{x_1 \in X'_1} \max_{x_2 \in X'_2} f(x_1, x_2) = \max_{x_2 \in X'_2} \min_{x_1 \in X'_1} f(x_1, x_2).$$

Proof From

$$\min_{x_1 \in X'_1} \max_{x_2 \in X'_2} f(x_1, x_2) \leq \max_{x_2 \in X'_2} f(x_{1m}, x_2) \leq f(x_{1m}, x_{2M})$$
$$\leq \min_{x_1 \in X'_1} f(x_1, x_{2M}) \leq \max_{x_2 \in X'_2} \min_{x_1 \in X'_1} f(x_1, x_2)$$

and Theorem 3.2.2 which closes the $\leq$-chain. ∎

Theorem 3.2.5 (*JM-commutativity*) *For a given $X \in I^*(\mathbb{R}^k)$, the joint validity of $\text{SDP}(f, X'_p, X'_i) \neq \emptyset$, $\text{SDP}(f, X'_i, X'_p) \neq \emptyset$ is equivalent to $f^*(X) = f^{**}(X)$; in this case,*

$$f^*(X) = f^{**}(X) = [\text{SDV}(f, X'_p, X'_i), \text{SDV}(f, X'_i, X'_p)].$$

Proof As

$$\text{SDV}(f, X'_p, X'_i) = \min_{x_p \in X'_p} \max_{x_i \in X'_i} f(x_p, x_i) = \max_{x_i \in X'_i} \min_{x_p \in X'_p} f(x_p, x_i),$$

$$\text{SDV}(f, X'_i, X'_p) = \min_{x_i \in X'_i} \max_{x_p \in X'_p} f(x_p, x_i) = \max_{x_p \in X'_p} \min_{x_i \in X'_i} f(x_p, x_i),$$

3.2 Semantic Functions

Fig. 3.3 Function $f(x_1, x_2) = x_1^2 + x_2^2$ in $X' = ([-1, 1]', [0, 2]')$

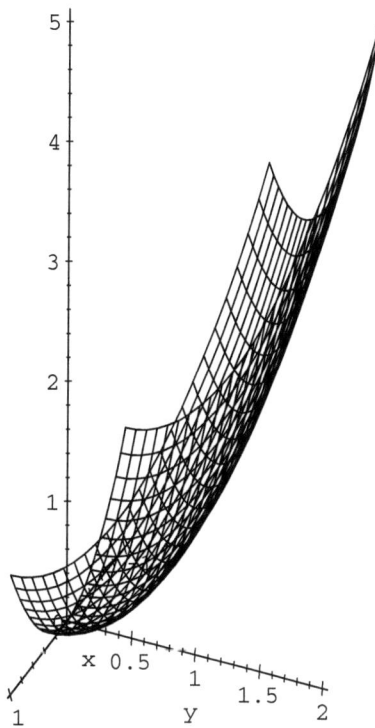

then

$$f^*(X) = \bigvee_{x_p \in X'_p} \bigwedge_{x_i \in X'_i} [f(x_p, x_i), f(x_p, x_i)]$$

$$= [\min_{x_p \in X'_p} \max_{x_i \in X'_i} f(x_p, x_i), \max_{x_p \in X'_p} \min_{x_i \in X'_i} f(x_p, x_i)]$$

$$= [\max_{x_i \in X'_i} \min_{x_p \in X'_p} f(x_p, x_i), \min_{x_i, X'_i} \max_{x_p \in X'_p} f(x_p, x_i)]$$

$$= \bigwedge_{x_i \in X'_i} \bigvee_{x_p \in X'_p} [f(x_p, x_i), f(x_p, x_i)] = f^{**}(X).$$

∎

Remark 3.2.3 The JM-commutativity of a function f implies the applicability of one of the two semantic theorems, the direct or its dual. Which one will depend only on the truncation's sense, outer or inner, of the computation of f^*.

Example 3.2.2 For the continuous real function $f(x_1, x_2) = x_1^2 + x_2^2$ the *-semantic and **-semantic extensions for $X = ([-1, 1], [2, 0])$ (see Fig. 3.3) are

Fig. 3.4 Function
$f(x_1, x_2) = x_1^2 + x_2^2$ in
$X' = ([-1, 1]', [-1, 1]')$

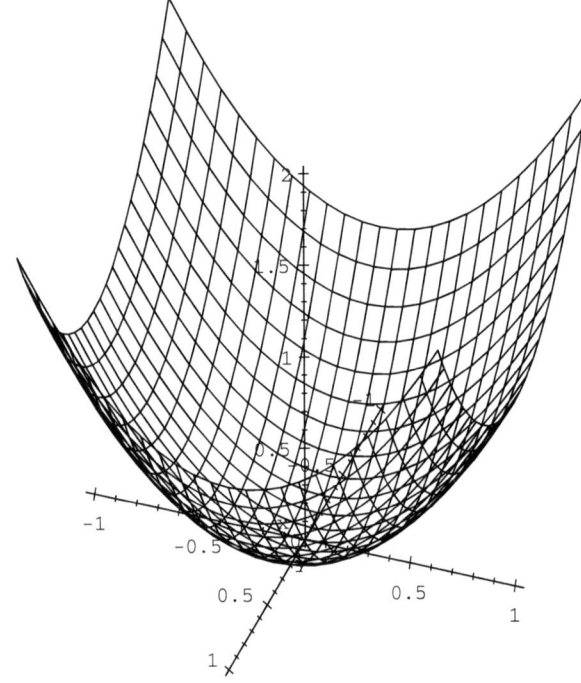

$$f^*([-1,1],[2,0]) = \bigvee_{x_1 \in [-1,1]'} \bigwedge_{x_2 \in [0,2]'} [x_1^2 + x_2^2, x_1^2 + x_2^2]$$

$$= \bigvee_{x_1 \in [-1,1]'} [x_1^2 + 4, x_1^2] = [4, 1],$$

$$f^{**}([-1,1],[2,0]) = \bigwedge_{x_2 \in [0,2]'} \bigvee_{x_1 \in [-1,1]'} [x_1^2 + x_2^2, x_1^2 + x_2^2]$$

$$= \bigwedge_{x_2 \in [0,2]'} [x_2^2, x_2^2 + 1] = [4, 1]$$

and

$$SDP(f, X'_p, X'_i) = SDP(f, [-1, 1]', [0, 2]') = \{(0, 2)\}$$
$$SDV(f, X'_p, X'_i) = SDV(f, [-1, 1]', [0, 2]') = 4$$
$$SDP(f, X'_i, X'_p) = SDP(f, [0, 2]', [-1, 1]') = \{(1, 0), (-1, 0)\}$$
$$SDV(f, X'_i, X'_p) = SDV(f, [0, 2]', [-1, 1]') = 1.$$

For the same function $f(x_1, x_2) = x_1^2 + x_2^2$, the *-semantic and **-semantic extensions for $X = ([-1, 1], [1, -1])$ (see Fig. 3.4) are

$$f^*([-1,1],[1,-1]) = \bigvee_{x_1 \in [-1,1]'} \bigwedge_{x_2 \in [-1,1]'} [x_1^2 + x_2^2, x_1^2 + x_2^2]$$

$$= \bigvee_{x_1 \in [-1,1]'} [x_1^2 + 1, x_1^2] = [1,1],$$

$$f^{**}([-1,1],[1,-1]) = \bigwedge_{x_2 \in [-1,1]'} \bigvee_{x_1 \in [-1,1]'} [x_1^2 + x_2^2, x_1^2 + x_2^2]$$

$$= \bigwedge_{x_2 \in [-1,1]'} [x_2^2, x_2^2 + 1] = [1,1]$$

and

$$\text{SDP}(f, X'_p, X'_i) = \text{SDP}(f, [-1,1]', [-1,1]') = \{(0,1), (0,-1)\}$$
$$\text{SDV}(f, X'_p, X'_i) = \text{SDV}(f, [-1,1]', [-1,1]') = 1$$
$$\text{SDP}(f, X'_i, X'_p) = \text{SDP}(f, [-1,1]', [-1,1]') = \{(-1,0), (1,0)\}$$
$$\text{SDV}(f, X'_i, X'_p) = \text{SDV}(f, [-1,1]', [-1,1]') = 1.$$

3.3 Semantic Theorems

The values of the extensions f^* or f^{**} may not yield, without further thought, much clear meaning about the values of the real f on its domain. Two key theorems reverse this misimpression, uncovering completely the meaning of the interval results f^* and f^{**} and characterizing them as the key referents for the semantic interval extensions previously defined in logical terms.

Theorem 3.3.1 (*-semantic theorem) *Given a continuous real function $f : \mathbb{R}^k \to \mathbb{R}$ and a modal vector $A \in I^*(\mathbb{R}^k)$, whenever $F(A) \in I^*(\mathbb{R})$ exists, we have that the following are equivalent propositions:*

a) $f^*(A) \subseteq F(A)$,
b) $(\forall X' \in I(\mathbb{R}^k)) ((x \in X') \in \text{Pred}^*(A) \Rightarrow (z \in f(X')) \in \text{Pred}^*(F(A)))$,
c) $(\forall a_p \in A'_p) Q(z, F(A)) (\exists a_i \in A'_i) z = f(a_p, a_i)$

Proof If $A_1, \ldots, A_p$ are the proper components of A and $A_{p+1}, \ldots, A_k$ the improper ones, then

$$\text{Impr}(f(a_p, A'_i)) = \text{Dual}(f^*(a_p, A'_i))$$

$$= \text{Dual}(\bigvee_{a_{p+1} \in A'_{p+1}} \cdots \bigvee_{a_k \in A'_k} [f(a_1, \ldots, a_p, a_{p+1}, \ldots, a_k),$$

$$f(a_1, \ldots, a_p, a_{p+1}, \ldots, a_k)])$$

$$= \bigwedge_{a_{p+1} \in A'_{p+1}} \cdots \bigwedge_{a_k \in A'_k} [f(a_1, \ldots, a_p, a_{p+1}, \ldots, a_k),$$
$$f(a_1, \ldots, a_p, a_{p+1}, \ldots, a_k)]$$
$$= f^*(\boldsymbol{a}_p, \boldsymbol{A}_i).$$

Therefore, (a) implies (b):

$(\forall \boldsymbol{X}' \in I(\mathbb{R}^k)) (\boldsymbol{x} \in \boldsymbol{X}') \in \text{Pred}^*(\boldsymbol{A})$

$\Leftrightarrow \text{Impr}(\boldsymbol{X}') \subseteq \boldsymbol{A}$

$\Leftrightarrow (\exists \boldsymbol{a}_p \in \boldsymbol{A}'_p) (\boldsymbol{a}_p \in \boldsymbol{X}'_1, \boldsymbol{X}'_2 \supseteq \boldsymbol{A}'_i)$

//$((\boldsymbol{X}_1, \boldsymbol{X}_2)$ is the components' splitting corresponding to $(\boldsymbol{A}_p, \boldsymbol{A}_i)$.

$\Rightarrow (\exists \boldsymbol{a}_p \in \boldsymbol{A}'_p) f(\boldsymbol{X}'_1, \boldsymbol{X}'_2) \supseteq f(\boldsymbol{a}_p, \boldsymbol{A}'_i)$

//$f(\boldsymbol{X}'_1, \boldsymbol{X}'_2)$ or $f(\boldsymbol{X}')$ designates the united extension of f on $\boldsymbol{X}'$.

$\Leftrightarrow (\exists \boldsymbol{a}_p \in \boldsymbol{A}'_p) \text{Impr}(f(\boldsymbol{X}')) \subseteq f^*(\boldsymbol{a}_p, \boldsymbol{A}_i)$

$\Rightarrow \text{Impr}(f(\boldsymbol{X}')) \subseteq f^*(\boldsymbol{A}_p, \boldsymbol{A}_i)$

$\Rightarrow \text{Impr}(f(\boldsymbol{X}')) \subseteq F(\boldsymbol{A})$

//see the hypothesis a).

$\Leftrightarrow (z \in f(\boldsymbol{X}')) \in \text{Pred}^*(F(\boldsymbol{A}))).$

(b) implies (a): Let $\boldsymbol{a}_p$ be any point of $\boldsymbol{A}'_p$ and $\boldsymbol{X}'$ the interval $(\boldsymbol{a}_p, \boldsymbol{A}'_i)$,

$(\forall \boldsymbol{a}_p \in \boldsymbol{A}'_p) (\boldsymbol{x} \in (\boldsymbol{a}_p, \boldsymbol{A}'_i) \in \text{Pred}^*(\boldsymbol{A}))$

$\Rightarrow (\forall \boldsymbol{a}_p \in \boldsymbol{A}'_p) (z \in f(\boldsymbol{a}_p, \boldsymbol{A}'_i)) \in \text{Pred}^*(F(\boldsymbol{A}))$

//Particularization of the hypothesis b).

$\Leftrightarrow (\forall \boldsymbol{a}_p \in \boldsymbol{A}'_p) \text{Impr}(f(\boldsymbol{a}_p, \boldsymbol{A}'_i)) \subseteq F(\boldsymbol{A})$

$\Leftrightarrow (\forall \boldsymbol{a}_p \in \boldsymbol{A}'_p) f^*(\boldsymbol{a}_p, \boldsymbol{A}_i) \subseteq F(\boldsymbol{A})$

$\Leftrightarrow f^*(\boldsymbol{A}_p, \boldsymbol{A}_i) \subseteq F(\boldsymbol{A}).$

(a) is equivalent to (c):

$f^*(\boldsymbol{A}) \subseteq F(\boldsymbol{A})$

$\Leftrightarrow (\forall \boldsymbol{a}_p \in \boldsymbol{A}'_p) f^*(\boldsymbol{a}_p, \boldsymbol{A}_i) \subseteq F(\boldsymbol{A})$

$\Leftrightarrow (\forall \boldsymbol{a}_p \in \boldsymbol{A}'_p) (z \in f(\boldsymbol{a}_p, \boldsymbol{A}'_i)) \in \text{Pred}^*(F(\boldsymbol{A}))$

3.3 Semantic Theorems

$$\Leftrightarrow (\forall a_p \in A'_p) \, Q(z, F(A)) \, z \in f(a_p, A'_i)$$
$$\Leftrightarrow (\forall a_p \in A'_p) \, Q(z, F(A)) \, (\exists a_i \in A'_i) \, z = f(a_p, a_i).$$

∎

Remark 3.3.1 The *-semantic theorem allows interpreting universal intervals as "regulating or feedback ranges", and existential intervals as "fluctuation or autonomous ranges" for the system consisting of the interval data and result $(A_p, A_i, F(A))$, and the analytical connection $z = f(a_p, a_i)$.

Example 3.3.1 For the continuous real function $f(x, y) = x + y$, from the definition of f^*,

$$f^*([x_1, x_2], [y_1, y_2]) = \underset{(x,[x_1,x_2])}{\Omega} \underset{(y,[y_1,y_2])}{\Omega} [x + y, x + y] = [x_1 + y_1, x_2 + y_2].$$

as will be proved in Sect. 5.3.1. For $X = [1, 2]$ and $Y = [2, 3]$ and since the result is $Z = [3, 5]$, we may write $[1, 2] + [2, 3] = [3, 5]$, with the meaning

$$(\forall x \in [1, 2]') \, (\forall y \in [2, 3]') \, (\exists z \in [3, 5]') \, x + y = z.$$

Similarly, for $X = [1, 2]$ and $Y = [4, 1]$ the result is $Z = [1, 2] + [4, 1] = [5, 3]$ which means, in this case,

$$(\forall x \in [1, 2]') \, (\forall z \in [3, 5]') \, (\exists y \in [1, 4]') \, x + y = z.$$

And so on, for $X = [2, 1]$ and $Y = [1, 4]$ the result is $Z = [2, 1] + [1, 4] = [3, 5]$, which means

$$(\forall y \in [1, 4]') \, (\exists x \in [1, 2]') \, (\exists z \in [3, 5]') \, x + y = z;$$

for $X = [2, 1]$ and $Y = [3, 2]$ the result is $Z = [2, 1] + [3, 2] = [5, 3]$ with the interpretation

$$(\forall z \in [3, 5]') \, (\exists x \in [1, 2]') \, (\exists y \in [2, 3]') \, x + y = z.$$

Moreover, these interval statements (or interpretations of the modal functional relation $f^*(X, Y) = Z$) are robust to "modal outer rounding" of the result, as is shown for example in the replacement of the Z-value $[3, 5]$ by $[2.9, 5.1] \supseteq [3, 5]$, or of $[5, 3]$ by $[4.9, 3.1] \supseteq [5, 3]$, the latter being equivalent to a set-theoretical inner rounding of $[3, 5]'$.

Example 3.3.2 Let us apply the result about the function $f(x, y) = x + y$ to a naturalistic context. Suppose we have two cable reels of lengths $a = 10$ and $b = 20$ units. When connected, they can cover an overall length $c = 30$. This most elementary situation can be expressed for all that computationally matters by the algebraic expression $a + b = c$.

Consider the parallel but more realistic interval-situation where the first reel of cable has a length a known only to lie in a range bounded by the interval $A' = [10, 20]'$, i.e., $a \in [10, 20]'$; about the second reel we know that $b \in B' = [10, 25]'$. Let us consider the connection between both reels and let us apply the *-Semantic Theorem 3.3.1 for f^* restricted to the function of addition $f(a, b) = a + b$.

Case 1: $[10, 20] + [25, 10] = [35, 30]$ means

$$(\forall a \in [10, 20]') \, (\forall c \in [30, 35]') \, (\exists b \in [10, 25]') \, c = a + b$$

that is, a determined length of a wider regulating interval $[25, 10]$ can be selected to get some, in principle, unknown but determinable length c lying within the improper interval $C = [35, 30]$, in spite of the value a belonging to the proper operand $A = [10, 20]$ being understood as coming out of some general random selection process.

Case 2: $[10, 20] + [10, 10] = [20, 30]$,
$\phantom{\text{Case 2: }}[10, 20] + [17, 17] = [27, 37]$,
$\phantom{\text{Case 2: }}[10, 20] + [25, 25] = [35, 45]$,

the variable b taking fixed values 10, 17 or 25 in the interval-set $B' = [10, 25]'$, the indeterminacy of $A = [10, 20]$ is carried to the interval C by the relation $c = a + b$ so that the value of c will range randomly and in parallel with a on one of the intervals $[20, 30]$, $[27, 37]$ or $[35, 45]$. The quantified statement (for example for the first equality) is, if b is bounded to the only value of the point interval $[10, 10]$,

$$(\forall a \in [10, 20]') \, (\exists c \in [20, 30]') \, c = a + 10.$$

Case 3: $[10, 20] + [10, 25] = [20, 45]$ means

$$(\forall a \in [10, 20]') \, (\forall b \in [10, 25]') \, (\exists c \in [20, 45]') \, c = a + b,$$

so that c will show the joint full indeterminacy coming from a and b.

Case 4: $[20, 10] + [25, 10] = [45, 20]$ will be interpreted by

$$(\forall c \in [20, 45]') \, (\exists a \in [10, 20]') \, (\exists b \in [10, 25]') \, c = a + b.$$

Case 5: $[10, 20] + [20, 15] = [30, 35]$ means

$$(\forall a \in [10, 20]') \, (\exists c \in [30, 35]') \, (\exists b \in [15, 20]') \, c = a + b$$

that is, with the same autonomous interval $A = [10, 20]$ and a narrower regulating interval $B = [20, 15]$, a determined length of $b \in B' = [15, 20]'$ should be selected (a regulation operation) just to get some length c lying within the domain of the proper interval $C = [30, 35]'$.

3.3 Semantic Theorems

A dual feedback semantics for proper and improper modal intervals is established by the following Dual Semantic theorem.

Theorem 3.3.2 (**-semantic theorem) *Given a continuous real functions $f : \mathbb{R}^k \to \mathbb{R}$ and a modal vector $A \in I^*(\mathbb{R}^k)$, whenever $F(A) \in I^*(\mathbb{R})$ exists, we have that the following are equivalent propositions:*

a) $f^{**}(A) \supseteq F(A)$,
b) $(\forall X' \in I(\mathbb{R}^k))\, ((x \notin X') \in \text{Copred}^*(A) \Rightarrow (z \notin f(X')) \in \text{Copred}^*(F(A)))$,
c) $(\forall a_i \in A'_i)\, Q(z, \text{Dual}(F(A)))\, (\exists a_p \in A'_p)\, z = f(a_p, a_i)$.

Proof From the definitions of $f^*(X)$ and $f^{**}(X)$ we obtain

$$\text{Dual}(f^*(X)) = f^{**}(\text{Dual}(X)).$$

Applying Theorem 3.3.1 to $f^*(\text{Dual}(A)) \subseteq \text{Dual}(F(A))$, Theorem 3.3.2 follows. ∎

Example 3.3.3 For the function $f(x, y) = xy$ and $X = [-1, 2]$, $Y = [5, 3]$ the values of f^* and f^{**} are $f^*([-1, 2], [5, 3]) = f^{**}([-1, 2], [5, 3]) = [-3, 6]$ (see Chap. 5). Then, in accordance with both semantic theorems,

$$(\forall x \in [-1, 2]')\, (\exists z \in [-3, 6]')\, (\exists y \in [3, 5]')\, z = xy,$$
$$(\forall y \in [3, 5]')\, (\forall z \in [-3, 6]')\, (\exists x \in [-1, 2]')\, z = xy.$$

Remark 3.3.2 The Semantic Theorems show that:

- The semantic decision to apply the *-semantic theorem or the **-semantic theorem is made when one of the modal roundings, outer or inner, is selected.
- The functions $f^*(X)$ and $f^{**}(X)$ are semantically optimal for each semantic theorem.
- The effective computation of a modal extension $F(A)$ is not indicated by these two theorems which give only modal bounds, $f^*(X)$ or $f^{**}(X)$ according to the chosen rounding, to any modal extension $F(A)$.

Example 3.3.4 The solution of the equation $[3, 7] * X = [4, 6]$ is (see Chap. 5)

$$X = [4, 6]/\text{Dual}[3, 7] = [4, 6]/[7, 3] = [4/3, 6/7].$$

As an inner rounding of X is $[1.334, 0.857]$, then

$$[3, 7] * [1.334, 0.857] \subseteq [4, 6]$$

and the *-semantic theorem gives a meaning to this result

$$(\forall a \in [3, 7]')(\exists b \in [4, 6]')(\exists x \in [1.334, 0.857]')\, ax = b.$$

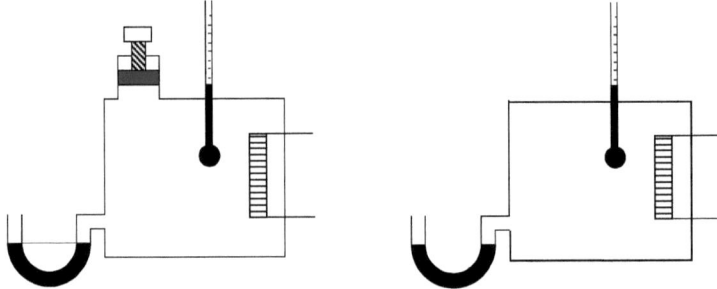

Fig. 3.5 Gas containers

The interpretation of the proper and improper intervals provided by the Semantic Theorems opens a wide field of technical applications for the theory of modal intervals, as the following suggestive example illustrates.

Example 3.3.5 Let us consider the volume v of a gas container, of which the temperature t takes values inside certain intervals. Assuming the validity of the equation

$$v = kt/p,$$

where k is the ideal gas constant, let us suppose the following intervals of variation

$$k \in K' = [0.00366, 0.00367]', \ t \in T' = [263, 283]', \ p \in P' = [0.99, 1.01]'.$$

The problem is to determine the volume, keeping the pressure p within certain pre-established bounds, that is

$$(\forall k \in [0.00366, 0.00367]') \ (\forall t \in [263, 283]') \ Q(v, V) \ (\exists p \in [0.99, 1.01]') \ v = kt/p.$$

This semantic is equivalent to the interval inclusion

$$v^*(K, T, P) \subseteq V$$

with K and T proper intervals and P an improper one. Computing v^*, the result is

$$v^*([0.00366, 0.00367], [263, 283], [1.01, 0.99]) = [0.97\ldots, 1.02\ldots] \subseteq [0.97, 1.03]$$

which means that for every value of k and t there exists a volume v between 0.97 and 1.03, depending on k and t, which makes the pressure within the desired limits. The container is to be built with a feedback valve to allow its volume to be regulated within the computed bounds, to keep the stated conditions, as the left graph of Fig. 3.5 illustrates.

3.4 Syntactic Functions

Allowing the pressure to vary within the domain $P' = [0.9, 1.1]'$, the resulting interval for the volume is

$$v^*([0.00366, 0.00367], [263, 283], [1.1, 0.9]) = [1.06\ldots, 0.94\ldots] \subseteq [1.06, 0.95].$$

In accordance with the *-semantic theorem, this result means that for every k and t and every volume v between 0.95 and 1.06, the pressure falls within the limits. The container can be built without any feedback valve because, for any volume between the bounds, the pressure is inside the stated conditions, as the right graph of Fig. 3.5 illustrates.

3.4 Syntactic Functions

The two applications of the meet–join operators to a continuous function f from $\mathbb{R}^k$ to $\mathbb{R}$, define the two semantic extensions f^* and f^{**}. From now on only real continuous functions with syntactic tree will be considered, so the existence of a syntactic tree for any function f is assumed and not explicitly repeated.

Looking at a syntactic tree of the continuous real function f, where the nodes are the operators, the leaves are the variables, and the branches define the domain of each operator, f can also be operationally extended to a syntactical function fR from $I^*(\mathbb{R}^k)$ to $I^*(\mathbb{R})$, by using the computational program implicitly defined by the syntactic tree of the expression defining the function.

3.4.1 Syntactic Extensions

Definition 3.4.1 (Modal syntactic *-extension) The function fR^* from $I^*(\mathbb{R}^k)$ to $I^*(\mathbb{R})$, called the *Modal syntactic *-extension* of f, is defined by the computational program indicated by a syntactic tree of the real function f from $\mathbb{R}^k$ to $\mathbb{R}$, when the real operators are transformed into their *-semantic extensions.

Definition 3.4.2 (Modal syntactic **-extension) The function fR^{**}, called *the Modal syntactic **-extension* of f, is defined similarly to fR^*, but with the operators transformed into their **-semantic extensions.

Example 3.4.1 For the continuous real function $f(x_1, x_2) = x_1 x_2 + g(x_1, x_2)$, with the operator $g(x_1, x_2) = (x_1 + x_2)^2$, syntactic trees of f, fR^* and fR^{**} are

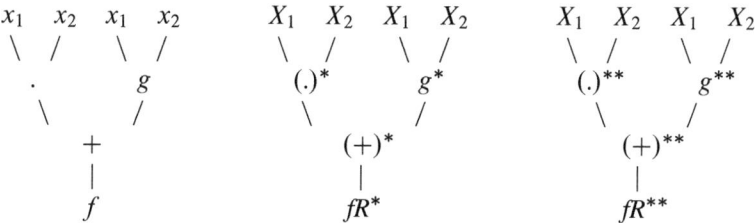

If $X_1 = [-1, 1]$, $X_2 = [1, -1]$, fR^* and fR^{**} are computed as follows. For the $x_1 x_2$ operator:

$$*\text{-extension}: \bigvee_{x_1 \in [-1,1]'} \bigwedge_{x_2 \in [-1,1]'} [x_1 x_2, x_1 x_2] = [0, 0],$$

$$**\text{-extension}: \bigwedge_{x_2 \in [-1,1]'} \bigvee_{x_1 \in [-1,1]'} [x_1 x_2, x_1 x_2] = [0, 0].$$

For the $g(x_1, x_2)$ operator:

$$*\text{-extension}: \bigvee_{x_1 \in [-1,1]'} \bigwedge_{x_2 \in [-1,1]'} [(x_1 + x_2)^2, (x_1 + x_2)^2] = [1, 0],$$

$$**\text{-extension}: \bigwedge_{x_2 \in [-1,1]'} \bigvee_{x_1 \in [-1,1]'} [(x_1 + x_2)^2, (x_1 + x_2)^2] = [0, 1].$$

Therefore,

$$fR^*([-1, 1], [1, -1]) = \bigvee_{y_1 \in [0,0]'} \bigwedge_{y_2 \in [0,1]'} [y_1 + y_2, y_1 + y_2] = [1, 0],$$

$$fR^{**}([-1, 1], [1, -1]) = \bigvee_{y_1 \in [0,0]'} \bigvee_{y_2 \in [0,1]'} [y_1 + y_2, y_1 + y_2] = [0, 1].$$

Lemma 3.4.1 (Duality relation)

$$\text{Dual}(fR^*(X)) = fR^{**}(\text{Dual}(X)).$$

Proof If Ψ is the computational program indicated by a syntactic tree of f and w_i are the operators, then

$$\text{Dual}(fR^*(X)) = \text{Dual}(\Psi(w_i^*, X)) = \Psi(w_i^{**}, \text{Dual}(X)) = fR^{**}(\text{Dual}(X)).$$

∎

Definition 3.4.3 (Modal syntactic operator) A modal syntactic operator is any continuous function f from $\mathbb{R}^k$ to $\mathbb{R}$ that is *JM*-commutable.

Definition 3.4.4 (Modal syntactic function) A modal syntactic function fR is a function defined similarly to fR^* or fR^{**}, but with all of its operators being *JM*-commutable, that is, modal syntactic operators.

This definition will extend considerably the framework of the four rational operators of real analysis $\{+, -, *, /\}$, since the constructive aspect which supports the four rational operators loses its interest within the numerical context where, obviously, all the operators are calculated with controlled deviations up to a certain degree.

A modal syntactic function will be, consequently, any function with the form of a continuous real function in which all its operators are modal syntactic operators, and where the functional correspondence *Arguments* → *Values* is obtained by the computational program indicated by the syntactic tree of the function.

3.4 Syntactic Functions

Later results will indicate the considerable repertoire of modal syntactic functions: $\mathrm{Abs}(x)$, $\mathrm{power}(x,n)$, $\log_a(x)$, $\mathrm{root}(x,n)$ are modal syntactic operators, continuous and unary, and consequently possible nodes of the syntactic tree of a modal syntactic function.

At this stage, it is necessary to fix some notations to make easier the discussion of the problems presented by modal syntactic functions:

1. For a function f the symbols f^* and f^{**} indicate the semantic functions $f^* : I^*(\mathbb{R}^k) \to I^*(\mathbb{R})$ and $f^{**} : I^*(\mathbb{R}^k) \to I^*(\mathbb{R})$ defined by the correspondences $X \to f^*(X)$ and $X \to f^{**}(X)$, which do not depend on any syntactic tree of f.
2. $fR(x)$ and $fR(X)$ indicate the functions $fR : \mathbb{R}^k \to \mathbb{R}$ and $fR : I^*(\mathbb{R}^k) \to I^*(\mathbb{R})$ established by the computational program indicated by the syntactic tree with which these functions are indicated, where $fR(X)$ exists when the operators of the syntactic tree of f are modal syntactic and the computation of $fR(\mathrm{Prop}(X))$ does not include any division by intervals containing zero.
3. Contrary to the equality $f(x) = fR(x)$ on $\mathbb{R}$, not only do the equalities between $f^*(X)$, $f^{**}(X)$ and $fR(X)$ not hold in general, but the forms of functions which are equivalent on $\mathbb{R}$, say f_1 and f_2, in the sense $f_1 R(x) = f_2 R(x)$, do not necessarily maintain this same equality on $I^*(\mathbb{R})$.

Example 3.4.2 The expressions

$$f_1(x) = \frac{1}{1-x} + \frac{1}{1+x}$$

$$f_2(x) = \frac{2}{1-x^2}$$

define the same continuous real function, for $x > 1$. Nevertheless, their syntactic extensions to the interval $X = [2, 3]$ are

$$f_1 R([2,3]) = \frac{1}{1-[2,3]} + \frac{1}{1+[2,3]} = [-3/4, -1/6]$$

$$f_2 R([2,3]) = \frac{2}{1-[2,3]^2} = [-2/3, -1/4],$$

which are different.

Theorem 3.4.1 (Inclusivity of the modal syntactic functions) *The modal syntactic extensions fR^* and fR^{**} (fR if it is the case) of a continuous real function f from $\mathbb{R}^k$ to $\mathbb{R}$, are inclusion-isotonic.*

Proof If $X \subseteq Y$, Ψ is a syntactic tree of f and w_i are its operators, for any w_i the implication $X \subseteq Y \Rightarrow w_i(X) \subseteq w_i(Y)$ holds, and therefore

$$fR^*(X) = \Psi(w_i^*, X) \subseteq \Psi(w_i^*, Y) = fR^*(Y).$$

The same reasoning holds for fR^{**}. ∎

3.4.2 Modal Syntactic Operators

The definition of modal syntactic operator extends the list of the real operators. Now, let us identify the most important classes of modal syntactic operators which will be the best interval operators for the syntactic tree of a modal syntactic extension.

Theorem 3.4.2 (One-variable operators) *Every one-variable continuous function is JM-commutable, and therefore a modal syntactic operator.*

Proof There is no commutation problem between the meet and join operations. ∎

Remark 3.4.1 The interesting operators are the monotonic operators or other easily programable ones like $abs(x)$, $power(x, n)$, $log(x)$ or $root(x, n)$ described in Chap. 5.

For the JM-commutativity of operators with two or more variables, the following definitions play an important role.

Definition 3.4.5 (Uniform monotonicity) A k-variable continuous function $f(x, y)$ is x-uniformly monotonic on a domain $(X', Y') \subseteq (\mathbb{R}, \mathbb{R}^{k-1})$, if it is monotonic for x on X', and it is unary or keeps the same sense of monotonicity for all the values y on Y'.

Definition 3.4.6 (Partial monotonicity) A k-variable continuous function $f(x, y)$ is x-partially monotonic on $(X', Y') \subseteq (\mathbb{R}, \mathbb{R}^{k-1})$ if it may increase with x for some y-values, and may decrease with x for the rest of the y-values on the domain Y'.

Example 3.4.3 The functions xy and x/y are partially monotonic. Uniformly monotonic functions, which are monotonic increasing or monotonic decreasing for each component, include, for example $x + y$, $x - y$, $\min(x, y)$ $\max(x, y)$.

Theorem 3.4.3 (Two-variable operators) *Every two-variable continuous function $f(x, y)$ which is (x, y)-partially monotonic on a domain (X', Y'), is JM-commutable for the corresponding interval arguments (X, Y).*

Proof If X and Y share the same modality, $f(x, y)$ is bounded by its values in the vertex of the domain (X', Y'). Otherwise, the possible cases, depending on the sign of the X, Y-bounds, are characterized by the behaviour of $f(x, y)$ on the borders of the two-dimensional interval domain, where the existence of two saddle-points is, case-by-case, easily assured by means of the continuity of f. These points are in the set of vertices of (X', Y'), or in some point of this domain. ∎

Figure 3.6 illustrates a case of this reasoning, showing an interval domain, arrows indicating the sense of monotonicity of some two-variables operator, and the corresponding saddle point, which coincide with the origin for these senses of monotonicity.

Remark 3.4.2 This is the case of the operators $x + y$, $x - y$, $x * y$ or x/y, described in detail in Chap. 5, and the interesting operators $\max(x, y)$ or $\min(x, y)$.

3.4 Syntactic Functions

Fig. 3.6 Saddle point for a two-variables operator

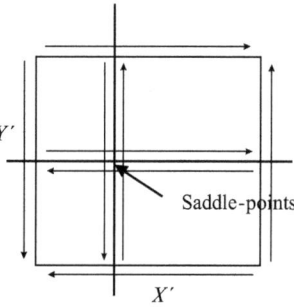

Fig. 3.7 Saddle points for addition

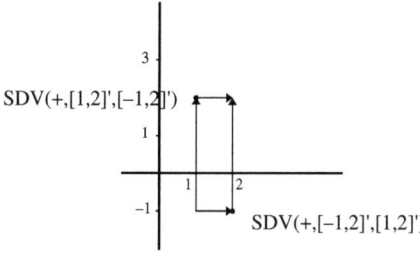

Example 3.4.4 For the addition of a proper interval and an improper one,

$$[1, 2] + [2, -1] = [SDV(+, [1, 2]', [-1, 2,]'), SDV(+, [-1, 2]', [1, 2]')]$$
$$= [\min_{x \in [1,2]'} \max_{y \in [-1,2]'} (x + y), \min_{y \in [-1,2]'} \max_{x \in [1,2]'} (x + y)] = [3, 1]$$

In Fig. 3.7 the sense of monotonicity of the sum in this interval and the saddle points are represented, where the arrows indicate the sense of monotonicity in the rectangular domain.

If both intervals are proper,

$$[1, 2] + [-1, 3] = [SDV(+, ([1, 2]', [-1, 3,]'), \emptyset), SDV(+, \emptyset, ([-1, 3]', [1, 2]'))]$$
$$= [\min_{x \in [1,2]'} \min_{y \in [-1,3]'} x + y, \max_{y \in [1,2]'} \max_{x \in [-1,3]'} x + y]$$
$$= [0, 5]$$

If both intervals are improper,

$$[1, -1] + [1, -2] = [SDV(+, \emptyset, ([-1, 1]', [-2, 1]')), SDV(+, ([-1, 1]', [-2, 1]'), \emptyset)]$$
$$= [\max_{x \in [-1,1]'} \max_{y \in [-2,1]'} x + y, \min_{y \in [-1,1]'} \min_{x \in [-2,1]'} x + y]$$
$$= [2, -3]$$

Fig. 3.8 Saddle points for the product function

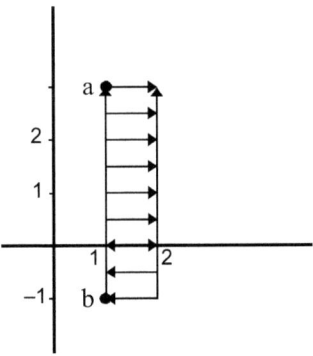

For the product, the main problem appears when the domain intersects the axis. For $[1, 2] * [3, -1]$,

$$[1, 2] * [3, -1] = [\text{SDV}(\cdot, [1, 2], [3, -1]), \text{SDV}(\cdot, [3, -1], [1, 2])]$$
$$= [\min_{x \in [1,2]'} \max_{y \in [-1,3]'} xy, \min_{y \in [-1,3]'} \max_{x \in [1,2]'} xy]$$
$$= [3, -1]$$

Figure 3.8 depicts the sense of monotonicity of the product in this interval and the saddle points are represented.

Remark 3.4.3 Partial monotonicity does not guarantee the *JM*-commutativity for more than two variables, as is seen in the case of the function $f(x, y, z) = x(y + z)$.

Theorem 3.4.4 (Uniform monotonicity) *Every uniformly monotonic continuous function $f(x, y)$, with $(x, y) \in \mathbb{R}^k$, x-monotonic increasing and y-monotonic decreasing on (X', Y'), is JM-commutable for (X, Y) and*

$$fR(X, Y) = f^*(X, Y) = f^{**}(X, Y)$$
$$= [f(\text{Inf}(X), \text{Sup}(Y)), f(\text{Sup}(X), \text{Inf}(Y))],$$

where

$$\text{Inf}(X) = (\text{Inf}(X_1), \ldots, \text{Inf}(X_m))$$
$$\text{Sup}(X) = (\text{Sup}(X_1), \ldots, \text{Sup}(X_m)),$$

and so on for Y.

Proof As f is uniformly monotonic,

$$X - \text{increasing} : \begin{cases} X_j \text{ proper} & \Rightarrow \text{ the minimum of } f \text{ is in Inf}(X_j) \\ X_j \text{ improper} & \Rightarrow \text{ the maximum of } f \text{ is in Inf}(X_j) \end{cases}$$

3.4 Syntactic Functions

$$Y - \text{decreasing}: \begin{cases} Y_j \text{ proper} \Rightarrow \text{ the minimum of } f \text{ is in Sup}(Y_j) \\ Y_j \text{ improper} \Rightarrow \text{ the maximum of } f \text{ is in Sup}(Y_j). \end{cases}$$

Therefore

$$\min_{(x_p, y_p) \in (X_p, Y_p)'} \max_{(x_i, y_i) \in (X_i, Y_i)'} f(\boldsymbol{x}_p, \boldsymbol{x}_i, \boldsymbol{y}_p, \boldsymbol{y}_i) = f(\text{Inf}(X), \text{Sup}(Y))$$

$$= \max_{(x_i, y_i) \in (X_i, Y_i)'} \min_{(x_p, y_p) \in (X_p, Y_p)'} f(\boldsymbol{x}_p, \boldsymbol{x}_i, \boldsymbol{y}_p, \boldsymbol{y}_i)$$

and, analogously,

$$\min_{(x_i, y_i) \in (X_i, Y_i)'} \max_{(x_p, y_p) \in (X_p, Y_p)'} f(\boldsymbol{x}_p, \boldsymbol{x}_i, \boldsymbol{y}_p, \boldsymbol{y}_i) = f(\text{Sup}(X), \text{Inf}(Y))$$

$$= \max_{(x_p, y_p) \in (X_p, Y_p)'} \min_{(x_i, y_i) \in (X_i, Y_i)'} f(\boldsymbol{x}_p, \boldsymbol{x}_i, \boldsymbol{y}_p, \boldsymbol{y}_i).$$

So, $f^*(X, Y) - f^{**}(X, Y)$ and

$$f^*(X, Y) = [f(\text{Inf}(X), \text{Sup}(Y)), f(\text{Sup}(X), \text{Inf}(Y))].$$

■

Example 3.4.5 The function $f(x, y, z) = (x - y)/(z + y)$ in the domain $X = [0, 2]$, $Y = [4, 3]$, $Z = [2, 1]$ is x-monotonic increasing, y-monotonic decreasing, and z-monotonic increasing, as can be shown from the constancy of the signs of $(x - y)$ and of $(z + y)$ on the particular interval domain involved in this example. Then

$$fR(X, Y, Z) = f^{**}(X, Y, Z) = f^*(X, Y, Z)$$
$$= [f(\text{Inf}(X), \text{Sup}(Y), \text{Inf}(Z)), f(\text{Sup}(X), \text{Inf}(Y), \text{Sup}(Z))]$$
$$= [(0 - 3)/(2 + 3), (2 - 4)/(1 + 4)]$$
$$= [-3/5, -2/5],$$

Example 3.4.6 A specially interesting example of continuous uniformly monotonic increasing operator is the function "limited identity", that is,

LID : $\mathbb{R}^3 \longrightarrow \mathbb{R}$

$$(t, x, y) \longrightarrow \text{LID}(t, x, y) = \begin{cases} \min(x, y) & \text{if } t \leq \min(x, y) \\ t & \text{if } \min(x, y) \leq t \leq \max(x, y) \\ \max(x, y) & \text{if } \max(x, y) \leq t \end{cases}$$

This operator is *JM*-commutable on its domain because it is a uniformly monotonic continuous function. Therefore

$$\text{LID}^{**}(T, X, Y) = \text{LID}^*(T, X, Y)$$
$$= [\text{LID}(\text{Inf}(T), \text{Inf}(X), \text{Inf}(Y)), \text{LID}(\text{Sup}(T), \text{Sup}(X), \text{Sup}(Y))]$$

If T is the interval $[-\infty, +\infty]$, then

$$\text{LID}^{**}([-\infty, +\infty], X, Y) = \text{LID}^*([-\infty, +\infty], X, Y)$$
$$= [\min(\text{Inf}(X), \text{Inf}(Y)), \max(\text{Sup}(X), \text{Sup}(Y))]$$
$$= X \vee Y$$

If T is the interval $[+\infty, -\infty]$, then

$$\text{LID}^{**}([+\infty, -\infty], X, Y) = \text{LID}^*([+\infty, -\infty], X, Y)$$
$$= [\max(\text{Inf}(X), \text{Inf}(Y)), \min(\text{Sup}(X), \text{Sup}(Y))]$$
$$= X \wedge Y$$

After admitting these T-arguments we see that LID incorporates, among the modal syntactic operators, the $\subseteq$-lattice operators "meet" and "join".

Actually, the enlargement of the set of modal syntactic operators from the classic one $(+, -, *, /)$, is quite important for applications, for example for control and approximation problems, mainly given the limitations imposed by the easy loss of information originating in multi-incidence and not-optimal syntactic trees of real expressions. In the computational context of $I^*(\mathbb{R})$, the classical rational operators $(+, -, *, /)$ have obviously no particular privilege over other programmable ones and holding the essential properties of continuity and *JM*-commutativity, since all of them are to be computed through the use of a suitably approximated arithmetic. Coming back to the "meet" and "join" operators, their use will only require an additional remark: the application of the semantic theorems will demand the consideration of the implicit t-variables they introduce.

Example 3.4.7 The operator

$$\text{CLIP}(t, x, y) = t - \text{LID}(t, x, y)$$

is uniformly t-monotonic increasing and (x, y)-monotonic decreasing and its program is

$$\text{CLIP}(t, x, y) = \begin{cases} t - \min(x, y) & \text{if } t \leq \min(x, y) \\ 0 & \text{if } \min(x, y) \leq t \leq \max(x, y) \\ t - \max(x, y) & \text{if } \max(x, y) \leq t \end{cases}$$

following a similar reasoning.

3.4 Syntactic Functions

Theorem 3.4.5 (Partial uniform monotonicity) *If $f(x, y)$ is a x-uniform monotonic continuous function in the domain (X', Y') and $X = (U, V)$, $Y = (Y_p, Y_i)$ is the component-splitting of X and Y into their proper and improper components, then*

$$f^*(X, Y) = (f^*(\mathrm{Inf}(U), \mathrm{Inf}(V), Y) \vee f^*(\mathrm{Sup}(U), \mathrm{Inf}(V), Y))$$
$$\wedge$$
$$(f^*(\mathrm{Inf}(U), \mathrm{Sup}(V), Y) \vee f^*(\mathrm{Sup}(U), \mathrm{Sup}(V), Y)). \qquad (3.1)$$

Proof Taking into account Definition 3.2.1 of *-semantic function, the associativity of the meet and join operators, and the x-monotonicity of f:

1) If $V = \emptyset$, $X = U$ is uni-modal proper and

$$f^*(X, Y) = \bigvee_{u \in U'(y, Y)} \Omega \, f(u, y) = \bigvee_{u \in U'} f^*(u, Y)$$
$$= f^*(\mathrm{Inf}(U), Y) \vee f^*(\mathrm{Sup}(U), Y).$$

2) If $U = \emptyset$, $X = V$ is uni-modal improper, as

$$f^*(X, Y) = [\min_{y_p \in Y'_p} \max_{y_i \in Y'_i} \max_{v \in V'} f(v, y_p, y_i), \max_{y_p \in Y'_p} \min_{y_i \in Y'_i} \min_{v \in V'} f(v, y_p, y_i)],$$

if f is v-monotonic increasing

$$f^*(X, Y) = [\min_{y_p \in Y'_p} \max_{y_i \in Y'_i} f(\mathrm{Inf}(V), y_p, y_i), \max_{y_p \in Y'_p} \min_{y_i \in Y'_i} f(\mathrm{Sup}(V), y_p, y_i)],$$

and if f is v-monotonic decreasing

$$f^*(X, Y) = [\min_{y_p \in Y'_p} \max_{y_i \in Y'_i} f(\mathrm{Sup}(V), y_p, y_i), \max_{y_p \in Y'_p} \min_{y_i \in Y'_i} f(\mathrm{Inf}(V), y_p, y_i)].$$

On the other hand, as

$$f^*(\mathrm{Inf}(V), Y) \wedge f^*(\mathrm{Sup}(V), Y)$$
$$= [\min_{y_p \in Y'_p} \max_{y_i \in Y'_i} f(\mathrm{Inf}(V), y_p, y_i), \max_{y_p \in Y'_p} \min_{y_i \in Y'_i} f(\mathrm{Inf}(V), y_p, y_i)]$$
$$\wedge$$
$$[\min_{y_p \in Y'_p} \max_{y_i \in Y'_i} f(\mathrm{Sup}(V), y_p, y_i), \max_{y_p \in Y'_p} \min_{y_i \in Y'_i} f(\mathrm{Sup}(V), y_p, y_i)],$$

if f is v-monotonic increasing

$$f^*(\text{Inf}(V), Y) \wedge f^*(\text{Sup}(V), Y)$$
$$= [\min_{y_p \in Y'_p} \max_{y_i \in Y'_i} f(\text{Inf}(V), y_p, y_i), \max_{y_p \in Y'_p} \min_{y_i \in Y'_i} f(\text{Sup}(V), y_p, y_i)]$$
$$= f^*(X, Y),$$

and if f is v-monotonic decreasing

$$f^*(\text{Inf}(V), Y) \wedge f^*(\text{Sup}(V), Y)$$
$$= [\min_{y_p \in Y'_p} \max_{y_i \in Y'_i} f(\text{Sup}(V), y_p, y_i), \max_{y_p \in Y'_p} \min_{y_i \in Y'_i} f(\text{Inf}(V), y_p, y_i)]$$
$$= f^*(X, Y).$$

3) If $X = (U, V)$ is multi-modal, applying successively (2) and (1), we obtain (3.1). ∎

Example 3.4.8 The function $f(x, y, z) = xy + 1/(x + y) + z - z^2$ in the domain $X = [0, 1]$, $Y = [2, 1]$, $Z = [4, 2]$ is x-monotonic increasing and z-monotonic decreasing. Its *-semantic extension

$$f^*(X, Y, Z) = \bigvee_{x \in [0,1]'} \bigwedge_{y \in [1,2]'} \bigwedge_{z \in [2,4]'} [xy + 1/(x+y) + z - z^2, xy + 1/(x+y) + z - z^2]$$

is not easily computable. But following Theorem 3.4.5, as

$$f^*(\text{Inf}(X), Y, \text{Inf}(Z)) = f^*([0, 0], [2, 1], [4, 4])$$
$$= \bigwedge_{y \in [1,2]'} [1/y - 12, 1/y - 12] = [-11, -11.5]$$

$$f^*(\text{Sup}(X), Y, \text{Inf}(Z)) = f^*([1, 1], [2, 1], [4, 4])$$
$$= \bigwedge_{y \in [1,2]'} [y + 1/(1 + y) - 12, y + 1/(1 + y) - 12]$$
$$= [-9.66\ldots, -10.5]$$

$$f^*(\text{Inf}(X), Y, \text{Sup}(Z)) = f^*([0, 0], [2, 1], [2, 2])$$
$$= \bigwedge_{y \in [1,2]'} [1/y - 2, 1/y - 2] = [-1, -1.5]$$

$$f^*(\text{Sup}(X), Y, \text{Sup}(Z)) = f^*([1, 1], [2, 1], [2, 2])$$
$$= \bigwedge_{y \in [1,2]'} [y + 1/(1 + y) - 2, y + 1/(1 + y) - 2]$$
$$= [0.33\ldots, -0.5].$$

3.4 Syntactic Functions

then

$$f^*(X,Y,Z) = \left\{ \begin{array}{c} ([-11,-11.5] \vee [-9.66\ldots,-10.5]) \\ \wedge \\ ([-1,-1.5] \vee [0.33\ldots,-0.5]) \end{array} \right\} = [-1,-10.5].$$

Theorem 3.4.6 (k-**Uniform monotonicity**) *Every continuous function $f(x, y)$, with $x \in \mathbb{R}$ and $y = (u, v) \in \mathbb{R}^{k-1}$, which is uniformly monotonic for the y-arguments on a domain (X', Y'), u-monotonic increasing and v-monotonic decreasing, is JM-commutable for the corresponding interval arguments (X, Y).*

Proof The continuity of $f(x, y_0)$ on X', for any $y_0 \in \mathbb{R}^{k-1}$, implies the existence on X' of the x-minimum and x-maximum of $f(x, y_0)$. This allows showing the existence of the saddle-points, which means its *JM*-commutativity.

Let us denote by (x, u, v) the split of (x, y), and by (x, u_m, v_M) the coordinates where $f(x, u, v)$ reaches a u-minimum and a v-maximum irrespective of the value for x. If

$$f(x_m, u_m, v_M) = \min_{x \in X'} f(x, u_m, v_M),$$

then $(x_m, u_m, v_M) \in \text{SDP}(f, (X', U'), V')$ because

$$(\forall x \in X')(\forall u \in U')(\forall v \in V')$$
$$(f(x_m, u_m, v) \leq f(x_m, u_m, v_M) \leq f(x, u_m, v_M) \leq f(x, u, v_M))$$

Let (x, u_M, v_m) be the coordinates where $f(x, u, v)$ reaches a u-maximum and a v-minimum irrespective of the value for x. If

$$f(x_M, u_M, v_m) = \max_{x \in X'} f(x, u_M, v_m),$$

then $(x_M, u_M, v_m) \in \text{SDP}(f, V', (X', U'))$ because

$$(\forall x \in X')(\forall u \in U')(\forall v \in V')$$
$$(f(x, u, v_m) \leq f(x, u_M, v_m) \leq f(x_M, u_M, v_m) \leq f(x_M, u_M, v)).$$

Therefore, following the proof of the previous theorem,

$$f^{**}(X, Y) = f^*(X, Y)$$

= if X proper then

$$[\min_{x \in X'} f(x, \text{Inf}(U), \text{Sup}(V)), \max_{x \in X'} f(x, \text{Sup}(U), \text{Inf}(V))]$$

if X improper then

$$[\max_{x \in X'} f(x, \text{Inf}(U), \text{Sup}(V)), \min_{x \in X'} f(x, \text{Sup}(U), \text{Inf}(V))].$$

∎

Example 3.4.9 The function $f(x, y, z) = (x - y)/(z + y)$ in the domains of $X = [0, 3]$, $Y = [2, 4]$, $Z = [3, 1]$ is x-monotonic increasing, y-monotonic decreasing and it is not monotonic in z. Then

$$f^*(X, Y, Z) = f^{**}(X, Y, Z) = fR(X, Y, Z)$$
$$= [\max_{z \in [1,3]'} f(\text{Inf}(X), \text{Sup}(Y), z), \min_{z \in [1,3]'} f(\text{Sup}(X), \text{Inf}(Y), z)]$$
$$= [\max_{z \in [1,3]'} ((0 - 4)/(z + 4)), \min_{z \in [1,3]'} ((3 - 2)/(z + 2))]$$
$$= [-4/7, 1/5].$$

Remark 3.4.4 The theorem is not essentially modified when x has more than one component if the condition of uni-modality is imposed on $\mathbf{X}$.

Remark 3.4.5 Actually, those functions of the type $f(\mathbf{x}, \mathbf{y})$, which are uniformly monotonic for $\mathbf{y} \in \mathbb{R}^k$ and JM-commutable for $\mathbf{x} \in \mathbb{R}^m$ on any vertex of the $\mathbf{y}$-prism defined by the $\mathbf{y}$-interval-arguments, can be admitted to the repertoire of modal syntactic operators. In fact,

$$f(\mathbf{x}_m, \mathbf{y}_M, \mathbf{u}_m, \mathbf{v}_M) = \min_{\mathbf{x} \in \mathbf{X}'} \max_{\mathbf{y} \in \mathbf{Y}'} f(\mathbf{x}, \mathbf{y}, \mathbf{u}_m, \mathbf{v}_M)$$

is a saddle-value of the $(\mathbf{x}, \mathbf{y})$-function $f(\mathbf{x}, \mathbf{y}, \mathbf{u}_m, \mathbf{v}_M)$ because for every $\mathbf{x} \in \mathbf{X}'$, $\mathbf{y} \in \mathbf{Y}'$, $\mathbf{u} \in \mathbf{U}'$ and $\mathbf{v} \in \mathbf{V}'$

$$f(\mathbf{x}_m, \mathbf{y}, \mathbf{u}_m, \mathbf{v}) \leq f(\mathbf{x}_m, \mathbf{y}_M, \mathbf{u}_m, \mathbf{v}_M) \leq f(\mathbf{x}, \mathbf{y}_M, \mathbf{u}_m, \mathbf{v}_M) \leq f(\mathbf{x}, \mathbf{y}_M, \mathbf{u}, \mathbf{v}_M).$$

Remark 3.4.6 Anyway, the operators of the syntactic trees should be as simple as possible for actual practice, in spite of constituting a larger family than the classical ones for real functions.

3.4.3 Modal Syntactic Computations with Rounding

A modal syntactic computation with outer or inner rounding is defined by the syntax of $fR(X)$ where the interval value of every component and the exact value of every operator are replaced by their modal inner or outer rounding.

Definition 3.4.7 (Outer-rounding computation of $fR^*(X)$) The outer-rounding computation $\text{Out}(fR^*(X))$ is the function defined by the computational program of $fR^*(X)$, in which the value of every X-component is replaced by its modal outer rounding $\text{Out}(X_i) \supseteq X_i$, and also the exact value of every operator $\omega^*(X_i, \ldots)$ is replaced by its computed actual outer-rounding $\text{Out}(\omega^*(X_i, \ldots)) \supseteq \omega^*(X_i, \ldots)$.

Definition 3.4.8 (Inner-rounding computation of $fR^{**}(X)$) The inner-rounding computation $\text{Inn}(fR^{**}(X))$ is the function defined by the program of $fR^{**}(X)$, in which every X-component X_i is replaced by $\text{Inn}(X_i) \subseteq X_i$, and every exact value $\omega^{**}(X_i,\ldots)$ by $\text{Inn}(\omega^{**}(X_i,\ldots)) \subseteq \omega^{**}(X_i,\ldots)$.

Remark 3.4.7 In a hypothetically ideal "real" arithmetic, the Out and Inn operators would reduce to the identity operator. If the real-arithmetics' rounding supporting Out and Inn is supposed to be $\leq$-monotonic increasing and the elements of the corresponding digital scale are applied to themselves, it is usual to speak of an optimal rounding.

Lemma 3.4.2 (Duality relation)

$$\text{Dual}(\text{Out}(fR^*(X))) = \text{Inn}(fR^{**}(\text{Dual}(X))).$$

Proof

$$\text{Dual}(\text{Out}(fR^*(X))) = \text{Inn}(\text{Dual}(fR^*(X))) = \text{Inn}(fR^{**}(\text{Dual}(X))).$$

∎

Theorem 3.4.7 (Inclusivity of the modal syntactic extensions) *The rounded modal syntactic extensions* $\text{Out}(fR^*(X))$ *and* $\text{Inn}(fR^{**}(X))$ *(or,* $\text{Out}(fR(X))$ *and* $\text{Inn}(fR(X))$*, if such be the case) of a continuous real function* f *from* $\mathbb{R}^k$ *to* $\mathbb{R}$*, are inclusion-monotonic increasing, if the supporting interval rounding of the arguments and of the operators are also inclusion-monotonic increasing.*

Proof This property holds for the modal syntactic extensions $fR^*(X)$ and $fR^{**}(X)$ (or $fR(X)$ if such be the case). For computations with rounding the result may be obtained by considering the different roundings as ordinary inclusion-monotonic increasing operators interposed into the syntactic tree of fR. ∎

Theorem 3.4.8 (Dual computing process) *If* $fR(X)$ *is a modal syntactic function, then*

$$\text{Inn}(fR(X)) = \text{Dual}(\text{Out}(fR(\text{Dual}(X)))).$$

Proof From Lemma 3.4.1. ∎

Remark 3.4.8 Computations with modal intervals do not need a double arithmetic, with inner and outer rounding. This theorem allows the implementation of only the outer rounding interval arithmetic. Note the application of the Out operator to $\text{Dual}(X)$ in the second term: Dual is not a modal syntactic operator and the information about X implied by this expression will be $\text{Inn}(X)$.

3.5 Concluding Remarks

From the operational point of view, the most outstanding characteristic of the system of modal intervals $I^*(\mathbb{R})$ is the following: in a similar way that real numbers are associated in pairs having the same absolute value but opposite signs, the modal intervals are associated in pairs too, each member corresponding to the same closed interval of the real line but having each one of the opposite selection modalities, existential or universal.

From the regularity point of view, $I^*(\mathbb{R})$ is a decisive improvement over $I(\mathbb{R})$ since $I^*(\mathbb{R})$ is not only the structural completion of $I(\mathbb{R})$, but also solves the referential character of interval computations to the isomorphic ones on the real line.

The system $I^*(\mathbb{R})$ provides a lot of properties immediately and consistently related with an informational approach to numeric data, as they arise from the procedures of measurement and digital computing. These properties are not obtained by using an additional over-imposed model, as in the models supported by probability, but are built on the inherent logic of the practical possibilities of the use of numbers. In fact, $I^*(\mathbb{R})$ is not a "model" for the numeric information, but the indispensable logical and operational frame for any geometrical model using numeric information. From the viewpoint of the technical constitution of the system $I^*(\mathbb{R})$, the main points are the following:

1. Association of each interval $A \in I^*(\mathbb{R})$ to the set Pred(A) of the predicates P on the real line that A accepts, that is, those by which the modally-quantified statement $Q(x, A) P(x)$ is made true. This step brings out the particular set-theoretical character of the inclusion of modal intervals, and supplies the important theorem about the mutual transfers of information between the "exact" result of a naturally analytical relation and the outer and inner-rounding of its interval computation.

2. The logical re-formulation of the "poor interval extensions" of continuous real functions, allowing of defining the "modal interval extensions" of these functions, with their all-important *- and **-semantic theorems, supports the application to modal intervals of the appealing intuition tied to the notions of "regulating" and "autonomous" ranges, and indicates the dependence between semantics and interval rounding. Starting from "poor interval extensions" selects also as meaningful only two of the different interval extensions of continuous functions that could be built if only the lattice completion of $I(\mathbb{R})$ was considered, a decision which would lead to an, in principle, different extension for each ordering of the meet and join operators.

3. The theory of interval modal syntactic functions clarifies the somewhat complicated relationship between the syntactic structure of the functions and their semantics, defined by their corresponding *- and **-semantic extensions. This is the key which solves the critical question of the dependence between computational process and the meaning of the computed results.

Chapter 4
Interpretability and Optimality

4.1 Introduction

The Semantic Theorems show that $f^*(X)$ and $f^{**}(X)$ are optimal from a semantic point of view, and clarify which $\subseteq$-sense of rounding is the right one when *-semantic or **-semantic are to be applied. They provide, therefore, a general norm that computational functions F from $I^*(\mathbb{R}^k)$ to $I^*(\mathbb{R})$ must satisfy to conform to the f^* or the f^{**}-semantic, but this is still not a general procedure by which these functions may be effectively computed. These procedures will be provided by the modal syntactic extension of continuous real functions, as far as they satisfy certain suitability conditions.

4.2 Interpretability and Optimality

The problem with the semantic extensions f^* and f^{**} is that they are not generally computable. A modal syntactic extension fR is computable from a syntactic tree of f but the result is hardly interpretable in reference to the original continuous real function. In order to remedy this lack of computability for f^* and f^{**} and the lack of meaning for the modal syntactic extensions fR, this chapter provides some relations between them, under some conditions.

The interpretation problem for the modal syntactic functions fR, which are the core of numerical computing, consists in relating them to the corresponding semantic functions by means of inclusion relations which are interpretable in accordance with the Semantic Theorems 3.3.1 and 3.3.2. In this case, if for $X \in I^*(\mathbb{R}^k)$ one of the relations

$$f^*(X) \subseteq fR(X) \quad \text{or} \quad f^{**}(X) \supseteq fR(X)$$

is true, then the computation $fR(X)$ is called *interpretable*.

On the other hand, the lack of computability of f^* and f^{**} can be remedied by means of modal syntactic computations which are inner or outer approximations, although in many cases this will involve a loss of information. To avoid that, it will be necessary to find criteria such that, in an ideal arithmetic without rounding,

$$f^*(X) = fR(X) = f^{**}(X).$$

In this case $fR(X)$ is called *optimal*, i.e., fR is an optimal computation on X.

Definition 4.2.1 (Optimality) A modal syntactic function fR is said to be *optimal* if for every $X \in I^*(\mathbb{R}^k)$ for which $fR(Prop(X))$ is defined, we have $f^*(X) = fR(X) = f^{**}(X)$.

Similarly we can speak of optimality on a given interval-domain: If this property holds particularly for an $A \in I^*(\mathbb{R}^k)$, we will say that fR is optimal in the domain A'.

Remark 4.2.1 A single operator (function) whose argument-places are allowed to be occupied by distinct variables, or by the same variable, can define a syntactic function (the most elementary one); this justifies writing $gR(X)$ instead of $g^*(X)$ for any *JM*-commutable function g used as an operator.

In case of only the equality $fR(X) = f^*(X)$ being meant, without any previous supposition about the equality of $f^*(X)$ and $f^{**}(X)$, we will speak of **-optimality*; and similarly we will speak of ***-optimality* for the case of $fR(X) = f^{**}(X)$.

4.2.1 Uni-incidence and Multi-incidence

An important role in obtaining these relations of inclusion and equality is played by the incidence of the involved variables.

Definition 4.2.2 (Uni-incidence and multi-incidence) A component x_i of x is *uni-incident* in a continuous real function f if it occupies only one leaf of the syntactic tree of f. Otherwise x_i is *multi-incident* in $f(x)$. A vector x is *uni-incident* in $f(x)$ if each of its components has this property.

Example 4.2.1 In the function $f : \mathbb{R}^2 \to \mathbb{R}$ defined by

$$f(x_1, x_2) = x_2 + \frac{x_1^2}{x_2}$$

the variable x_1 is uni-incident and x_2 multi-incident.

Remark 4.2.2 Beware of the fact that the concepts of uni-incidence and multi-incidence only have meaning when the discourse is about the definition of a modal syntactic function, which points to a definite computing program. Multi-incidence is not multi-incidence of values, but variables, and always deals with the repetition of the same variable in different leaves of the syntactic tree of a function. A continuous real function can, anyway, be considered as a pure function, defined by the classic correspondence between its arguments' values and the function's value, as well as a computing program. Correspondingly, naming *-variables the variables of any real function when considered as a pure function, and R-variables when their different places in the syntactic tree of the computing program, we may say that, for any continuous real function, *-variables are always uni-incident, and that each R-variable has an order of incidence equal to the number of leaves it occupies in the syntactic tree of the function.

4.2.2 Interpretability in the Uni-incidence Case

A first result will relate the modal semantic extension f^* with the modal semantic extensions of their operators. For that, and considering a function as a composition of their operators, it is possible to consider successive applications of the semantic theorems as the following lemmas suggest.

Lemma 4.2.1 (*-interpretability of one-step links) *If f, g_i, and h_j are continuous real functions in a suitably large domain, G_i, X, Y are existential interval vectors, H_j, U, V are universal interval vectors, $F \in I^*(\mathbb{R})$ satisfies $f^*(G_i, H_j) \subseteq F$, $g_i^*(X, U) \subseteq G_i$, $h_j^*(Y, V) \subseteq H_j$, and the components of u and v, corresponding to the universal vectors U and V do not have common components in the set of the g_i and h_j lists, then*

$$(f \circ (g_i, h_j))^*(X, Y, U, V) \subseteq F,$$

where $f \circ (g_i, h_j)(x, y, u, v) = f(g_i(x, u), h_j(y, v))$.

Proof This result is to be obtained by resolution with the prenex forms of the conjunctions of the corresponding logical formulas provided by the application of the semantic theorem to the analytical relations $f^*(G_i, H_j) \subseteq F$, $g_i^*(X, U) \subseteq G_i$ and $h_j^*(Y, V) \subseteq H_j$. From the *-Semantic Theorem 3.3.1:

a) $h_j^*(Y, V) \subseteq H_j \Leftrightarrow (\forall y \in Y') (\forall h_j \in H_j') (\exists v \in V') \, h_j = h_j(y, v)$;
b) $g_i^*(X, U) \subseteq G_i \Leftrightarrow (\forall x \in X') (\exists g_i \in G_i') (\exists u \in U') \, g_i = g_i(x, u)$;
c) $f^*(G_i, H_j) \subseteq F \Leftrightarrow (\forall g_i \in G_i') \, Q(f \in F) (\exists h_j \in H_j') \, f = f(g_i, h_j)$.

From the conjunction of (b) and (c) we obtain

$$(\forall x \in X') \, Q(f, F) (\exists h_j \in H_j') (\exists u \in U') \, f = f(g_i(x, u), h_j)$$

and with (a)

$$(\forall y \in Y') \, (\forall x \in X') \, Q(f, F) \, (\exists u \in U') \, (\exists v \in V') \, f = f(g_i(x, u), h_j(y, v)),$$

And from the *-semantic theorem we have $(f \circ (g_i, h_j))^*(X, Y, U, V) \subseteq F$. ∎

Remark 4.2.3 In this proof, if $\text{Mod}(F) = \forall$, the prefix $Q(f, F)$ can be commuted with $(\forall g_i \in G'_i)$ and extracted from the parenthesis before any other quantifier prefix and, after extracting the rest, it can be placed in the center of the list of quantifiers; if $\text{Mod}(F) = \exists$, it can be commuted with $(\exists h_j \in H'_j)$ and extracted from the parenthesis after any other quantifier prefix; after extracting the rest it can also be placed in the center of the list of quantifiers. The prefixes of the lists $(\forall x \in X')$ and $(\forall y \in Y')$ can be extracted first, irrespective of whether x or y are repeated or not in the lists of variables of g and h.

Lemma 4.2.2 (**-interpretability of one-step links) *If f, g_i, h_j are continuous real functions in a suitably large domain, G_i, X, Y are universal interval vectors, H_j, U, V are existential interval vectors, $F \in I^*(\mathbb{R})$ satisfies $f^{**}(G_i, H_j) \supseteq F$, $g_i^{**}(X, U) \supseteq G_i$, $h_j^{**}(Y, V) \supseteq H_j$, and the components of u and v which correspond to the existential vectors U and V do not have common components in the set of the g_i and h_j lists, then*

$$(f \circ (g, h))^{**}(X, Y, U, V) \supseteq F,$$

where $f \circ (g, h)(x, y, u, v) = f(g_i(x, u), h_j(y, v))$.

Proof The proof is the dual to the one of the previous Lemma 4.2.1. ∎

Both lemmas introduce the basic results for obtaining the conditions of interpretation for the rounded modal syntactic computations $\text{Out}(fR(X))$ and $\text{Inn}(fR(X))$.

Theorem 4.2.1 (*-interpretability of modal syntactic functions) *If the improper components of X are uni-incident in $fR^*(X)$ and $\text{Out}(fR^*(\text{Prop}(X)))$ exists, then*

$$f^*(X) \subseteq \text{Out}(fR^*(X)).$$

Proof Since f is the composition of its operators, we can use Lemma 4.2.1, which is insensitive to the multi-incidence of the proper components of X. The computational aspect is supported by the inclusion-isotony of the operators and of the outer-rounding. ∎

Remark 4.2.4 The existence of $\text{Out}(fR^*(\text{Prop}(X)))$ guarantees the existence of the operands and of the operators implied by $\text{Out}(fR^*(X))$.

This theorem states that if the continuity of the functions on the implied domains by $\text{Out}(fR(\text{Prop}(X)))$ is assured, and if the outer rounding is used for the digitalization of the data of X and all the elementary operations defining fR, then the result $fR(X)$ is interpretable in terms of the *-semantic theorem.

4.2 Interpretability and Optimality

Theorem 4.2.2 (-interpretability of modal syntactic functions)** *If the proper components of X are uni-incident in $f^{**}R(X)$ and $\text{Out}(fR^*(\text{Prop}(X)))$ exists, then*

$$\text{Inn}(fR^{**}(X)) \subseteq f^{**}(X).$$

Proof Through a dual demonstrative march of the one of Theorem 4.2.1, by means of Lemma 4.2.2.

According to this theorem, if the domains implied in the definition of $f^{**}(X)$ and the definition and computation of $fR(\text{Prop}(X))$ are well defined, because they do not comprise any divide by 0, then the result of the computation $\text{Inn}(fR(X))$ is acceptable from the point of view of the **-semantic theorem and can be interpreted according to this same theorem.

Theorem 4.2.3 (Interpretability of modal syntactic functions) *If all the variables are uni-incident in fR, then with an ideal arithmetic*

$$f^*(X) \subseteq fR(X) \subseteq f^{**}(X)$$

or else

$$f^*(X) \subseteq \text{Out}(fR(X)) \quad \text{and} \quad \text{Inn}(fR(X)) \subseteq f^{**}(X).$$

Proof From the previous lemmas. ∎

Example 4.2.2 To illustrate the necessity of the condition of uni-incidence for the improper components of X in $fR(X)$, let's examine the case of the function $f(x) = x - x$ for the interval-value $X = [1, 2]$:

$$f^*(X) = [0, 0] \subseteq fR(X) = [1, 2] - [1, 2] = [-1, 1]$$

But in the case of the value $X = [2, 1]$, we would obtain

$$f^*(X) = [0, 0] \nsubseteq fR(X) = [2, 1] - [2, 1] = [1, -1].$$

Example 4.2.3 It should be stressed that there are real functions f with all fR-variables uni-incident, but which are not globally JM-commutable: this is the case with

$$f(a, b, c, d) = (a + b)(c + d),$$

which satisfies $f^*(X) \subseteq fR(X) \subseteq f^{**}(X)$. For the intervals $A = [-2, 2]$, $B = [1, -1]$, $C = [-1, 1]$ and $D = [2, -2]$

$$fR(A, B, C, D) = ([-2, 2] + [1, -1]) * ([-1, 1] + [2, -2]) = [-1, 1] * [1, -1]$$
$$= [0, 0]$$

$$f^*(A,B,C,D) = [\min_{a\in A'}\min_{c\in C'}\max_{b\in B'}\max_{d\in D'}((a+b)(c+d)),$$

$$\max_{a\in A'}\max_{c\in C'}\min_{b\in B'}\min_{d\in D'}((a+b)(c+d))]$$

$$= [3/2, -3/2]$$

$$f^{**}(A,B,C,D) = [\max_{b\in B'}\max_{d\in D'}\min_{a\in A'}\min_{c\in C'}((a+b)(c+d)),$$

$$\min_{b\in B'}\min_{d\in D'}\max_{a\in A'}\max_{c\in C'}((a+b)(c+d))]$$

$$= [-3/2, 3/2]$$

So $[3/2, -3/2] \subseteq [0,0] \subseteq [-3/2, 3/2]$ and it is not JM-globally commutable.

4.2.3 Optimality in the Uni-incidence Case

In this section we will find criteria which characterize, assuming that all the fR-variables are uni-incident and ideal computations (without rounding), when the program fR is such that

$$f^*(X) = fR(X) = f^{**}(X).$$

Theorem 4.2.4 (Optimality and uni-incidence) *If in fR(X) all arguments are uni-incident and f is globally JM-commutable, then*

$$f^*(X) = fR(X) = f^{**}(X).$$

Proof This follows from Theorem 4.2.3, taking into account that the JM-commutability means $f^*(X) = f^{**}(X)$. ∎

Remark 4.2.5 In particular, if all of the X-components are uni-incident and have the same modality, $f^*(X) = fR(X) = f^{**}(X)$.

Now we will construct the fundamental class of uni-incident optimal modal syntactic functions. The uni-incidence hypothesis is assumed but usually not explicitly repeated.

Lemma 4.2.3 (Left monotonic associativity) *If g is a monotonic operator of one variable and fR(X) is optimal, then gR(fR(X)) is also optimal.*

Proof If g is for instance monotonic increasing, then

$$(g \circ f)^*(X) = \bigvee_{x_p\in X'_p}\bigwedge_{x_i\in X'_i}[g(f(x_p,x_i)), g(f(x_p,x_i))]$$

$$= [\min_{x_p\in X'_p}\max_{x_i\in X'_i} g(f(x_p,x_i)), \max_{x_p\in X'_p}\min_{x_i\in X'_i} g(f(x_p,x_i))]$$

4.2 Interpretability and Optimality

$$= [g(\min_{x_p \in X'_p} \max_{x_i \in X'_i} f(x_p, x_i)), g(\max_{x_p \in X'_p} \min_{x_i \in X'_i} f(x_p, x_i))]$$

$$= [g(\text{Inf}(f^*(X))), g(\text{Sup}(f^*(X)))]$$

// in the case of g monotonic decreasing, the bounds of this interval
// would be obtained in the reverse order.

$$= gR(f^*(X))$$

Similarly $(g \circ f)^{**}(X) = g^{**}(f^{**}(X))$. Hence, as g is an one-variable operator and f is optimal,

$$(g \circ f)^*(X) = gR(f^*(X)) = gR(f^{**}(X)) = (g \circ f)^{**}(X)$$

and

$$(g \circ f)R(X) = gR(fR(X)) = gR(f^*(X)) = (g \circ f)^*(X). \quad \blacksquare$$

Remark 4.2.6 A more precise and only a little more verbose statement of Lemma 4.2.3 would be: if g is a one variable monotonic operator and if $fR(X)$ does exist,

$$(g \circ f)^*(X) = g^*(f^*(X))$$

and in case $fR(X)$ is optimal

$$(g \circ f)^*(X) = (g \circ f)R(X) = gR(fR(X)).$$

Example 4.2.4 Let us consider the function $h(x, y) = e^{x+y}$ composed of the operators $f(x, y) = x + y$ and $g(z) = e^z$. As fR is optimal and g is one-variable and monotonic, then $h^*(X, Y) = hR(X, Y) = e^{X+Y}$. For $h(x, y) = (x + y)^2$, composed of $f(x, y) = x + y$ and $g(z) = z^2$ and fR is optimal but g is one-variable and not monotonic. So the result is not applicable.

Lemma 4.2.4 (Right unary associativity) *If $g_1(x_1), \ldots, g_k(x_k)$ are continuous operators of one variable and $fR(X)$ is optimal, then $fR(g_1 R(X_1), \ldots, g_k R(X_k))$ is also optimal.*

Proof Let X_p and X_i be the proper and improper components, respectively, of $X = (X_1, \ldots, X_k)$

$$(f \circ (g_1, \ldots, g_k))^*(X_p, X_i) = \bigvee_{x_p \in X'_p} \bigwedge_{x_i \in X'_i}$$

$$[f(g_1(x_1), \ldots, g_k(x_k)), f(g_1(x_1), \ldots, g_k(x_k))].$$

Let be $y_1 = g_1(x_1), \ldots, y_k = g_k(x_k)$ and $Y_1 = g_1 R(X_1), \ldots, Y_k = g_k R(X_k)$. Then

$$f^*(Y_p, Y_i) = \bigvee_{y_p \in Y'_p} \bigwedge_{y_i \in Y'_i} [f(y_p, y_i), f(y_p, y_i)].$$

Since X_j and Y_j, for $j = 1, \ldots, k$, have the same modality and the corresponding meet and join operators originate the bounds of the interval function, obtained step-by-step from $(f \circ (g_1, \ldots, g_k))(x)$ and $f(y_1, \ldots, y_k)$, to range over equal domains, therefore

$$(f \circ (g_1, \ldots, g_k))^*(X_p, X_i) = f^*(g_p R(X_p), g_i R(X_i)).$$

Similarly we would obtain $(f \circ (g_1, \ldots, g_k))^{**}(X_p, X_i) = f^{**}(g_p R(X_p), g_i R(X_i))$. From the optimality of f we obtain the intended result. ∎

Example 4.2.5 For right associativity, the right one-variable operators do not need monotonicity; but the left one-variable operators do. Thus,

a) $f(x, y) = x^2 + y^2$ is optimal because it is composed of $g_1 R(x) = x^2$, $g_2 R(y) = y^2$, which are not monotonic, and

$$fR(X, Y) = X + Y \text{ is optimal}$$

$$g_1 R \text{ is optimal (one-variable and continuous)}$$

$$g_2 R \text{ is optimal (one-variable and continuous).}$$

Therefore, by Theorem 4.2.4

$$fR(g_1 R(x), g_2 R(y)) = g_1 R(x) + g_2 R(y) = X^2 + Y^2$$

is optimal. If $X = [4, 3]$ and $Y = [-1, 5]$, then

$$f^*([4, 3], [-1, 5]) = fR([4, 3], [-1, 5]) = [4, 3]^2 + [-1, 5]^2 = [16, 34]$$

b) $f(x, y, z, t) = (x + y)/(|z| + |t|)$ is also optimal because it is composed of $g_1 R(x) = x$, $g_2 R(y) = y$, $g_3 R(z) = |z|$, $g_4 R(t) = |t|$ which are continuous and $fR(X, Y, Z, T) = (X + Y)/(Z + T)$ optimal.

Lemma 4.2.5 (Uniformly monotonic left-associativity)

a) If $g(x_1, \ldots, x_k)$ is a uniformly monotonic operator, and $f_1 R(Y_1), \ldots, f_k R(Y_k)$ are optimal, then $gR(f_1 R(Y_1), \ldots, f_k R(Y_k))$ is also optimal.
b) If g has the form $g(z, x_1, \ldots, x_k)$, it is uniformly monotonic for $x_1, \ldots, x_k$, and $f_1 R(Y_1), \ldots, f_k R(Y_k)$ are optimal, then $gR(Z, f_1 R(Y_1), \ldots, f_k R(Y_k))$ is optimal too.

4.2 Interpretability and Optimality

Proof Under the hypotheses, the manifold of saddle-points of $f_1(y_1), \ldots, f_k(y_k)$ provides a saddle-point of $g(f_1(y_1), \ldots, f_k(y_k))$.

a) Let $(x_1, x_2) = (x_1, \ldots, x_k)$ be such that $g(x_1, x_2)$ is x_1-monotonic increasing and x_2-monotonic decreasing. From Theorem 3.4.4

$$g^*(X_1, X_2) = [g(\text{Inf}(X_1), \text{Sup}(X_2)), g(\text{Sup}(X_1), \text{Inf}(X_2))].$$

Suppose $X_1 = f_1^*(Y_{1p}, Y_{1i})$ and $X_2 = f_2^*(Y_{2p}, Y_{2i})$; if

$$\text{Inf}(X_1) = \min_{y_{1p} \in Y'_{1p}} \max_{y_{1i} \in Y'_{1i}} f_1(y_{1p}, y_{1i}) = \text{SDV}(f_1, Y'_{1p}, Y'_{1i}) = f_1(y_{1pm}, y_{1iM})$$

and

$$\text{Sup}(X_2) = \max_{y_{2p} \in Y'_{2p}} \min_{y_{2i} \in Y'_{2i}} f_2(y_{2p}, y_{2i}) = \text{SDV}(f_2, Y'_{2i}, Y'_{2p}) = f_2(y_{2pM}, y_{2im}),$$

then

$$(\forall y_{1p} \in Y'_{1p})\,(\forall y_{1i} \in Y'_{1i})\,(f_1(y_{1pm}, y_{1i}) \leq f_1(y_{1pm}, y_{1iM}) \leq f_1(y_{1p}, y_{1iM}))$$

and

$$(\forall y_{2p} \in Y'_{2p})\,(\forall y_{2i} \in Y'_{2i})\,(f_2(y_{2p}, y_{2im}) \leq f_2(y_{2pM}, y_{2im}) \leq f_2(y_{2pM}, y_{2i})).$$

Therefore, since $g(x_1, x_2)$ is uniformly x_1-monotonic increasing and x_2-monotonic decreasing,

$$(\forall y_{1p} \in Y'_{1p})\,\forall(y_{1i} \in Y'_{1i})\,(\forall y_{2p} \in Y'_{2p})\,(\forall y_{2i} \in Y'_{2i})$$
$$g(f_1(y_{1pm}, y_{1i}), f_2(y_{2pM}, y_{2i})) \leq g(f_1(y_{1pm}, y_{1iM}), f_2(y_{2pM}, y_{2im}))$$
$$\leq g(f_1(y_{1p}, y_{1iM}), f_2(y_{2p}, y_{2im})).$$

Since $g(\text{Inf}(X_1), \text{Sup}(X_2)) = \text{SDV}(g \circ (f_1, f_2), (Y'_{1p}, Y'_{2p}), (Y'_{1i}, Y'_{2i}))$, we have

$$\text{Inf}(g^*(X_1, X_2)) = \text{Inf}(g \circ (f_1, f_2))^*(Y_{1p}, Y_{2p}, Y_{1i}, Y_{2i}))$$

and similarly

$$\text{Sup}(g^*(X_1, X_2)) = \text{Sup}(g \circ (f_1, f_2))^*(Y_{1p}, Y_{2p}, Y_{1i}, Y_{2i})).$$

In consequence

$$g^*(f_1^*(Y_1), f_2^*(Y_2)) = (g \circ (f_1, f_2))^*(Y_1, Y_2)$$

and similarly for g^{**} and $g \circ (f_1, f_2)^{**}$.
b) See Theorem 3.4.6 and Lemma 4.2.4. ∎

Remark 4.2.7 If $f_0 R(Y_0)$ is any continuous one-variable operator, then $gR(f_0 R(Y_0), f_1 R(Y_1), \ldots, f_k R(Y_k))$ is also optimal.

Example 4.2.6 For the function $h(x_1, x_2, x_3, x_4) = x_1 x_2 + x_3 x_4$ composed by the operator $g(z_1, z_2) = z_1 + z_2$, which is uniformly monotonic, and $f_1(x_1, x_2) = x_1 x_2$ y $f_2(x_3, x_4) = x_3 x_4$, then as $f_1 R$ y $f_2 R$ are optimal, hR is optimal too.

For the function $h(x_1, x_2, x_3, x_4) = z(x_1 x_2 + x_3 x_4)$ composed by the operator $g(z, f_1, f_2) = z(f_1 + f_2)$, which is uniformly monotonic with respect to f_1 and f_2, but not with respect to z, and $f_1(x_1, x_2) = x_1 x_2$ y $f_2(x_3, x_4) = x_3 x_4$, then as $f_1 R$ y $f_2 R$ are optimal, hR is optimal too.

4.2.4 Tree-Optimality

The following concept leads to important results about optimality.

Definition 4.2.3 (Tree-optimal modal syntactic functions) The modal syntactic function $fR(X)$ is tree-optimal if, given any one of its non-uniformly monotonic elementary branches, the fR-tree is followed upwards only by one-variable operators.

Remark 4.2.8 In this definition the idea of a branch develops from the elementary formal connections between any operator and each one of its ordered immediate operands, and of these operands (or the final or root result) with the immediately following operator.

Theorem 4.2.5 (Optimality of tree-optimal modal syntactic functions) If $fR(X)$ is tree-optimal and X is uni-incident in fR, then $fR(X)$ is optimal (that is, $f^*(X) = fR(X) = f^{**}(X)$ whenever X is uni-incident in fR).

Proof By Lemmas 4.2.3–4.2.5. ∎

Remark 4.2.9 In the case where $X_1, \ldots, X_k$ are uni-incident and uni-modal, the optimality in the sense of $f^*(X) = fR(X) = f^{**}(X)$ holds independently of the syntactic structure of the fR tree (be aware of the need to deal with the fR-tree without any operator built upon the duality transformation). In as much as this condition means a restriction upon the modalities of the arguments $X_1, \ldots, X_k$, this case does in fact correspond to the notion of conditioned optimality to be introduced later on.

4.2 Interpretability and Optimality

Remark 4.2.10 Checking the tree-optimality of $fR(X)$ can be restricted to the sub-tree defined by its non-uniformly monotonic variables.

Example 4.2.7 The $\mathbb{R}^4$ to $\mathbb{R}$ continuous function defined by $f(x, y, z, u) = xy + zu$ is tree-optimal and therefore an optimal modal syntactic function for every $(X, Y, Z, U) \in I^*(\mathbb{R}^4)$:

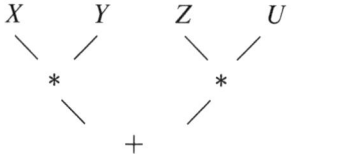

non-uniformly monotonic operator

uniformly monotonic operator

The $\mathbb{R}^4$ to $\mathbb{R}$ continuous function defined by $g(x, y, z, u) = (x + y)(z + u)$ is neither tree-optimal nor optimal for some (X, Y, Z, U)

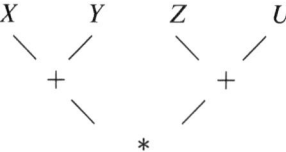

non-unary operator

non-uniformly monotonic operator

and, for example:

$$g^*([-2, 2], [1, -1], [-1, 1], [2, -2]) = [1.5, -1.5],$$
$$g^{**}([-2, 2], [1, -1], [-1, 1], [2, -2]) = [-1.5, 1.5],$$
$$gR([-2, 2], [1, -1], [-1, 1], [2, -2]) = [0, 0].$$

But it is optimal in other domains, for example

$$g^*([1, 3], [0, 3], [4, 2], [3, 1]) = [7, 18],$$
$$g^{**}([1, 3], [0, 3], [4, 2], [3, 1]) = [7, 18],$$
$$gR([1, 3], [0, 3], [4, 2], [3, 1]) = [7, 18].$$

Actually, it cannot be confusing to speak of tree-optimality on a determinate interval-domain of $\mathbb{R}^k$

4.2.5 Interpretability in the Multi-incidence Case

Until this point we have been dealing with uni-incident variables and when the syntactic extensions are interpretable or optimal. Now we are going to provide theorems to handle the case of modal syntactic functions with multi-incident variables and how to transform the different interval incidences in order to obtain interpretable or optimal computations.

The following theorems provide two important results in case there are multi-incident components.

Theorem 4.2.6 (Coercion to *-interpretability) *If $fR(X)$ has multi-incident improper components, and if Xt^* is obtained from X by substituting all the incidences of each multi-incident improper component into a point-wise interval defined for any point of its domain, then $f^*(X) \subseteq fR(Xt^*)$ (the existence requirements being presupposed).*

Proof For $X = (X_p, X_i)$, let X_{i1} be the improper components which were uni-incident in $fR(X)$, and denote the multi-incident improper components by X_{i2}. Let x_{i2} be any point of X'_{i2} and $Xt^* = (X_p, X_{i1}, [x_{i2}, x_{i2}])$ which does not have multi-incident improper components. Then from Theorem 4.2.3,

$$f^*(X_p, X_i) \subseteq f^*(Xt^*) \subseteq fR(Xt^*). \qquad \blacksquare$$

Example 4.2.8 The *-semantic extension of the continuous function $f(x, y) = y - xy$ to the intervals $X_1 = [2, 3], X_2 = [4, 3]$ is

$$f^*([2,3], [4,3]) = \bigvee_{x \in [2,3]'} \bigwedge_{y \in [3,4]'} [y - xy, y - xy] = [-6, -4].$$

For the modal syntactic extensions $fR(Xt^*)$ we have

$$fR_1([2, 3], [4, 4]) = [4, 4] - [2, 3] * [4, 4] = [-8, -4]$$
$$fR_2([2, 3], [3, 3]) = [3, 3] - [2, 3] * [3, 3] = [-6, -3]$$

and $f^*(X) \subseteq fR_1(Xt^*)$, $f^*(X) \subseteq fR_2(Xt^*)$.

Theorem 4.2.7 (Coercion to **-interpretability) *If $fR(X)$ has multi-incident proper components, and if Xt^{**} is obtained from X by substituting all the incidences of each multi-incident proper component into a point-wise interval defined for any point of its domain, then $fR(Xt^{**}) \subseteq f^{**}(X)$ (the existence requirements being presupposed).*

Proof The proof is the dual of the previous one. $\qquad \blacksquare$

Theorem 4.2.8 (Interval coercion to *-interpretability) *If $fR(X)$ has multi-incident improper components, and if XT^* is obtained from X by transforming, for every multi-incident improper component, all incidences but one into their duals, then*

$$f^*(X) \subseteq fR(XT^*)$$

(the existence requirements being presupposed).

4.2 Interpretability and Optimality

Proof Let us suppose that $X = (X_p, X_i)$ has only one j-incident improper X_m. Let us define $XT^* = (X_p, X_{m1}, \text{Dual}(X_{m2}), \ldots, \text{Dual}(X_{mj}), X_{iu})$, where

- X_p are the proper components,
- X_{iu} are the uni-incident improper components,
- X_{m1} are the improper incidences of X_m which maintain its original improper modality.
- $X_{m2}, \ldots, X_{mj}$ are the other incidences of X_m, which have been dualized, and $x_{m2}, \ldots, x_{mj}$ their corresponding variables.

Since in XT^* there are no multi-incident improper components, in accordance with Theorem 4.2.3 we have

$$f^*(XT^*) \subseteq fR(XT^*) = [a, b]$$

$$\Updownarrow$$

$$(\forall x_p \in X'_p)\,(\forall x_{m2} \in X'_m) \cdots (\forall x_{mj} \in X'_m)\,Q(z, [a, b])\,(\exists x_{m1} \in X'_m)\,(\exists x_{iu} \in X'_{iu})$$
$$f(x_p, x_{m1}, x_{m2}, \ldots, x_{mj}, x_{iu}) = z.$$

Since f is continuous on all its domain $\text{Prop}(X)$, given x going over the domain X'_m and choosing $x_{m2} = \ldots = x_{mj} = x$, for any x_p and for any (or some) z, the logical formula

$$(\forall x_p \in X'_p)\,(\forall x \in X')\,Q(z, [a, b])\,(\exists x_{m1} \in X'_m)\,(\exists x_{iu} \in X'_{iu})$$
$$f(x_p, x_{m1}, x, \ldots, x, x_{iu}) = z.$$

proves that there exists a continuous function (because of the continuity of f)

$$\varphi : X'_m \to X'_m$$
$$x \to x_{m1} = \varphi(x)$$

Applying the Fixed Point Theorem to φ, there exists a point $x_m \in X'_m$ such that $x_{m1} = x_m$ and the logical formula implies that

$$(\forall x_p \in X'_p)\,Q(z, [a, b])\,(\exists x_m \in X'_m)\,(\exists x_{iu} \in X'_{iu})\,f(x_p, x_m, x_{iu}) = z$$

which is equivalent to

$$f^*(X_p, X_i) \subseteq [a, b] = fR(XT^*).$$

The more general case, with more than one multi-incident components in X, is not essentially different. ∎

Theorem 4.2.9 (Interval coercion to **-interpretability) *If $fR(X)$ has multi-incident proper components, and if XT^{**} is obtained from X by transforming, for every multi-incident proper component, all incidences but one into its dual, then*

$$fR(XT^{**}) \subseteq f^{**}(X)$$

(the existence requirements being presupposed).

Proof The proof is parallel to that of the previous theorem. ∎

Example 4.2.9 For $f(x_1, x_2) = x_2 - x_1 x_2$, and $X = ([2,3], [4,3])$,

$$f^*(X) = \bigvee_{x_1 \in [2,3]'} \bigwedge_{x_2 \in [3,4]'} [x_2 - x_1 x_2, x_2 - x_1 x_2]$$

$$= \bigvee_{x_1 \in [2,3]'} [3 - 3x_1, 4 - 4x_1] = [-6, -4],$$

$$fR(XT^*) = [4,3] - [2,3] * [3,4] = [-8, -3],$$

or

$$fR(XT^*) = [3,4] - [2,3] * [4,3] = [-6, -4],$$

where, for both computations, the relation $f^*(X) \subseteq fR(XT^*)$ holds. For the same function and $X = ([1,3], [3,4])$

$$f^{**}(X) = f^*(X) = \bigvee_{x_2 \in [3,4]'} \bigvee_{x_1 \in [1,3]'} [x_2 - x_1 x_2, x_2 - x_1 x_2]$$

$$= \bigvee_{x_2 \in [3,4]'} [x_2 - 3x_2, x_2 - x_2] = [-8, 0]$$

$$fR(XT^{**}) = [3,4] - [1,3] * [4,3] = [-6, 0],$$

or

$$fR(XT^{**}) = [4,3] - [1,3] * [3,4] = [-8, 0]$$

and $fR(XT^{**}) \subseteq f^{**}(X)$. (The reason for the coincidence, when such is the case, will be found later on).

The computation of an interpretable syntactic interval program fR, tree-optimal or not, may result in a loss of information, even when all its arguments are uni-incident. This is a heavy drawback for every computation meant to serve an actual use, since its loss of information will be usually far more important than the one produced by the common numerical rounding. This loss can be cancelled or reduced in some cases: for example if there exist multi-incident improper components

4.2 Interpretability and Optimality

and $fR(Xt_1^*), \ldots, fR(Xt_n^*)$ are n results obtained transforming each one of these multi-incidences in point-wise intervals, in several different ways, then

$$f^*(X) \subseteq fR(Xt_1^*), \ldots, f^*(X) \subseteq fR(Xt_n^*) \Rightarrow f^*(X)$$
$$\subseteq fR(Xt_1^*) \wedge \ldots \wedge fR(Xt_n^*) = B$$

and B will possibly be a better result than every $fR(Xt_i^*)$. Similarly, for the case of multi-incident proper components,

$$fR(Xt_1^{**}) \vee \ldots \vee fR(Xt_n^{**}) \subseteq f^{**}(X).$$

One has the same results when using the XT^* or XT^{**} transformations.

Example 4.2.10 For the function f from $\mathbb{R}^2$ to $\mathbb{R}$ given by

$$f(x_1, x_2) = x_1 + x_2 - x_1 x_2,$$

and the interval $X = ([1, 3], [5, 2])$,

$$f^*(X) = \bigvee_{x_1 \in [1,3]'} \bigwedge_{x_2 \in [2,5]'} [x_1 + x_2 - x_1 x_2, x_1 + x_2 - x_1 x_2]$$

$$= \bigvee_{x_1 \in [1,3]'} [-x_1 + 2, -4x_1 + 5] = [-1, 1]$$

$$f^{**}(X) = \bigwedge_{x_2 \in [2,5]'} \bigvee_{x_1 \in [1,3]'} [x_1 + x_2 - x_1 x_2, x_1 + x_2 - x_1 x_2]$$

$$= \bigwedge_{x_2 \in [2,5]'} [-2x_2 + 3, 1] = [-1, 1].$$

As

$$fR(XT_1^*) = fR([1, 3], [2, 5], [5, 2]) = [1, 3] + [2, 5] - [1, 3] * [5, 2]$$
$$= [3, 8] - [5, 6] = [-3, 3]$$
$$fR(XT_2^*) = fR([1, 3], [5, 2], [2, 5]) = [1, 3] + [5, 2] - [1, 3] * [2, 5]$$
$$= [6, 5] - [2, 15] = [-9, 3]$$
$$fR(XT_1^{**}) = fR([1, 3], [3, 1], [5, 2]) = [1, 3] + [5, 2] - [3, 1] * [5, 2]$$
$$= [6, 5] - [15, 2] = [4, -10]$$
$$fR(XT_2^{**}) = fR([3, 1], [1, 3], [5, 2]) = [3, 1] + [5, 2] - [1, 3] * [5, 2]$$
$$= [8, 3] - [5, 6] = [2, -2],$$

we have

$$f^*(X) = [-1,1] \subseteq \begin{cases} fR(XT_1^*) = [-3,3] \\ fR(XT_2^*) = [-9,3] \end{cases} \subseteq [-3,3] \wedge [-9,3] = [-3,3]$$

$$f^{**}(X) = [-1,1] \supseteq \begin{cases} fR(XT_1^{**}) = [4,-10] \\ fR(XT_2^{**}) = [2,-2] \end{cases} \supseteq [4,-10] \vee [2,-2] = [2,-2].$$

Similarly,

$$fR(Xt_1^*) = fR([1,3],[2,2]) = [1,3] + [2,2] - [1,3] * [2,2]$$
$$= [3,5] - [2,6] = [-3,3]$$
$$fR(Xt_2^*) = fR([1,3],[5,5]) = [1,3] + [5,5] - [1,3] * [5,5]$$
$$= [6,8] - [5,15] = [-9,3]$$
$$fR(Xt_1^{**}) = fR([1,1],[5,2]) = [1,1] + [5,2] - [1,1] * [5,2]$$
$$= [6,3] - [5,2] = [4,-2]$$
$$fR(Xt_2^{**}) = fR([3,3],[5,2]) = [3,3] + [5,2] - [3,3] * [5,2]$$
$$= [8,5] - [15,6] = [2,-10],$$

so

$$f^*(X) = [-1,1] \subseteq \begin{cases} fR(Xt_1^*) = [-3,3] \\ fR(Xt_2^*) = [-9,3] \end{cases} \subseteq [-3,3] \wedge [-9,3] = [-3,3]$$

$$f^{**}(X) = [-1,1] \supseteq \begin{cases} fR(Xt_1^{**}) = [4,-2] \\ fR(Xt_2^{**}) = [2,-10] \end{cases} \supseteq [4,-2] \vee [2,-10] = [2,-2].$$

The fact that there can be a semantic loss of information by the computation of the program of an modal syntactic function, for which only the relations $f^*(X) \subseteq fR(X) \subseteq f^{**}(X)$ can be assured, leads to the immediate problem of obtaining criteria ensuring the relation of optimality $f^*(X) = fR(X) = f^{**}(X)$.

Definition 4.2.4 (Total monotonicity) A continuous real function f is x-totally monotonic for a multi-incident variable $x \in \mathbb{R}$ if it is uniformly monotonic for this variable and for each one of its incidences (considering each leaf of the syntactic tree as an independent variable). Any uni-incident uniformly monotonic variable is totally monotonic too.

Remark 4.2.11 A modal syntactic function fR will be described as totally monotonic for a multi-incident component X if

4.2 Interpretability and Optimality

1) f is uniformly monotonic for the variable x (this sense of monotonicity of f for x is called global on the tree fR) and
2) f is uniformly monotonic for each incidence x_j of x, considered as independent (this sense of monotonicity for this incidence of x_j in x is called local on the tree fR).

Example 4.2.11 The function

$$f(x, y) = \frac{x}{x + y}$$

in the domain $(X, Y) = ([1, 3]', [2, 4]')$ is uniformly monotonic for the variable x,

$$\frac{\partial f}{\partial x} = \frac{y}{(x + y)^2} \geq 0$$

and for each one of its incidences,

$$\frac{\partial f}{\partial x_1} = \frac{1}{x + y} \geq 0 \qquad \frac{\partial f}{\partial x_2} = \frac{-x}{(x + y)^2} \leq 0.$$

So f is x-totally monotonic and fR is X-totally monotonic.

The following two lemmas do not provide any computational procedure, but they will support the proofs of some subsequent propositions.

Lemma 4.2.6 *Let $X = (Y, Z)$ be an interval vector and fR, defined in the domain $Prop(X)$, be totally monotonic for the subset Z of their multi-incident components. Let (Y, ZD) be the enlarged vector of X such that each incidence of every multi-incident component of Z is included as an independent component, but transformed into its dual if fR has a local monotonicity sense contrary to the global one of the corresponding Z-component. Then*

$$f^*(Y, Z) = f^*(Y, ZD).$$

Proof Let us suppose that Z consists only of one multi-incident proper component and f, for example is z-monotonic increasing. Let $ZD = (ZD_+, ZD_-)$ split into ZD_+, the incidences for which f is monotonic in the same sense as that of Z, and ZD_-, the incidences for which f is monotonic in the opposite sense. Then

$$f^*(Y, Z)$$
$$= \bigvee_{y_p \in Y'_p} \bigvee_{z \in Z'} \bigwedge_{y_i \in Y'_i} [f(y_p, y_i, z), f(y_p, y_i, z)]$$

// from Theorem 3.4.6

$$= \bigvee_{y_p \in Y'_p} \bigwedge_{y_i \in Y'_i} \bigvee_{z \in Z'} [f(y_p, y_i, z), f(y_p, y_i, z)]$$

// from the z-total monotonicity

$$= \bigvee_{y_p \in Y'_p} \bigwedge_{y_i \in Y'_i} \bigvee_{zd_+ \in ZD'_+} \bigwedge_{zd_- \in ZD'_-} [f(\mathbf{y}_p, \mathbf{y}_i, zd_+, zd_-), f(\mathbf{y}_p, \mathbf{y}_i, zd_+, zd_-)]$$

// with a demonstration fully parallel to that of Theorem 3.4.6.

$$= \bigvee_{y_p \in Y'_p zd_+ \in ZD'_+} \bigvee_{y_i \in Y'_i zd_- \in ZD'_-} \bigwedge [f(\mathbf{y}_p, \mathbf{y}_i, zd_+, zd_-), f(\mathbf{y}_p, \mathbf{y}_i, zd_+, zd_-)]$$

$$= f^*(Y, ZD).$$

The general case of possessing more multi-components proper for Z is not essentially different. ∎

Lemma 4.2.7 *Let $X = (Y, Z)$ be an interval vector and fR, defined in the domain $Prop(X)$, be totally monotonic for the subset Z of their multi-incident components. Let (Y, ZD) be the enlarged vector of X such that each incidence of every multi-incident component of Z is included as an independent component, but transformed into its dual if fR has a local monotonicity sense contrary to the global one of the corresponding Z-component. Then*

$$f^{**}(Y, Z) = f^{**}(Y, ZD).$$

Proof The proof is parallel to that of the previous lemma. ∎

Theorem 4.2.10 (*-partially optimal coercion) *Let X be an interval vector, and suppose fR is defined in the domain $Prop(X)$ and is totally monotonic for a subset Z of multi-incident components. Let XDt^* be the enlarged vector of X, such that:*

1) *each incidence of every multi-incident component of Z is included in XDt^* as an independent component, but transformed into its dual if f has a local monotonicity sense contrary to the global one of the corresponding Z-component;*
2) *for the rest, every multi-incident improper component is transformed into a point-wise interval defined by any point of its domain in every of its incidences.*

Then

$$f^*(X) \subseteq fR(XDt^*).$$

Proof Let $X = (Y, Z)$ be a splitting of X in such a way that f is totally monotonic for the components of Z and $XDt^* = (Yt^*, ZD)$. From Lemma 4.2.6 and Theorem 4.2.6 of coercion to interpretability,

$$f^*(X) = f^*(Y, Z) = f^*(Y, ZD) \subseteq fR(Yt^*, ZD) = fR(XDt^*). \qquad \blacksquare$$

4.2 Interpretability and Optimality

Theorem 4.2.11 (-partially optimal coercion)** *Let X be an interval vector, and let fR be defined on the domain $Prop(X)$ and totally monotonic for a subset Z of its multi-incident components. Let XDt^{**} be the enlarged vector of X satisfying:*

1) *each incidence of every multi-incident component of Z is included in XDt^{**} as an independent component, but transformed into its dual if f has a local monotonicity sense contrary to the global one of the corresponding Z-component;*
2) *for the rest, every multi-incident proper component is transformed into a pointwise interval defined by any point of its domain in every of its incidences.*

Then

$$fR(XDt^{**}) \subseteq f^{**}(X).$$

Proof The proof is the dual of that of Theorem 4.2.10. ∎

Theorem 4.2.12 (Interval *-partially optimal coercion) *Let X be an interval vector, and suppose fR is defined in the domain $Prop(X)$ and is totally monotonic for a subset Z of multi-incident components. Let XDT^* be the enlarged vector of X satisfying:*

1) *each incidence of every multi-incident component of Z is included in XDT^* as an independent component, but transformed into its dual if f has a local monotonicity sense contrary to the global one of the corresponding Z-component;*
2) *for the rest, every multi-incident improper component is transformed into its dual in all but one of its incidences.*

Then

$$f^*(X) \subseteq fR(XDT^*).$$

Moreover if $fR(X)$ is tree-optimal,

$$fR(XDT^*) \subseteq fR(XT^*)$$

provided that the multi-incident components not belonging to Z undergo in XT^* the same transformation as in XDT^*.

Proof Let $X = (Y, Z)$ be a splitting of X such that f is totally monotonic for the components of Z, and $XDT^* = (YT^*, ZD)$. From Lemma 4.2.6 and the theorem of interval coercion to *-interpretability (Theorem 4.2.8),

$$f^*(X) = f^*(Y, Z) = f^*(Y, ZD) \subseteq fR(YT^*, ZD) = fR(XDT^*).$$

Given the independence of the components of $fR(YT^*, ZD)$,

$$fR(XDT^*) = fR(YT^*, ZD) = f^*(YT^*, ZD)$$
$$= f^*(YT^*, Z) \subseteq fR(YT^*, ZT^*) = fR(XT^*). \quad ∎$$

Remark 4.2.12 It is interesting to note that in the resulting transformation of ZD into ZT^*, all the components proper in Z either keep their proper modality or they come from improper to proper through the local component change. The improper components in Z go to proper modality in ZT^* for each incidence of Z except one. If this latter has the same monotonicity sense as the global one, it keeps its modality in the ZD to ZT^* transformation; if the contrary case happens, the total monotonicity prevents this "contracting effect" from prevailing in the final $\subseteq$-relation.

Example 4.2.12 Let us consider the continuous function f from $\mathbb{R}^2$ to $\mathbb{R}$ defined by $f(x, y) = xy + \dfrac{1}{x+y}$ with $X = [10, 5]$ and $Y = [2, -1]$. Its *-semantic extension is

$$f^*([10, 5], [2, -1]) = [20.08\overline{3}, -9.\overline{8}]$$

Due to the theorem of coercion to *-interpretability,

$$fR(XT_1^*) = \mathrm{Dual}(X) * \mathrm{Dual}(Y) + \frac{1}{X+Y} \subseteq [-9.75, 20.0834]$$

$$fR(XT_2^*) = X * \mathrm{Dual}(Y) + \frac{1}{\mathrm{Dual}(X)+Y} \subseteq [-4.8889, 10.1429]$$

which includes into $f^*(X)$. The derivatives prove that f is y-totally monotonic because f is y-uniformly monotonic, monotonic increasing for the first incidence and monotonic decreasing for the second one. As X is improper, there exist two possibilities for the computation of $fR(XDT^*)$:

$$fR(XDT_1^*) = X * Y + \frac{1}{\mathrm{Dual}(X)+\mathrm{Dual}(Y)} \subseteq [20.0833, -9.75]$$

$$fR(XDT_2^*) = \mathrm{Dual}(X) * Y + \frac{1}{X+\mathrm{Dual}(Y)} \subseteq [10.1428, -4.888]$$

which includes to $f^*(X)$. Moreover, as fR is tree-optimal, $fR(XDT_1^*)$ and $fR(XDT_2^*)$ are contained in $fR(XT_2^*)$ and $fR(XT_1^*)$

Theorem 4.2.13 (Interval **-partially optimal coercion) *Let X be an interval vector, and suppose fR is defined in the domain $Prop(X)$ and totally monotonic for a subset Z of multi-incident components. Let XDT^{**} be the enlarged vector of X satisfying*

1) *each incidence of every multi-incident component of Z is included in XDT^{**} as an independent component, but transformed into its dual if f has a local monotonicity sense contrary to the global one of the corresponding Z-component;*
2) *for the rest, every multi-incident proper component is transformed into its dual in all but one of its incidences.*

4.2 Interpretability and Optimality

Then
$$fR(XDT^{**}) \subseteq f^{**}(X).$$

Moreover if $fR(X)$ is tree-optimal,
$$fR(XT^{**}) \subseteq fR(XDT^{**}),$$

provided that the multi-incident components not belonging to Z suffer in XT^{**} the same transformation as in XDT^{**}.

Proof The proof is the dual to that of Theorem 4.2.12. ∎

Theorem 4.2.14 (Partially optimal coercion) *Let X be an interval vector, and suppose fR is defined on the domain $Prop(X)$ and is totally monotonic for all its multi-incident components. Let XD be the enlarged vector of X such that each incidence of every multi-incident component is included in XD as an independent component, but transformed into its dual if the corresponding incidence-point has a monotonicity-sense contrary to the global one of the corresponding X-component. Then*
$$f^*(X) \subseteq fR(XD) \subseteq f^{**}(X).$$

Proof In this case, all the multi-incident components are totally monotonic, therefore $XDT^* = XD$ and we come to this result using Theorem 4.2.12 about *-partially optimal coercion. ∎

4.2.6 Optimality in the Multi-incidence Case

This section gives criteria to characterize the optimality of the program fR, assuming only ideal computations (without rounding), in the general case of multi-incident variables.

Theorem 4.2.15 (Coercion to optimality) *Let X, fR and XD be defined under the hypotheses of Theorem 4.2.14, and let fR be tree-optimal on the domain $Prop(X)$. In this case,*
$$f^*(X) = fR(XD) = f^{**}(X).$$

Proof From Lemmas 4.2.6 and 4.2.7 and Theorem 4.2.14,
$$f^*(XD) = f^*(X) \subseteq fR(XD) \subseteq f^{**}(X) = f^{**}(XD),$$

due to the tree-optimality of $fR(XD)$ and the uni-incidence of the XD components, $f^*(XD) = f^{**}(XD)$. Therefore $f^*(X) = fR(XD) = f^{**}(X)$. ∎

Example 4.2.13 1. for $f(x) = x - x$, one has $fR(XD) = X - \text{Dual}(X)$ or $fR(XD) = \text{Dual}(X) - X$,
2. for $f(x) = x/x$, and $fR(XD) = X/\text{Dual}(X)$ or $fR(XD) = \text{Dual}(X)/X$, whenever $0 \notin X'$,
3. for $f(x) = 1/(1+x) + 1/(1-x)$ and $X = [1/4, 1/2]$, and also

$$fR(XD) = \frac{1}{1+\text{Dual}(X)} + \frac{1}{1-X} = \frac{1}{1+[1/2,1/4]} + \frac{1}{1-[1/4,1/2]},$$

because the f-tree is optimal in this case for $X' \subseteq [0, 1)'$.

Example 4.2.14 The function

$$f(x, y) = xy + \frac{1}{x+y}$$

for $(X, Y) = ([5, 10], [2, 1])$ is x-monotonic increasing and y-monotonic increasing. Therefore

$$XD = (X, Y, \text{Dual}(X), \text{Dual}(Y)) \Rightarrow fR(XD) = X * Y + \frac{1}{\text{Dual}(X) + \text{Dual}(Y)}$$

So

$$fR(XD) = [5, 10] * [2, 1] + \frac{1}{[10, 5] + [1, 2]} = [10, 10] + \frac{1}{[11, 7]} = [71/7, 111/11]$$

and

$$f^*(X, Y) = [\min_{x \in [5,10]'} \max_{y \in [1,2]'} (xy + \frac{1}{x+y}), \max_{x \in [5,10]'} \min_{y \in [1,2]'} (xy + \frac{1}{x+y})]$$

$$= [\min_{x \in [5,10]'} (2x + \frac{1}{x+2}), \max_{x \in [5,10]'} (x + \frac{1}{x+1})]$$

$$= [71/7, 111/11]$$

$$f^{**}(X, Y) = [\max_{y \in [1,2]'} \min_{x \in [5,10]'} (xy + \frac{1}{x+y}), \min_{y \in [1,2]'} \max_{x \in [5,10]'} (xy + \frac{1}{x+y})]$$

$$= [\max_{y \in [1,2]'} (5y + \frac{1}{5+y}), \min_{y \in [1,2]'} (10y + \frac{1}{10+y})]$$

$$= [71/7, 111/11]$$

We remark that the syntactic tree of f is optimal for $x, y \in I^*([1, +\infty))$.

4.2 Interpretability and Optimality

Optimality means JM-commutativity and equality between the semantical and the syntactical interval extensions, i.e., with an ideal arithmetic without rounding,

$$f^*(X) = fR(X) = f^{**}(X).$$

But in the case of rounding computations only Out($fR(X)$) or Inn($fR(X)$) are reachable and the optimality relations become

$$\text{Out}(fR(X)) \supseteq f^*(X) = fR(X) = f^{**}(X) \supseteq \text{Inn}(fR(X)).$$

Example 4.2.15 Let us consider the electrical circuit where the voltage v across the resistor r is $v = \dfrac{er}{\rho + r + s}$. The derivative of v with respect to the multi-incident variable r is

$$\frac{dv}{dr} = \frac{e(\rho + s)}{(\rho + r + s)^2},$$

which is positive for any positive domain of the variables. The derivative with respect to the first incidence of r is positive and the derivative with respect to the second incidence of r is negative. The theorem of coercion to optimality proves that for any positive intervals E, R, R_0, and S, the modal syntactic extension

$$V = \frac{E * R}{R_0 + \text{Dual}(R) + S} \tag{4.1}$$

is optimal, i.e., $v^*(E, R, R_0, S) = V$.

Let us suppose that the applied voltage source e, the internal resistance of the generator ρ, and the resistance r are inside the intervals $E' = [9, 11]'$, $R'_0 = [1.5, 2.5]'$ and $R' = [1, 3]'$ respectively. The regulation problem is to find an interval S such that if the resistance s takes values in the interval S', the voltage v takes values of the given interval V'.

If the voltage v is inside the domain $V' = [2, 4]'$, isolating S from the circuit equation (4.1), following the rules of interval arithmetic in $I^*(\mathbb{R})$, the result is

$$S = \frac{\text{Dual}(E * R)}{V} - \text{Dual}(R_0) - R = \left[\frac{33}{4} - 4.5, \frac{9}{2} - 3.5\right] \supseteq [3.8, 1]$$

(considering the inner rounding to a decimal digit). Applying the *-semantic theorem to the inclusion

$$v^*([9, 11], [1, 3], [1.5, 2.5], [3.8, 1]) = [9, 11] * [1, 3]/([1.5, 2.5] + [3, 1] + [3.8, 1])$$
$$\subseteq [2, 4]$$

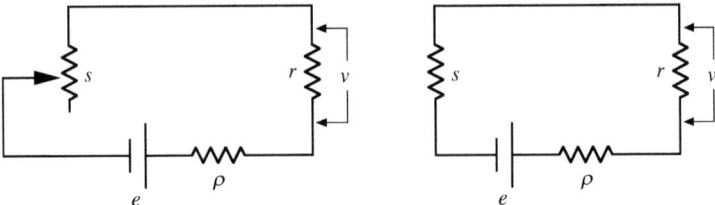

Fig. 4.1 Electrical circuits

with S an improper interval, i.e., a regulating interval, the equivalent logical formula is

$$(\forall e \in [9, 11]')\ (\forall r \in [1, 3]')\ (\forall \rho \in [1.5, 2.5]')\ (\exists s \in [1, 3.8]')\ (\exists v \in [2, 4]')$$
$$v = er/(\rho + r + s)$$

which means that a rheostat of resistance varying between 1 and 3.8 is necessary to force the output voltage to be inside the given interval $[2, 4]'$ (left part of the Fig. 4.1).

Suppose now that the voltage v can vary inside a wider interval, for example $V' = [2, 7]'$. Analogous computations then yield $S \supseteq [0.22, 1]$. Applying the *-semantic theorem to the inclusion

$$v^*([9, 11], [1, 3], [1.5, 2.5], [0.22, 1]) = [9, 11] * [1, 3]/([1.5, 2.5] + [3, 1] + [0.22, 1])$$
$$\subseteq [2, 7]$$

with S a proper interval, i.e., a fluctuation interval, the semantic result is

$$(\forall e \in [9, 11]')\ (\forall r \in [1, 3]')\ (\forall \rho \in [1.5, 2.5]')\ (\forall s \in [0.22, 1]')\ (\exists v \in [2, 7]')$$
$$v = er/(\rho + r + s)$$

which means that for any resistance between 0.22 and 1, the output voltage is inside the given domain $[2, 7]'$ and no control is necessary in this case (right part of the Fig. 4.1).

Definition 4.2.5 (Equivalent modal syntactic functions) Two modal syntactic functions fR and gR are equivalent when f and g coincide as continuous real functions, independently of the form of their syntactic trees.

Theorem 4.2.16 (Equivalent optimality) *Let $fR(X)$ be any uni-incident modal syntactic function with an equivalent modal syntactic function gR which has an optimal computation for X. Then $fR(X)$ is optimal for X.*

4.2 Interpretability and Optimality

Proof From the hypothesis

$$f^*(X) \subseteq fR(X) \subseteq f^{**}(X), \ f^*(X) = g^*(X), \ f^{**}(X) = g^{**}(X).$$

As $gR(X)$ is optimal,

$$g^*(X) = gR(X) = g^{**}(X) \Rightarrow f^*(X) = fR(X) = f^{**}(X). \qquad\blacksquare$$

Example 4.2.16 The $\mathbb{R}^3$ to $\mathbb{R}$ continuous real functions $f(a, b, c) = a(b + c)$ and $g(a, b, c) = ab + ac$ are equivalent and fR is a uni-incident modal syntactic function.

1. For $A = [-1, 1]$, $B = [1, 2]$ and $C = [3, 1]$, g is uniformly monotonic for the variable a and monotonic increasing for the two incidences of a. The computation $gR(A, B, C) = A * B + A * C = [-1, 1] * [1, 2] + [-1, 1] * [3, 1] = [-3, 3]$ is optimal. Theorem 4.2.16 implies the optimality of the computation

$$fR(A, B, C) = A*(B+C) = [-1, 1]*([1, 2]+[3, 1]) = [-1, 1]*[4, 3] = [-3, 3].$$

2. For $A = [-1, 1]$, $B = [3, 4]$ and $C = [-1, -3]$, g is uniformly monotonic for the variable a and for the two incidences of a. The computation

$$gR(A, B, C) = A * B + A * C = [-1, 1] * [3, 4] + [-1, 1] * [-1, -3] = [-5, 5]$$

is neither optimal nor interpretable, but

$$gR(A, B, C) = A * B + \mathrm{Dual}(A) * C$$
$$= [-1, 1] * [3, 4] + \mathrm{Dual}([-1, 1]) * [-1, -3] = [-1, 1]$$

is coerced to optimal. Theorem 4.2.16 implies the optimality of the computation

$$fR(A, B, C) = A * (B + C)$$
$$= [-1, 1] * ([3, 4] + [-1, -3]) = [-1, 1] * [2, 1] = [-1, 1].$$

3. For $A = [-1, 1]$, $B = [1, 2]$ and $C = [0, -4]$, g is only partially monotonic for the variable a. Then there is no criterion for an acceptable coercion on gR and

$$fR(A, B, C) = A * (B + C)$$
$$= [-1, 1] * ([1, 2] + [0, -4]) = [-1, 1] * [1, -2] = [0, 0].$$

can be different from f^*, from f^{**}, or from both.

4.3 Conditional Optimality

We will now introduce new conditions to assess the syntactic optimality of modal syntactic functions, which depend on the modalities of their arguments. The simplest examples are provided by the case of uni-modal syntactic functions whose modal syntactic computation is optimal, as far as its arguments are uni-incident.

Theorem 4.3.1 (Coercion to optimality for uni-modal arguments) *Let X, fR and XD satisfy the conditions of Theorem 4.2.14, and let X be a uni-modal interval vector. Then*

$$f^*(X) = fR(XD) = f^{**}(X).$$

Proof From Theorem 4.2.14 and the equality $f^*(X) = f^{**}(X)$ due to the uni-modality. ∎

Example 4.3.1 The $\mathbb{R}^2$ to $\mathbb{R}$ continuous function defined by

$$f(x_1, x_2) = x_1(x_2 + x_1)$$

for $X_1 = [-2, -1]$, $X_2 = [4, 8]$ is uni-modal; f is x_1-totally monotonic increasing since the partial derivative with respect to x_1 is positive, the partial derivative respect to the first incidence of x_1 is positive, and the partial derivative with respect to the second incidence of x_1 is negative. In accordance with Theorem 4.3.1

$$f^*(X) = f^{**}(X) = X_1 * (X_2 + \text{Dual}(X_1)) = [-2, -1] * ([4, 8] + [-1, -2])$$
$$= [-12, -3].$$

Definition 4.3.1 (Uni-incident list of vectors) $(x_1, \ldots, x_k)$ is a uni-incident list of vectors when every x_i has only uni-incident variable components and the different x_i share no common component.

Remark 4.3.1 $f^*(g_1(X_1), \ldots, g_k(X_k))$ will designate $f^*(X_1, \ldots, X_k)$, where f is the function $f_0 \circ (g_1, \ldots, g_k)$, and f_0 the main operator of the f-syntactic tree which, in what follows, always is supposed a two-variable partially monotonic function (exceptions to this general case will be indicated whenever the need arises).

Lemma 4.3.1 (Conditional *-optimality of two-variables partially monotonic operators) *If X is a proper vector and $f(g(x), h(y, v))$ is a uni-incident, continuous, and h-partially monotonic function, then*

$$f^*(g(X), h(Y, V)) = f_0^*(g^*(X), h^*(Y, V)),$$

where Y and V are the proper and improper arguments of h.

4.3 Conditional Optimality

Proof Let us define

$$G' = \{g(x) \mid x \in X'\},$$
$$G'_+ = \{g \in G' \mid f(g, h) \text{ is } h\text{-monotonic increasing }\},$$
$$G'_- = \{g \in G' \mid f(g, h) \text{ is } h\text{-monotonic decreasing}\}.$$

We have

$$f^*(g(X), h(Y, V)) = \bigvee_{x \in X'} \bigvee_{y \in Y'} \bigwedge_{v \in V'} [f_0(g(x), h(y, v)), f_0(g(x), h(y, v))]$$

$$= \bigvee_{g \in G'} \bigvee_{y \in Y'} \bigwedge_{v \in V'} [f_0(g, h(y, v)), f_0(g, h(y, v))]$$

$$= (\bigvee_{g \in G'_+} \bigvee_{y \in Y'} \bigwedge_{v \in V'} [f_0(g, h(y, v)), f_0(g, h(y, v))])$$

$$\vee (\bigvee_{g \in G'_-} \bigvee_{y \in Y'} \bigwedge_{v \in V'} [f_0(g, h(y, v)), f_0(g, h(y, v))])$$

// from the associativity of the lattice operators.

$$= (\bigvee_{g \in G'_+} [f_0(g, \min_{y \in Y'} \max_{v \in V'} h(y, v)), f_0(g, \max_{y \in Y'} \min_{v \in V'} h(y, v))])$$

$$\vee (\bigvee_{g \in G'_-} [f_0(g, \max_{y \in Y'} \min_{v \in V'} h(y, v)), f_0(g, \min_{y \in Y'} \max_{v \in V'} h(y, v))])$$

// $f_0(g, .)$ is uniformly monotonic for $g \in G'_+$ and $g \in G'_-$

$$= \bigvee_{g \in G'} f_0^*(g, h^*(Y, V)) = f_0^*(g^*(X), h^*(Y, V)).$$ ∎

Remark 4.3.2 The only conditions on $g(x)$ and $h(y, v)$ are continuity and uni-incidence (which supposes only the independence between the components of x and (y, v)).

Remark 4.3.3 If $g^*(X)$ and $h^*(Y, V)$ have *-optimal computations, denoted by $gR^*(X)$ and $hR^*(Y, V)$, then

$$f_0^*(g(X), h(Y, V)) = f_0^*(gR^*(X), hR^*(Y, V)),$$

denoted by $fR^*(X, Y, V)$ or $fR^*(g(X), h(Y, V))$. In this case, fR^* is *-optimal too.

Some examples can show the relevance of the hypotheses of this Lemma.

Example 4.3.2 The $\mathbb{R}^3$ to $\mathbb{R}$ continuous function $f(x, y, v) = x(y + v)$ can be expressed as $f_0(g, h) = gh$ with $g(x) = x$ and $h(y, v) = y + v$. For $X = [-1, 1]$, $Y = [1, 2]$, $V = [0, -4]$ f_0 is h-partially monotonic. Then, it is true that

$$f^*(g(X), h(Y, V)) = f_0^*(gR^*(X), hR^*(Y, V)) = f_0^*([-1, 1], [1, -2]) = [0, 0].$$

Example 4.3.3 The condition that X have proper uni-modality is necessary: with the same functional structure of the latter example, for $X = [1, -1]$, $Y = [0, 4]$, $V = [-2, -3]$, we will obtain

$$f^*(g(X), h(Y, V)) = [0.5, -0.5]$$

different from

$$f_0^*(gR^*(X), hR^*(Y, V)) = f_0^*([1, -1], [-2, 1]) = [0, 0].$$

Example 4.3.4 The modality of $h^*(Y, V)$ has no influence on the hypothetical *-optimality of fR^*: for the same system of functions of the Example 4.3.2 and for $X = [-3, 1]$, $Y = [-1, 2]$, $V = [1, -1]$, it results that

$$h^*(Y, V) = h^*([-1, 2], [1, -1]) = [0, 1]$$

which is a proper interval, but f is not JM-commutative since $f^*(g(X), h(Y, V)) = [-3, 1]$ is different from

$$f^{**}(g(X), h(Y, V)) = [-3, 2.25].$$

Anyway, as f_0 is h-partially monotonic and X is proper, as is guaranteed by Lemma 4.3.1,

$$f^*(g(X), h(Y, V)) = f_0^*(gR^*(X), hR^*(Y, V)) = f_0^*([-3, 1], [0, 1]) = [-3, 1].$$

Example 4.3.5 The g-monotonicity of f_0 is not necessary for *-optimality; for the $\mathbb{R}^3$ to $\mathbb{R}$ continuous function $f(x, y, v) = x^2(y + v)$, for $X = [-3, 1]$, $Y = [-1, 2]$, $V = [1, -1]$ we obtain

$$f^*(X, Y, V) = f^*([-3, 1], [-1, 2], [1, -1]) = [0, 9].$$

Expressed in the form $f_0(g, h) = g^2 h$ with $g(x) = x$ and $h(y, v) = y + v$ (f is not g-partially monotonic) is

$$f_0^*(gR^*(X), hR^*(Y, V)) = f_0^*([-3, 1], [0, 1]) = [0, 9].$$

Expressed in the form $f_0(g, h) = gh$ with $g(x) = x^2$ and $h(y, v) = y + v$ (f is g-partially monotonic) is

$$f_0^*(gR^*(X), hR^*(Y, V)) = f_0^*([0, 9], [0, 1]) = [0, 9].$$

4.3 Conditional Optimality

Example 4.3.6 Nevertheless, the h-monotonicity of f_0 is necessary for *-optimality; for the $\mathbb{R}^3$ to $\mathbb{R}$ continuous function $f(x, y, v) = x(y + v)^2$, for $X = [1, 2]$, $Y = [-4, 1]$, $V = [2, 0]$ we obtain

$$f^*(X, Y, V) = f^*([1, 2], [-4, 1], [2, 0]) = [1, 8].$$

Expressed in the form $f_0(g, h) = gh^2$ with $g(x) = x$ and $h(y, v) = y + v$ (f is not h-partially monotonic) is

$$f_0^*(gR^*(X), hR^*(Y, V)) = f_0^*([1, 2], [-2, 1]) = [0, 8]$$

different from $f^*(X, Y, V)$.

Lemma 4.3.2 (Conditional **-optimality of two-variables partially monotonic operators) *If U is an improper vector and $f(g(u), h(y, v))$ is a uni-incident, continuous and h-partially monotonic function, then*

$$f^{**}(g(U), h(Y, V)) = f_0^{**}(g^{**}(U), h^{**}(Y, V)),$$

where Y and V are the proper and improper arguments of h.

Proof The proof is the dual of that of Lemma 4.3.1 ∎

Remark 4.3.4 If $g^{**}(U)$ and $h^{**}(Y, V)$ do have **-optimal computations, denoted by $gR^{**}(U)$ and $hR^{**}(Y, V)$, then

$$f_0^{**}(g(U), h(Y, V)) = f_0^{**}(gR^{**}(U), hR^{**}(Y, V))$$

denoted by $fR^{**}(U, Y, V)$ or $fR^{**}(g(U), h(Y, V))$. In this case, f is **-optimal too.

Example 4.3.7 As we see in Example 4.3.3, for the $\mathbb{R}^3$ to $\mathbb{R}$ continuous function $f(x, y, v) = x(y + v)$, expressed as $f_0(g, h) = gh$ with $g(x) = x$ and $h(y, v) = y + v$ for $X = [1, -1]$, $Y = [0, 4]$, $V = [-2, -3]$, fR is not *-optimal; however it is **-optimal, since

$$f^{**}(g(X), h(Y, V)) = [0, 0]$$

and

$$f_0^{**}(gR^{**}(X), hR^{**}(Y, V)) = f_0^{**}([1, -1], [-2, 1]) = [0, 0].$$

Definition 4.3.2 (Split modality) The function $f(g_1(x_1), \ldots, g_k(x_k))$ satisfies the condition of split modality when f is JM-commutable and $X_1, \ldots, X_k$ are unimodal vectors.

Particularly the structure $f(g(x), h(v))$ satisfies the condition of split modality when f_0 is (g, h)-partially monotonic, X is a proper uni-modal vector, and V an improper uni-modal vector.

Theorem 4.3.2 (Split optimality) *If $f(g_1(x_1), \ldots, g_k(x_k))$ is a continuous function satisfying the split modality condition, then*

$$f^*(g_1(X_1), \ldots, g_k(X_k)) = f_0^*(g_1^*(X_1), \ldots, g_k^*(X_k))$$
$$= f_0^{**}(g_1^{**}(X_1), \ldots, g_k^{**}(X_k))$$
$$= f^{**}(g_1(X_1), \ldots, g_k(X_k)).$$

Proof The proof is essentially similar to that of Lemma 4.2.4 (right unary associativity). When $k = 2$, X_1 is a uni-modal proper vector, and X_2 is a uni-modal improper vector, the theorem is a direct consequence of Lemmas 4.3.1 and 4.3.2 (the lemmas of conditional *-optimality and **-optimality). ∎

Example 4.3.8 If $A_1, \ldots, A_k \in I^*(\mathbb{R})$, then the k-dimensional product $A_1 * \ldots * A_k$ is optimal. Effectively, let us suppose $A = (A_1, \ldots, A_p, A_{p+1}, \ldots, A_k)$ with $A_1, \ldots, A_p$ proper and $A_{p+1}, \ldots, A_k$ improper. Then

$$f^*(A) = f_0^*(g^*(A_1, \ldots, A_p), h^*(A_{p+1}, \ldots, A_k))$$
$$= (A_1 * \ldots * A_p) * (A_{p+1} * \ldots * A_k),$$

since the condition of split modality is verified and g and h are JM-commutable (they are uni-modal).

Lemma 4.3.3 (First lemma of lateral optimality) *Let $f(g(u), h(y, v))$ be a continuous function, h-uniformly monotonic in (Y, V). If U is an improper vector, if Y and V are, respectively, the proper and improper components of the h arguments, and if U and V have no common components, then f is *-optimal, i.e.,*

$$f^*(g(U), h(Y, V)) = f_0^*(g^*(U), h^*(Y, V)).$$

Proof Developing the right-hand side

$$f_0^*(g^*(U), h^*(Y, V)) = f_0^{**}(g^*(U), h^*(Y, V)) = \bigwedge_{g \in G'} f^*(g, h^*(Y, V)),$$

since f is h-uniformly monotonic function and consequently JM-commutable (see Theorem 3.4.4). Developing the left-hand side

$$f^*(g(U), h(Y, V)) = \bigvee_{y \in Y'} \bigwedge_{u \in U'} \bigwedge_{v \in V'} [f(g(u), h(y, v)), f(g(u), h(y, v))].$$

4.3 Conditional Optimality

Since $f(g, h)$ is h-uniformly monotonic, for example h-uniformly monotonic increasing, then

$$\bigvee_{y \in Y'} \bigwedge_{g \in G'} [f(g, \max_{v \in V'} h(y, v)), f(g, \min_{v \in V'} h(y, v))].$$

As $f(g(u), h_1(y))$ and $f(g(u), h_2(y))$, with

$$h_1(y) = \max_{v \in V'} h(y, v),$$

$$h_2(y) = \min_{v \in V'} h(y, v),$$

are split modality functions in (G, Y), they are JM-commutable. Consequently

$$f^*(g(U), h(Y, V)) = \bigwedge_{g \in G'} \bigvee_{y \in Y'} [f(g, \max_{v \in V'} h(y, v)), f(g, \min_{v \in V'} h(y, v))$$

$$= \bigwedge_{g \in G'} [f(g, \min_{y \in Y'} \max_{v \in V'} h(y, v)), f(g, \max_{y \in Y'} \min_{v \in V'} h(y, v))]$$

$$= \bigwedge_{g \in G'} f^*(g, h^*(Y, V)).$$

Analogous reasoning takes care of the case when $f(g, h)$ is h-uniformly monotonic decreasing. ∎

Lemma 4.3.4 (Second lemma of lateral optimality) *Let $f(g(x), h(y, v))$ be a function continuous and h-uniformly monotonic in (Y, V). If X is a proper vector, if Y and V are, respectively, the proper and improper components of the h arguments, and if X and Y have no common components, then f is **-optimal, i.e.,*

$$f^{**}(g(X), h(Y, V)) = f_0^{**}(g^{**}(X), h^{**}(Y, V)).$$

Proof The proof is the dual of that of Lemma 4.3.3. ∎

Theorem 4.3.3 (Lateral optimality) *Suppose $f(g(x), h(y, v))$ is a continuous function, h-uniformly monotonic in (Y, V). Then if X is a uni-modal vector, if Y and V are, respectively, the proper and the improper components of the h-argument, and if X, Y and V have no common component, then*

$$f^*(g(X), h(Y, V)) = f_0^*(g^*(X), h^*(Y, V))$$
$$= f_0^{**}(g^{**}(X), h^{**}(Y, V)) = f^{**}(g(X), h(Y, V)).$$

Proof The first and third equalities come from Lemmas 4.3.1 and 4.3.2 (conditioned *-optimality and **-optimality) and from Lemmas 4.3.3 and 4.3.4 (lateral optimality). The second equality comes from the fact that when X is uni-modal, $g^*(X) = g^{**}(X)$ is h JM-commutable because it is h-uniformly nonotonous, therefore $h^*(Y,V) = h^{**}(Y,V)$, and from Theorem 3.4.4 applied to f_0, $f_0^* = f_0^{**}$. ∎

Example 4.3.9 Consider the continuous real function $f(x,y,v) = x(y+v)$, expressed by $f_0(g,h) = gh$ with $g(x) = x$ and $h(y,v) = y+v$ for $X = [1,0]$, $Y = [-4,0]$, $V = [2,1]$, this function satisfies the lateral optimality condition since it is h-uniformly monotonic and X is uni-modal. Then

$$f^*(g[1,0], h([-4,0],[2,1])) = f^{**}(g[1,0], h([-4,0],[2,1])) = [0,0].$$

Remark 4.3.5 Whenever $h^*(Y,V)$ had an optimal modal syntactic computation $hR(Y,V)$, we could write $fR(g(X), h(Y,V)) = f_0^*(gR(X), hR(Y,V))$.

Remark 4.3.6 The functions g and h can be vectors, as long as all the arguments of every $g_i(X_i)$ have the same modality, f is uniformly monotonic for each h_i, and overall uni-incidence holds.

4.3.1 Conditional Tree-Optimality

Definition 4.3.3 (Modally conditioned optimal modal syntactic operators) A modal syntactic function $fR(X)$ is called a c-optimal modal syntactic operator when f satisfies one of the following conditions

1. f is of split modality,
2. f is laterally optimal.

Definition 4.3.4 (Conditionally tree-optimal modal syntactic functions) A modal syntactic function $fR(X)$ is c-tree optimal if any of its non-uniformly monotonic sub-trees is optimal or c-optimal.

Remark 4.3.7 The restrictions required by c-optimality propagate up the syntactic tree.

The c-tree-optimal condition extends the set of tree-optimal syntactic extensions to which theorems similar to Theorems 4.2.5 and 4.2.15 can be applied, as the following results show.

Theorem 4.3.4 (Conditional optimality) If $fR(X)$ is uni-incident and c-tree-optimal, then $fR(X)$ is optimal, i.e.,

$$f^*(X) = fR(X) = f^{**}(X).$$

Proof This follows from the previous lemmas and theorems. ∎

Example 4.3.10 Consider the continuous real function $f(x,y,v) = x(y+v)$ given by $f_0(g,h) = gh$ with $g(x) = x$ and $h(y,v) = y+v$ for $X = [1,0]$, $Y = [-4,0]$, $V = [2,1]$. This function is not tree-optimal. Nevertheless it is c-optimal, because it

4.3 Conditional Optimality

satisfies the condition of lateral optimality, see Example 4.3.9, and the conditioned optimal coercion theorem gives

$$f^*(X, Y, V) = f^{**}(X, Y, V) = fR(X, Y, V) = fR([1, 0], [-4, 0], [2, 1]) = [0, 0].$$

The syntactic tree of fR is

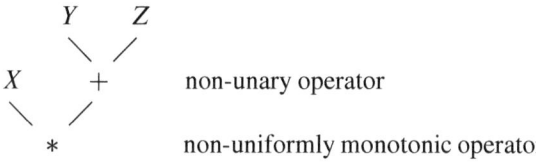

Theorem 4.3.5 (Coercion to conditional optimality) *Let X be an interval vector; let $fR(X D)$ be defined and c-tree-optimal on the domain $Prop(X)$ and totally monotonic for all its multi-incident components. Let $X D$ be defined as the enlarged vector of X such that each incidence of every multi-incident component is included in $X D$ as an independent component, but transformed into its dual if the corresponding incidence point has a monotonicity sense contrary to the global one of the corresponding X-component. Then*

$$f^*(X) = fR(X D) = f^{**}(X).$$

Proof The proof is a re-statement of that of Theorem 4.2.15, adjusted for the case of c-optimality. Note the role of $fR(X D)$ in the demonstration of this theorem. ∎

Example 4.3.11 The function $f(x, y, z) = x(y + z) - y$ for $X = [1, 0]$, $Y = [2, 3]$ and $Z = [-3, -1]$ is not tree-optimal; therefore, it is not possible to apply the optimal-coercion theorem. However, the modal syntactic function associated to f is c-tree-optimal (the product satisfies the conditions of Theorem 4.3.3) and hence it is possible to apply the conditioned optimal coercion Theorem 4.3.5. Then

$$f^*(X) = fR(X D) = X * (\text{Dual}(Y) + Z) - Y$$
$$= [1, 0] * ([3, 2] + [-3, -1]) - [2, 3] = [-3, -2].$$

The syntactic tree of fR is

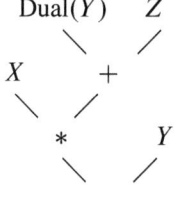

The equivalent function $g(x,y,z) = xy + xz - y$ does not have an optimal computation since it is not globally uniformly monotonic for the multi-incident variable x. The relevant difference from f is that x is not multi-incident in f, but it is multi-incident in g.

We may observe that the c-optimality cannot, in this example, be broken by the transformation from X to XD, since the branch conditioned to uni-modality contains the only variable X.

Remark 4.3.8 In accordance with the corresponding definitions of optimality and tree-optimality conditions, both imply c-conditional optimality. Therefore Theorems 4.3.5 and 4.3.4 provide a more extensive class of optimal modal syntactic functions.

Definition 4.3.5 (Interval property) We call an interval property any property which, in case it holds on an interval-domain $(X'_1, \ldots, X'_k)$, it then also holds on any interval sub-domain $(Y'_1, \ldots, Y'_k) \subseteq (X'_1, \ldots, X'_k)$.

Theorem 4.3.6 *The classes of tree-optimal and c-tree-optimal modal syntactic functions do not exhaust all the cases of modal syntactic optimality.*

Proof Since tree-optimality and c-tree-optimality are interval properties; however JM-commutability and modal syntactic optimality are not, as the following example shows. ∎

Example 4.3.12 Consider the continuous function $f(x,y) = |x+y|$. For the intervals $(X,Y) = ([1,-1], [0,2])$, the modal syntactic function $fR(x,y)$ is optimal, since

$$f^*(X,Y) = f^{**}(X,Y) = fR(X,Y) = [1,1];$$

however, after reducing the interval domain $([-1,1]', [0,2]')$ to a sub-domain $([-1,1]', [0,1]')$, for $(X,Y) = ([1,-1], [0,1])$,

$$f^*(X,Y) = [1,0] \text{ and } f^{**}(X,Y) = [1,0.5]$$

and $fR(X,Y)$ is not optimal. If the reference domain is enlarged to $([-1,1]', [-1,2]')$, the JM-commutability is lost, since for $([1,-1], [-1,2])$

$$f^*(X,Y) = [1, 1.5] \text{ and } f^{**}(X,Y) = [0, 1.5].$$

Remark 4.3.9 The relation $f^*(X) \subseteq fR(X) \subseteq f^{**}(X)$ for uni-incident modal syntactic functions, makes optimality a subsidiary property of JM-commutability.

4.4 m-Dimensional Computations

The previous results about interpretability can be extended to systems of functions under certain hypotheses.

4.4 m-Dimensional Computations

If $X \in I^*(\mathbb{R}^m)$ and $f : \mathbb{R}^k \to \mathbb{R}^m$ defined by $f(x) = (f_1(x), \ldots, f_m(x))$ continuous in X' is such that the functions $f_1, \ldots, f_m$ do not share any improper variable, the *-semantic theorems can be applied separately to $(f_1^*(X), \ldots, f_m^*(X))$ so that the conjunction of the logical formulas obtained provide the semantics for $f^*(X) = (f_1^*(X), \ldots, f_m^*(X))$. Analogously for the semantic interpretation of $f^{**}(X)$.

Example 4.4.1 For $f : \mathbb{R}^3 \to \mathbb{R}^2$ defined by

$$f(x_1, x_2, x_3) = (x_1 + x_2 + x_3, x_1 x_2 - x_1)$$

with $X = ([1, 6], [2, 6], [9, -3])$, there are no common variables corresponding to improper intervals. As

$$f_1^*(X) = fR(X_1, X_2, X_3) = [1, 6] + [2, 6] + [9, -3] = [12, 9],$$

which is equivalent to

$$(\forall x_1 \in [1, 6]')(\forall x_2 \in [2, 6]') \; (\forall z_1 \in [9, 12]') \; (\exists x_3 \in [-3, 9]') \; z_1 = x_1 + x_2 + x_3$$

and f_2 satisfies the conditions of Theorem 4.2.15,

$$f_2^*(X) = fR(X D) = [1, 6] * [2, 6] - \mathrm{Dual}([1, 6]) = [1, 30],$$

which is equivalent to

$$(\forall x_1 \in [1, 6]') \; (\forall x_2 \in [2, 6]') \; (\exists z_2 \in [1, 30]') \; z_2 = x_1 x_2 - x_1,$$

then

$$f^*([1, 6], [2, 6], [9, -3]) = (f_1^*([1, 6], [2, 6], [9, -3]), f_2^*([1, 6], [2, 6]))$$
$$= ([12, 9], [1, 30])$$

is equivalent to

$$(\forall x_1 \in [1, 6]')(\forall x_2 \in [2, 6]') \; (\forall z_1 \in [9, 12]') \; (\exists z_2 \in [1, 30]') \; (\exists x_3 \in [-3, 9]')$$
$$(z_1 = x_1 + x_2 + x_3, z_2 = x_1 x_2 - x_1).$$

Example 4.4.2 Consider a physical system consisting of two connected tanks which contain saline solution. Tank 1 holds v_1 l and tank 2 holds v_2 l. Denoting by $x(t)g$ and $y(t)g$ the mass of salt in tanks 1 and 2, respectively, every second $u(t)g$ of salt is introduced into the tanks at a rate of k into tank 1 and $(1 - k)$ into the tank 2. Every second c_1 l of solution flow into tank 1 from tank 2 and c_2 l flow from tank 1 into tank 2. The mass of salt in each one of the two tanks at the initial instant are

Fig. 4.2 Two tanks system

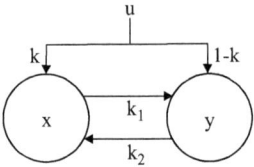

$x(0)g$ and $y(0)g$. The problem is to know the evolution of the mass of salt in each tank during the time of the simulation (see Fig. 4.2).

Taking into account the balance of masses of salt in both tanks, the discrete mathematical model in differences for this physical system is

$$x(t + \Delta_t) = (1 - \Delta_t k_1)x(t) + \Delta_t k_2 y(t) + \Delta_t k u(t)$$
$$y(t + \Delta_t) = (1 - \Delta_t k_2)y(t) + \Delta_t k_1 x(t) + \Delta_t (1 - k)u(t),$$

where $k_1 = c_1/v_1$, $k_2 = c_2/v_2$ and $\Delta_t < \min(1/k_1, 1/k_2)$, which provides the variation of the mass of salt in both tanks along the time. Thus, for

$$v_1 = 11\,l$$
$$v_2 = 21\,l$$
$$c_1 = 0.7\,l/s$$
$$c_2 = 0.7\,l/s$$
$$k = 0.2$$
$$u(t) = \begin{cases} 0.7\,g/s & \text{for } t < 20 \\ 0.0\,g/s & \text{for } t \geq 20 \end{cases}$$
$$x(0) = 7\,g$$
$$y(0) = 4\,g$$
$$\Delta_t = 1\,s,$$

the obtained results of the simulation are summarized in Table 4.1 which shows the evolution of the mass of salt in each tank.

Introducing interval uncertainties in the physical system, let K, K_1 and K_2 be intervals of variation for the parameters of the system, $X(t)$ and $Y(t)$ be intervals of variation for the state variables, and $X(0)$ and $Y(0)$ be intervals of variation for the initial states. Now the logical statement of the problem is to find intervals $X(t)$ and $Y(t)$, for $t = 0, \Delta_t, 2\Delta_t, \ldots, n\Delta_t$, such that

$$(\forall k_1 \in K_1)\,(\forall k_2 \in K_2)\,(\forall k \in K)\,(\forall u(t) \in U(t))$$
$$(\forall x(t) \in X(t))\,(\exists (y(t) \in Y(t))\,(\exists x(t + \Delta_t) \in X(t + \Delta_t))$$
$$x(t + \Delta_t) = (1 - \Delta_t k_1)x(t) + \Delta_t k_2 y(t) + \Delta_t k u(t)$$

4.4 m-Dimensional Computations

Table 4.1 Simulations

t	$x(t)$	$y(t)$
1	6.82788	4.87212
2	6.69578	5.70422
...	...	...
20	8.10952	16.8905
21	8.29648	17.4035
...	...	...
70	8.83074	16.8693
71	8.8311	16.8689
...	...	...

and

$$(\forall k_1 \in K_1)\,(\forall k_2 \in K_2)\,(\forall k \in K)\,(\forall u(t) \in U(t))$$
$$(\forall y(t) \in Y(t))\,(\exists (x(t) \in X(t))\,(\exists y(t + \Delta_t) \in Y(t + \Delta_t))$$
$$y(t + \Delta_t) = (1 - \Delta_t k_2) y(t) + \Delta_t k_1 x(t) + \Delta_t (1 - k) u(t)$$

which, by the *-Semantic Theorem 3.3.1, are equivalent to the following interval computations

$$X(t + \Delta_t) = (1 - \Delta_t K_1) * X(t) + \Delta_t K_2 * \text{Impr}(Y(t)) + \Delta_t K * U(t)$$
$$Y(t + \Delta_t) = (1 - \Delta_t K_2) * Y(t) + \Delta_t K_1 * \text{Impr}(X(t)) + \Delta_t (1 - K) * U(t),$$

for $t = 0, \Delta_t, 2\Delta_t, \ldots, n\Delta_t$. Thus, for

$$V_1 = [10, 12]\, l$$
$$V_2 = [20, 22]\, l$$
$$C_1 = [0.68, 0.72]\, l/s$$
$$C_2 = [0.68, 0.72]\, l/s$$
$$K = [0.18, 0.22]$$
$$U(t) = \begin{cases} [0.65, 0.75]\, g/s & \text{for } t < 20 \\ [0.0, 0.0]\, g/s & \text{for } t \geq 20 \end{cases}$$
$$X(0) = [6.5, 7.5]\, g$$
$$Y(0) = [3.5, 4.5]\, g$$
$$\Delta_t = 1\, s,$$

the results of the interval simulation are summarized in Table 4.2 which shows the uncertainty intervals of the mass of salt in each tank for several steps of the simulation.

Table 4.2 Interval simulations

t	$X(t)$	$Y(t)$
1	[6.28809, 7.366]	[4.306, 5.44391]
2	[6.12061, 7.26861]	[5.07539, 6.34339]
...	...	...
20	[7.17966, 9.06766]	[15.3723, 18.4603]
21	[7.35031, 9.27223]	[15.8398, 19.0217]
...	...	...
70	[7.91813, 9.75162]	[15.3604, 18.4539]
71	[7.91842, 9.752]	[15.36, 18.4536]
...	...	...

When there exist shared improper variables, the conjunction of the logical formulas resulting from the application of the Semantic Theorems to the different component functions is not possible due to the existential quantifiers affecting the same variable in different functions. In this case the following theorems are valid.

Theorem 4.4.1 (Interpretability of m-dimensional computations) *Let $X \in I^*(\mathbb{R}^m)$ and $f : \mathbb{R}^k \to \mathbb{R}^m$ defined by $f(x) = (f_1(x), \ldots, f_m(x))$ continuous in X'. Let $(Z_1, \ldots, Z_m)$ be a system of interpretable outer-rounded computations of*

$$(f_1^*(A_1), \ldots, f_m^*(A_m)),$$

where $(A_1, \ldots, A_m)$ are derived from X so as to have the same proper components as X, but with each multi-incident improper component of X transformed into a point-wise interval defined for any point of its domain. In this case,

$$(\forall x_p \in X'_p) \, Q^*(z, Z) \, (\exists x_i \in X'_i) \, z = f(x_p, x_i),$$

with $X = (X_p, X_i)$ and $Q^*(z, Z)$ being the sequence of the prefixes

$$Q(z_1, Z_1), \ldots, Q(z_m, Z_m),$$

with the ones corresponding to the universal quantifiers heading the sequence.

Proof Under these hypotheses, no improper component of X is repeated in more than one $f_i^*(X)$, because for every multi-incident universal variable, all its incidences have been transformed into a point-wise interval defined for any of the points of their domains (if this were not the case, multiple existential quantifiers over the same variable cannot be extracted to obtain the prenex form of the semantic theorem for the overall m-dimensional system). The result is obtained by the logical product of the separate semantics of each of the computations $fR_1(A_1), \ldots, fR_m(A_m)$. ∎

Theorem 4.4.2 (Dual interpretability of m-dimensional computations) *Let $X \in I^*(\mathbb{R}^k)$ and $f : \mathbb{R}^k \to \mathbb{R}^m$ be defined by $f(x) = (f_1(x), \ldots, f_m(x))$ continuous on X'. Let $(Z_1, \ldots, Z_m)$ be a system of interpretable inner-rounded computations of*

$$(f_1^{**}(A_1), \ldots, f_m^{**}(A_m)),$$

4.4 m-Dimensional Computations

where $(A_1, \ldots, A_m)$ *are derived from X so as to have the same improper components of X, but with each multi-incident proper component of X transformed into a point-wise interval defined for any point of its domain. In this case*

$$(\forall x_i \in X'_i)\, Q^*(z, \text{Dual}(Z))\, (\exists x_p \in X'_p)\, z = f(x_p, x_i),$$

with $X = (X_p, X_i)$ and $Q^*(z, \text{Dual}(Z))$ being the sequence of the prefixes

$$Q(z_1, \text{Dual}(Z_1)), \ldots Q(z_m, \text{Dual}(Z_m)),$$

with the ones corresponding to the universal quantifiers heading the sequence.

Proof Similar to the previous one. ∎

Example 4.4.3 For $f : \mathbb{R}^3 \to \mathbb{R}^2$ defined by

$$f(x_1, x_2, x_3) = (x_1 + x_2 + x_3, x_1 x_2 - x_1)$$

with $X = ([6, 1], [6, 2], [-3, 2])$, we may proceed as follows:

$$f_1(x_1, x_2, x_3) = x_1 + x_2 + x_3$$
$$f_2(x_1, x_2) = x_1 x_2 - x_1,$$
$$A_1 = ([1, 1], [2, 2], [-3, 2])$$
$$A_2 = ([1, 1], [2, 2])$$
$$Z_1 = [1, 1] + [2, 2] + [-3, 2] = [0, 5]$$
$$Z_2 = [1, 1] * [2, 2] - [1, 1] = [1, 1]$$

where Z_1 is an interpretable modal syntactic computation of $f_1^*(A_1)$ and Z_2 of $f_2^*(A_2)$ after the substitution of the multi-incident interval $[6, 1]$ by $[1, 1]$ and $[6, 2]$ by $[2, 2]$. Therefore, the interpretation of these results is

$$(\forall x_3 \in [-3, 2]')\, (\forall z_2 \in [1, 1]')\, (\exists z_1 \in [0, 5]')\, (\exists x_1 \in [1, 6]')\, (\exists x_2 \in [2, 6]')$$
$$(z_1 = x_1 + x_2 + x_3, z_2 = x_1 x_2 - x_1).$$

Similarly

$$A_1 = ([6, 6], [3, 3], [-3, 2])$$
$$A_2 = ([6, 6], [3, 3])$$
$$Z_1 = [6, 6] + [3, 3] + [-3, 2] = [6, 11]$$
$$Z_2 = [6, 6] * [3, 3] - [6, 6] = [12, 12]$$

after the substitution of the multi-incident interval [6, 1] by [6, 6] and [6, 2] by [3, 3]. Therefore, the interpretation of these results is

$$(\forall x_3 \in [-3, 2]')\, (\forall z_2 \in [12, 12]')\, (\exists z_1 \in [6, 11]')\, (\exists x_1 \in [1, 6]')\, (\exists x_2 \in [2, 6]')$$
$$(z_1 = x_1 + x_2 + x_3, z_2 = x_1 x_2 - x_1).$$

Theorem 4.4.3 (Interval interpretability of m-dimensional computations) *Let $X \in I^*(\mathbb{R}^k)$ and suppose $f : \mathbb{R}^k \to \mathbb{R}^m$ defined by $f(x) = (f_1(x), \ldots, f_m(x))$ is continuous in X'. Let $A_1, \ldots, A_m$ be the vectors containing the arguments of $f_1, \ldots, f_m$, derived from X so as to have the same proper components as X, but with all except one multi-incident improper component of X transformed into its dual in each X_j. Let $(Z_1, \ldots, Z_m)$ be a system of interpretable outer-rounded computations of*

$$(f_1^*(A_1), \ldots, f_m^*(A_m)),$$

i.e., $f_j R(A_j) \subseteq Z_j$ for $j = 1, \ldots, m$. In this case

$$(\forall x_p \in X_p')\, Q^*(z, Z)\, (\exists x_i \in X_i')\, z = f(x_p, x_i),$$

with $X = (X_p, X_i)$ and $Q^*(z, Z)$ being the sequence of the prefixes

$$Q(z_1, Z_1), \ldots, Q(z_m, Z_m),$$

with the ones corresponding to the universal quantifiers heading the sequence.

Proof Let us define the distance function

$$d(x, z) = (f_1(x) - z_1)^2 + \ldots + (f_m(x) - z_m)^2. \tag{4.2}$$

By definition, $A_1, \ldots, A_m$ are a T-transformation of X, so in accordance with Theorem 4.2.8,

$$d^*(X, \text{Dual}(Z_1), \ldots, \text{Dual}(Z_m)) \subseteq dR(A_1, \ldots, A_m, \text{Dual}(Z_1), \ldots, \text{Dual}(Z_m)).$$

The definition of Z_j is equivalent to

$$f_j R(A_j) - \text{Dual}(Z_j) \subseteq [0, 0],$$

for $j = 1, \ldots, m$, therefore from (4.2) and the inclusion-isotony of the operators

$$dR(A_1, \ldots, A_m, \text{Dual}(Z_1), \ldots, \text{Dual}(Z_m)) \subseteq [0, 0]$$

Consequently

$$d^*(X, \text{Dual}(Z_1), \ldots, \text{Dual}(Z_m)) \subseteq [0, 0].$$

4.4 m-Dimensional Computations

From the *-semantic theorem, applied to this inclusion,

$$(\forall x_p \in X'_p)\, Q^*(z, Z')\, (\exists x_i \in X'_i)\, (f_1(x_p, x_i) - z_1)^2 + \ldots + (f_m(x_p, x_i) - z_m)^2 = 0$$

equivalent to

$$(\forall x_p \in X'_p)\, Q^*(z, Z')\, (\exists x_i \in X'_i)\, (z_1 = f_1(x_p, x_i), \ldots, z_m = f_m(x_p, x_i)) = 0,$$

which is the conclusion of the theorem. ∎

Theorem 4.4.4 (Dual interval interpretability of m-dimensional computations) *Let $X \in I^*(\mathbb{R}^k)$ and suppose $f : \mathbb{R}^k \to \mathbb{R}^m$ defined by $f(x) = (f_1(x), \ldots, f_m(x))$ is continuous on X'. Let $A_1, \ldots, A_m$ be the vectors containing the arguments of $f_1, \ldots, f_m$, derived from X so as to have the same proper components as X, but with all except one multi-incident proper component of X transformed into its dual in each X_j. Let $(Z_1, \ldots, Z_m)$ be a system of interpretable inner-rounded computations of*

$$(f_1^{**}(A_1), \ldots, f_m^{**}(A_m)),$$

i.e., $Z_j = f_j R(A_j)$ for $j = 1, \ldots, m$. In this case

$$(\forall x_i \in X'_i)\, Q^*(z, \mathrm{Dual}(Z))\, (\exists x_p \in X'_p)\, z = f(x_p, x_i),$$

with $X = (X_p, X_i)$ and $Q^*(z, \mathrm{Dual}(Z))$ being the sequence of the prefixes

$$Q(z_1, \mathrm{Dual}(Z_1)), \ldots Q(z_m, \mathrm{Dual}(Z_m)),$$

with the ones corresponding to the universal quantifiers heading the sequence.

Example 4.4.4 For $f : \mathbb{R}^3 \to \mathbb{R}^2$ defined by

$$f(x_1, x_2, x_3) = (x_1 + x_2 + x_3, x_1 x_2 - x_1)$$

with $X = ([6, 1], [6, 2], [-3, 2])$, we may proceed as follows:

$$f_1(x_1, x_2, x_3) = x_1 + x_2 + x_3$$
$$f_2(x_1, x_2, x_3) = x_1 x_2 - x_1,$$
$$A_1 = ([1, 6], [6, 2], [-3, 2])$$
$$A_2 = ([6, 1], [2, 6])$$
$$Z_1 = [1, 6] + [6, 2] + [-3, 2] = [4, 10]$$
$$Z_2 = [6, 1] * [2, 6] - [1, 6] = [6, 5]$$

where Z_1 is an interpretable modal syntactic computation of $f_1^*(A_1)$ and Z_2 of $f_2^*(A_2)$. Therefore, the semantics of these results is

$$(\forall x_3 \in [-3, 2]')\ (\forall z_2 \in [5, 6]')\ (\exists z_1 \in [4, 10]')\ (\exists x_1 \in [1, 6]')\ (\exists x_2 \in [2, 6]')$$
$$(z_1 = x_1 + x_2 + x_3, z_2 = x_1 x_2 - x_1).$$

The arguments $(A_1, \ldots, A_m)$ can be possibly modified by coercion to interpretability of the corresponding ones of X (Theorems 4.2.8 and 4.2.9). So, in a computational program of the form

$$f_1^*(X) \subseteq Z_1, \ldots, f_m^*(X) \subseteq Z_m,$$

assuming the *-semantics, no improper component of X can be repeated, keeping its own modality in more than one $f_i^*(X)$, since multiple existential quantifiers over the same variable cannot be extracted to obtain the prenex form of the semantic theorem for the overall m-dimensional system. If this is not the case, for every multi-incident improper variable, all its incidences but one must be changed into its dual.

After imposing the previous condition, the functions $f_i^*(X)$ should be individually coerced to optimality to obtain component-wise optimality

Example 4.4.5 Consider f the continuous function from $\mathbb{R}^2$ to $\mathbb{R}^2$ defined by the formula $f(x_1, x_2) = (x_1 + x_2, x_1 x_2 - x_2)$. For $X = (X_1, X_2) = ([1, 3], [2, -1])$ there exist multi-incident improper components. To get an interpretable computation, according to the n-dimensional semantic theorem, we must transform some improper incidences into their dual. For example for

$$(X_1 + X_2, X_1 * \text{Dual}(X_2) - \text{Dual}(X_2)) = ([3, 2], [-5, 7])$$

the logical product of the semantic theorems for each component gives the overall semantics

$$(\forall x_1 \in [1, 3]')(\forall z_1 \in [3, 2]')(\exists z_2 \in [-5, 7]')(\exists x_2 \in [2, -1]')$$
$$(z_1 = x_1 + x_2, z_2 = x_1 x_2 - x_2).$$

To make an effective component-wise optimal computation, the system $(fR_i(X))$ would become

$$(X_1 + X_2, X_1 * \text{Dual}(X_2) - X_2) = ([3, 2], [-2, 4]),$$

since the second incidence of X_2 in the second equation of the system is of a monotonicity sense contrary to the one of the overall variable in this second component.

4.5 Additional Examples

The semantic theorems prove the equivalence between some modal interval inclusions concerning to *-semantic extensions, and some first order logical formulas, where the domains are intervals, the universal quantifiers precede the existential ones and there is a unique predicate involving a continuous real function.

Practical problems involving uncertainties can be stated by means of these types of formulas. In fact, the logical statement is the core of almost all technical problems dealing with uncertainties, hardly stated without their implicit or explicit logical formulation. When the uncertainty sets are intervals, MIA techniques are very suitable for solving them, following three steps:

- State the problem as a formula of first order logic.
- If this formula accomplishes the previous requirements, use the *-semantic theorem to reduce it to modal interval inclusions affecting to *-semantic extensions of the functions involved.
- Verify the inclusions computing the corresponding *-semantic extensions.

So, the problem is reduced to computing a *-semantic extension $f^*(X)$ of a continuous function f to a modal interval vector X. The difficulty of this computation depends on the function f. Under some monotonicity conditions, several results of MIA reduce the computation of f^* to simple modal interval arithmetic operations. When the function involved does not satisfy these conditions, some results can be partially applied to obtain better approximations by splitting the variable space by means of a branch-and-bound algorithm (see Chap. 7).

Several technical examples of application are developed in this section to illustrate this process. They contain simple functions and try to be a bridge between modal interval analysis and the world of an engineer. Their source is the technical report [88].

The first example shows the semantical ability of the modal intervals to give results which are logically sure when uncertainties are present, unlike other classical procedures such as classical error analysis.

Example 4.5.1 For an ideal lens, the distance g between an object and the lens, the focal length f, and the distance b between the lens and the image of the object, are related by

$$g = \frac{1}{\frac{1}{f} - \frac{1}{b}}.$$

Let us suppose that the focal length is $f = 20 \pm 1$ and the distance between the image and the lens is $b = 25 \pm 1$ (in centimeters). The problem is to find the distance between the object and the lens together with its uncertainty due to the uncertainties

Fig. 4.3 Lens

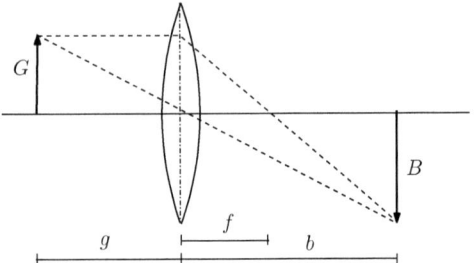

in the measurements (see Fig. 4.3). This question can be handled by means of an error estimation method: as $f_0 = 20$ and $b_0 = 25$, then

$$g_0 = \frac{1}{\frac{1}{20} - \frac{1}{25}} = 100;$$

as the absolute errors are $\Delta f = 1$ and $\Delta b = 1$, then the error estimate for g is

$$\Delta g = (1 - \frac{f_0}{b_0})^{-2} \Delta f + (\frac{b_0}{f_0} - 1)^{-2} \Delta b = (1 - \frac{20}{25})^{-2} + (\frac{25}{20} - 1)^{-2} = 41.$$

The answer would be $g = 100 \pm 41$. This variation is the interval $G' = [59, 141]'$, but this result is wrong, as the interval solution of the problem will show.

As $f \in [19, 21]'$ and $b \in [24, 26]'$, the logical statement of this problem is to find an interval G such that

$$(\forall f \in [19, 21]') \, (\forall b \in [24, 26]') \, (\exists g \in G') \, g = \frac{1}{\frac{1}{f} - \frac{1}{b}}.$$

which is equivalent to the modal inclusion

$$g^*([19, 21], [24, 26]) \subseteq G.$$

This function g is tree-optimal and uni-incident, and so

$$g^*([19, 21], [24, 26]) = \frac{1}{\frac{1}{[19,21]} - \frac{1}{[24,26]}} \subseteq [70.57, 168.01].$$

Hence $g \in [70.57, 168.01]'$ and the solution is the interval $G' = [70.57, 168.01]'$.

The following example illustrates the use of the Theorem of Coercion to compute the *-semantic extension for some types of functions.

Example 4.5.2 A procedure to determine the density of a fluid is to measure the weight of a body having volume v in the fluid, and to weigh it also in other fluids

4.5 Additional Examples

with known densities. A body submerged in a liquid of density ρ meets a lifting force equal to

$$-\rho v g$$

where g is the gravitational acceleration. Let us suppose that the body is a cube of edge length equal to 1 cm. Taking into account the uncertainties of the measurement processes, let us suppose that the cube's weights in grams are: in the air, $m_1 \in [9, 10]'$, submerged in water, $m_2 \in [0, 0.1]'$, and submerged in the fluid of unknown density, $m_3 \in [2.7, 3]'$. From

$$m_1 g - \rho_{H_2O} v g = m_2 g$$
$$m_1 g - \rho v g = m_3 g$$

the unknown density is

$$\rho = \rho_{H_2O} \frac{m_1 - m_3}{m_1 - m_2}.$$

Supposing $\rho_{H_2O} \in [0.99, 1.01]'$, the logical statement of this problem is to find an interval R such that

$$(\forall m_1 \in [9, 10]')\,(\forall m_2 \in [0, 0.1]')\,(\forall m_3 \in [2.7, 3]')\,(\forall \rho_{H_2O} \in [0.99, 1.01]')(\exists \rho \in R')$$
$$\rho = \rho_{H_2O} \frac{m_1 - m_3}{m_1 - m_2}.$$

This logical formula is equivalent to the modal inclusion

$$\rho^*([0.99, 1.01], [9, 10], [0, 0.1], [2.7, 3]) \subseteq R$$

As the function ρ is tree-optimal in the explicit domains, it has a multi-incident variable m_1 with respect to which it is totally monotonic (with a positive partial derivative with respect to m_1, a positive partial derivative with respect to the first incidence of m_1, and a negative partial derivative with respect to the second incidence of m_1), Theorem 4.2.15 ensures that

$$\rho^*([0.99, 1.01], [9, 10], [0, 0.1], [2.7, 3]) = [0.99, 1.01] * \frac{[9, 10] - [2.7, 3]}{[10, 9] - [0, 0.1]} \subseteq [0.66, 0.75]$$

and the solution is $R' = [0.66, 0.75]'$, which is the uncertainty interval for the density ρ due to the uncertainties in the measurements.

The third example concerns semantic interpretation when a vectorial function is involved.

Fig. 4.4 Circuit

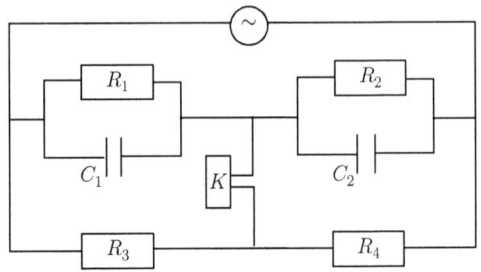

Example 4.5.3 To find the capacity of an unknown capacitor c_1 and the resistance of an unknown resistor r_1, let us balance the variable capacity of a capacitor c_2 and the variable resistance of a resistor r_2 until the tone of an earphone k, supposed with a negligible resistance, is minimal or vanishes, where r_3 and r_4 are two resistors with known resistances (see Fig. 4.4). At this moment, the equations for this physical system are

$$c_1 = \frac{r_4 c_2}{r_3}$$

$$r_1 = \frac{r_3 r_2}{r_4}$$

Supposing that $r_3 \in [9.9, 10.1]'$, $r_4 \in [6.8, 6.9]'$, according to the manufacturing specifications and, due to measurement uncertainties, $c_2 \in [40.2, 41.5]'$, $r_2 = [18.3, 19.8]'$, the logical statement of the problem is

$$(\forall r_2 \in [18.3, 19.8]') \, (\forall r_3 \in [9.9, 10.1]') \, (\forall r_4 \in [6.8, 6.9]') \, (\forall c_2 \in [40.2, 41.5]')$$
$$(\exists c_1 \in C_1') \, (\exists r_1 \in R_1') \, (c_1 = \frac{r_4 c_2}{r_3}, \, r_1 = \frac{r_3 r_2}{r_4}).$$

As the functions c_1 and r_1 do not share any existentially quantified variables, this formula is equivalent to the conjunction of

$$(\forall r_3 \in [9.9, 10.1]') \, (\forall r_4 \in [6.8, 6.9]') \, (\forall c_2 \in [40.2, 41.5]') \, (\exists c_1 \in C_1') \, c_1 = \frac{r_4 c_2}{r_3}$$

and

$$(\forall r_2 \in [18.3, 19.8]') \, (\forall r_3 \in [9.9, 10.1]') \, (\forall r_4 \in [6.8, 6.9]') \, (\exists r_1 \in R_1') \, r_1 = \frac{r_3 r_2}{r_4}.$$

Applying the *-Semantic Theorem, the first formula is equivalent to the inclusion

$$c_1^*([9.9, 10.1], [6.8, 6.9], [40.2, 41.5]) \subseteq C_1,$$

4.5 Additional Examples

with C_1 a proper interval. As the function c_1 is tree-optimal and all their variables are uni-incident, it is optimal and consequently

$$c_1^* = \frac{[6.8, 6.9] * [40.2, 41.5]}{[9.9, 10.1]} \subseteq [27.06, 28.93].$$

The second formula is equivalent to the inclusion

$$r_1^*([18.3, 19.8], [9.9, 10.1], [6.8, 6.9]) \subseteq R_1,$$

with R_1 a proper interval, which is optimal too. Therefore

$$r_1^* = \frac{[9.9, 10.1] * [18.3, 19.8]}{[6.8, 6.9]} \subseteq [26.25, 29.41].$$

The solution to the problem is $c_1 \in C_1 = [27.06, 28.93]'$ and $r_1 \in R_1 = [26.25, 29.41]'$.

Let us change the physical system, now supposing that r_3 represents a potentiometer of variable resistance in the interval [9.0, 11.0]. Now the logical statement of the problem becomes

$$(\forall r_2 \in [18.3, 19.8]') \ (\forall r_4 \in [6.8, 6.9]') \ (\forall c_2 \in [40.2, 41.5]')$$

$$(\exists r_3 \in [9.0, 11.0]') \ (\exists c_1 \in C_1') \ (\exists r_1 \in R_1') \ (c_1 = \frac{r_4 c_2}{r_3}, \ r_1 = \frac{r_3 r_2}{r_4}).$$

Both functions c_1 and r_1 share the existentially quantified variable r_3 and the logical formula can not be split in two. Following Theorem 4.4.1, now

$$f(r_2, r_3, r_4, c_2) = (\frac{r_4 c_2}{r_3}, \frac{r_3 r_2}{r_4})$$

with $X = ([18.3, 19.8], [11.0, 9.0], [6.8, 6.9], [40.2, 41.5])$ and we may proceed as follows:

$$c_1(r_2, r_3, r_4, c_2) = \frac{r_4 c_2}{r_3}$$

$$r_1(r_2, r_3, r_4, c_2) = \frac{r_3 r_2}{r_4},$$

$$A_1 = ([10.0, 10.0], [6.8, 6.9], [40.2, 41.5])$$

$$A_2 = ([18.3, 19.8], [10.0, 10.0], [6.8, 6.9])$$

$$c_1 R(A_1) = \frac{[6.8, 6.9] * [40.2, 41.5]}{[10.0, 10.0]} \subseteq [27.33, 28.64]$$

$$r_1 R(A_2) = \frac{[10.0, 10.0] * [18.3, 19.8]}{[6.8, 6.9]} \subseteq [26.52, 29.12]$$

$$B = ([27.33, 28.64], [26.52, 29.12]).$$

So, $c_1 \in C'_1 = [27.33, 28.64]'$, and $r_1 \in R'_1 = [26.52, 29.12]'$ is the solution to the problem.

Chapter 5
Interval Arithmetic

5.1 Introduction

In this chapter the semantic interval extensions of the simplest elementary functions: the arithmetic operators (addition, multiplication, division), logarithm, exponential functions (including the hyperbolic functions), power function, and the trigonometric functions and their inverses, are considered, together with their most important properties and their arithmetic implementations.

The results for the arithmetic operations will match with the definitions that Kaucher stated for the arithmetic operations in his extended interval space. This can induce to consider the Modal Interval Analysis as a particular extension of the Kaucher theory. Nevertheless, there are important differences between them. Kaucher started from a formal extension of classical intervals including intervals with arbitrary bounds, i.e., starting from $\mathbb{R}^2$, and searching "...to write formulas, theorems and proofs in a closed form..." and "...to shorten theorems and proofs..." [49] he defines an algebraic structure with arithmetic operators, which coincide with the previous results in an unavoidable way, if the Kaucher extension makes sense. Gardenyes started from classical intervals by adding a quantifier to an interval. By defining relations, equality, inclusion, etc, and lattice operators, he made it possible to consider functions over interval domains and, as a particular case, arithmetic operators which, owing to the main idea of associating a quantifier with an interval, have a semantic sense and meaning. Kaucher extended interval space and Gardenyes modal intervals coincide only in the implementation of the arithmetic operators and not in the logical properties, non-existent in the Kaucher theory. Thus, Modal Interval Analysis, is not only an algebraic completion of the classical intervals, but it provides an essential logical and operational framework for problems where computations semantically interpretable are involved.

5.2 One-Variable Function

Every one-variable continuous function is *JM*-commutable, that is both semantic extensions f^* and f^{**} are equal, because there is no commutation problem between the meet and join operations. The interesting operators are the monotonic operators or other easily programmable operators.

Remark 5.2.1 Following Remark 4.2.1, we define in this case $fR(X) = f^*(X)$ and, therefore, one-variable functions become optimal syntactic functions.

If $X = [x_1, x_2]$, for the logarithmic and exponential functions

$$\ln X = [\ln x_1, \ln x_2] \quad \text{whenever } X' > 0;$$

$$\exp(X) = [\exp(x_1), \exp(x_2)].$$

For the absolute value function:

$$|X| = \underline{\text{if}} \ x_1 \geq 0, \ x_2 \geq 0 \ \underline{\text{then}} \ [x_1, x_2],$$
$$\underline{\text{if}} \ x_1 < 0, \ x_2 < 0 \ \underline{\text{then}} \ [|x_2|, |x_1|],$$
$$\underline{\text{if}} \ x_1 < 0, \ x_2 \geq 0 \ \underline{\text{then}} \ [0, \max(|x_1|, |x_2|)],$$
$$\underline{\text{if}} \ x_1 \geq 0, \ x_2 < 0 \ \underline{\text{then}} \ [\max(|x_1|, |x_2|), 0].$$

For the operator $\text{power}(x, n)$:

$$X^n = \underline{\text{if}} \ n \text{ is odd } \underline{\text{then}} \ [x_1^n, x_2^n],$$
$$\underline{\text{if}} \ n \text{ is even } \underline{\text{then}}$$
$$\underline{\text{if}} \ (x_1 \geq 0, \ x_2 \geq 0) \ \underline{\text{then}} \ [x_1^n, x_2^n],$$
$$\underline{\text{if}} \ (x_1 < 0, \ x_2 < 0) \ \underline{\text{then}} \ [x_2^n, x_1^n],$$
$$\underline{\text{if}} \ (x_1 < 0, \ x_2 \geq 0) \ \underline{\text{then}} \ [0, \max(|x_1|^n, |x_2|^n)],$$
$$\underline{\text{if}} \ (x_1 \geq 0, \ x_2 < 0) \ \underline{\text{then}} \ [\max(|x_1|^n, |x_2|^n), 0].$$

For the operator $\text{root}(x, n)$:

$$\sqrt[n]{X} = [\sqrt[n]{x_1}, \sqrt[n]{x_2}] \ \text{if} \ x_1 \geq 0, \ x_2 \geq 0.$$

5.3 Arithmetic Operators

Interval arithmetic is obtained from the semantic extension of the arithmetic operators, which overcomes the most problematic difficulties of the domain of values approach.

5.3 Arithmetic Operators

Let $\omega \in \{+, -, *, /\}$ be an arithmetical operation on $\mathbb{R}$. Using the fact that the arithmetic operators are consistently programmable, the extension of the arithmetic operators ω to $I^*(\mathbb{R})$ is suitably defined by a semantic extension of the real continuous function $f(x, y) = x\omega y$, i.e.,

$$A \omega B = f^*(A, B) = f^{**}(A, B) = \underset{(x,A)}{\Omega} \underset{(y,B)}{\Omega} [x\omega y, x\omega y].$$

since they are *JM*-commutable two-variables operators.

In what follows we include their explicit computation programs and some of their most basic properties. The results are the same as those for the arithmetic operators in Kaucher's [48, 49] extended interval space and detailed proofs can be found in [11, 89].

Lemma 5.3.1 *For $A = [a_1, a_2]$, $B = [b_1, b_2]$,*

$$A + B = [a_1 + b_1, a_2 + b_2].$$

Proof For every $x \in A'$,

$$\underset{(y,B)}{\Omega} [x + y, x + y]$$

$$= \underline{\text{if}} \; B \; \text{proper} \; \underline{\text{then}} \; \underset{y \in B'}{\bigvee} [x + y, x + y] \; \underline{\text{else}} \; \underset{y \in B'}{\bigwedge} [x + y, x + y]$$

$$= \underline{\text{if}} \; B \; \text{proper} \; \underline{\text{then}} \; [\underset{y \in B'}{\min}(x + y), \underset{y \in B'}{\max}(x + y)]$$

$$\underline{\text{else}} \; [\underset{y \in B'}{\max}(x + y), \underset{y \in B'}{\min}(x + y)]$$

$$= \underline{\text{if}} \; B \; \text{proper} \; \underline{\text{then}} \; [x + \underset{y \in B'}{\min}(y), x + \underset{y \in B'}{\max}(y)]$$

$$\underline{\text{else}} \; [x + \underset{y \in B'}{\max}(y), x + \underset{y \in B'}{\min}(y)]$$

$$= [x + b_1, x + b_2].$$

Therefore

$$A + B = \underset{(x,A)}{\Omega} \underset{(y,B)}{\Omega} [x + y, x + y] = \underset{(x,A)}{\Omega} [x + b_1, x + b_2] = [a_1 + b_1, a_2 + b_2]. \blacksquare$$

Lemma 5.3.2 *For $A = [a_1, a_2]$, $B = [b_1, b_2]$,*

$$A - B = [a_1 - b_2, a_2 - b_1].$$

Proof The proof is similar to the case of addition. $\blacksquare$

Lemma 5.3.3 *For $A = [a_1, a_2]$, $r \in \mathbb{R}$,*

$$rA = \underset{(x,A)}{\Omega} [rx, rx] = \underline{\text{if }} r \geq 0 \underline{\text{ then }} [ra_1, ra_2]$$

$$\underline{\text{if }} r < 0 \underline{\text{ then }} [ra_2, ra_1].$$

Proof The proof is similar to the case of addition. ∎

The multiplication rule is a bit trickier, one must use Ratschek's function $\chi(A)$, see [72, 73], defined for intervals $A \neq [0, 0]$, defined by assigning the value which is equal to the bound of A of lesser absolute value divided by the one with greater absolute value. Now, here is the multiplication rule:

(a) If $A \supseteq [0, 0]$ or $B \subseteq [0, 0]$, or viceversa, then $A * B = [0, 0]$.
(b) If $A = [a_1, a_2]$ and $B = [b_1, b_2]$ are of the same modality, then $A * B$ is the interval of the same modality of A or B having as bounds the extreme left and right values of a_1b_1, a_1b_2, a_2b_1, a_2b_2.
(c) If $A = [a_1, a_2]$ and $B = [b_1, b_2]$ are of opposite modalities, then, first reorder A and B such that $\chi(A) \leq \chi(B)$ and second $A * B$ is the interval of the same modality of A built with the numerical values a_1b_1, a_1b_2, a_2b_1, a_2b_2 after removing the most extreme ones to the right and to the left.

These rules lead to the computation program for the product stated in the following lemma.

Lemma 5.3.4 *For $A = [a_1, a_2]$, $B = [b_1, b_2]$,*

$A * B = \underline{\text{if }} a_1 \geq 0,\ a_2 \geq 0,\ b_1 \geq 0,\ b_2 \geq 0 \underline{\text{ then }} [a_1b_1, a_2b_2]$,

$\underline{\text{if }} a_1 \geq 0,\ a_2 \geq 0,\ b_1 \geq 0,\ b_2 < 0 \underline{\text{ then }} [a_1b_1, a_1b_2]$,

$\underline{\text{if }} a_1 \geq 0,\ a_2 \geq 0,\ b_1 < 0,\ b_2 \geq 0 \underline{\text{ then }} [a_2b_1, a_2b_2]$,

$\underline{\text{if }} a_1 \geq 0,\ a_2 \geq 0,\ b_1 < 0,\ b_2 < 0 \underline{\text{ then }} [a_2b_1, a_1b_2]$,

$\underline{\text{if }} a_1 \geq 0,\ a_2 < 0,\ b_1 \geq 0,\ b_2 \geq 0 \underline{\text{ then }} [a_1b_1, a_2b_1]$,

$\underline{\text{if }} a_1 \geq 0,\ a_2 < 0,\ b_1 \geq 0,\ b_2 < 0 \underline{\text{ then }} [\max(a_2b_2, a_1b_1), \min(a_2b_1, a_1b_2)]$,

$\underline{\text{if }} a_1 \geq 0,\ a_2 < 0,\ b_1 < 0,\ b_2 \geq 0 \underline{\text{ then }} [0, 0]$,

$\underline{\text{if }} a_1 \geq 0,\ a_2 < 0,\ b_1 < 0,\ b_2 < 0 \underline{\text{ then }} [a_2b_2, a_1b_2]$,

$\underline{\text{if }} a_1 < 0,\ a_2 \geq 0,\ b_1 \geq 0,\ b_2 \geq 0 \underline{\text{ then }} [a_1b_2, a_2b_2]$,

$\underline{\text{if }} a_1 < 0,\ a_2 \geq 0,\ b_1 \geq 0,\ b_2 < 0 \underline{\text{ then }} [0, 0]$,

$\underline{\text{if }} a_1 < 0,\ a_2 \geq 0,\ b_1 < 0,\ b_2 \geq 0 \underline{\text{ then }} [\min(a_1b_2, a_2b_1), \max(a_1b_1, a_2b_2)]$,

$\underline{\text{if }} a_1 < 0,\ a_2 \geq 0,\ b_1 < 0,\ b_2 < 0 \underline{\text{ then }} [a_2b_1, a_1b_1]$,

$\underline{\text{if }} a_1 < 0,\ a_2 < 0,\ b_1 \geq 0,\ b_2 \geq 0 \underline{\text{ then }} [a_1b_2, a_2b_1]$, if $a_1 < 0$, $a_2 < 0$, $b_1 \geq 0$, $b_2 < 0$ then $[a_2 b_2, a_2 b_1]$,

if $a_1 < 0$, $a_2 < 0$, $b_1 < 0$, $b_2 \geq 0$ then $[a_1 b_2, a_1 b_1]$,

if $a_1 < 0$, $a_2 < 0$, $b_1 < 0$, $b_2 < 0$ then $[a_2 b_2, a_1 b_1]$.

For the operation of division, if $0 \notin [b_1, b_2]$,

$$1/[b_1, b_2] = [1/b_2, 1/b_1]$$

and

$$[a_1, a_2]/[b_1, b_2] = [a_1, a_2] * (1/[b_1, b_2]),$$

taking into account that $\chi(1/[b_1, b_2]) = \chi([b_1, b_2])$. So, the computation program for division is stated in the following lemma.

Lemma 5.3.5 *For $A = [a_1, a_2]$, $B = [b_1, b_2]$, $0 \notin B'$,*

$A/B = $ if $a_1 \geq 0$, $a_2 \geq 0$, $b_1 > 0$, $b_2 > 0$ then $[a_1/b_2, a_2/b_1]$,

if $a_1 \geq 0$, $a_2 \geq 0$, $b_1 < 0$, $b_2 < 0$ then $[a_2/b_2, a_1/b_1]$,

if $a_1 \geq 0$, $a_2 < 0$, $b_1 > 0$, $b_2 > 0$ then $[a_1/b_2, a_2/b_2]$,

if $a_1 \geq 0$, $a_2 < 0$, $b_1 < 0$, $b_2 < 0$ then $[a_2/b_1, a_1/b_1]$,

if $a_1 < 0$, $a_2 \geq 0$, $b_1 > 0$, $b_2 > 0$ then $[a_1/b_1, a_2/b_1]$,

if $a_1 < 0$, $a_2 \geq 0$, $b_1 < 0$, $b_2 < 0$ then $[a_2/b_2, a_1/b_2]$,

if $a_1 < 0$, $a_2 < 0$, $b_1 > 0$, $b_2 > 0$ then $[a_1/b_1, a_2/b_2]$,

if $a_1 < 0$, $a_2 < 0$, $b_1 < 0$, $b_2 < 0$ then $[a_2/b_1, a_1/b_2]$.

Remark 5.3.1 It is easy to observe that these operation rules coincide with the ones for the standard set-intervals when the operands are proper. Therefore, the algebra obtained for the completed intervals is an extension of the standard algebra for classical intervals.

5.3.1 Properties of the Arithmetic Operations

Their main properties are now summarized.

Properties of addition:

1. $A+B = B+A$
2. $(A+B)+C = A+(B+C)$

3. $A + \bigvee_{i \in I} B(i) = \bigvee_{i \in I} (A + B(i)) = A + \bigwedge_{i \in I} B(i) = \bigwedge_{i \in I} (A + B(i))$
4. $A \subseteq B, C \subseteq D \Rightarrow A + C \subseteq B + D$
5. $A \leq B, C \leq D \Rightarrow A + C \leq B + D$
6. $\text{Dual}(A + B) = \text{Dual}(A) + \text{Dual}(B)$
7. $A + [0, 0] = A$
8. The opposite of any interval $A = [a_1, a_2]$ exists and it is

$$\text{Opp}(A) = [-a_1, -a_2]$$

with $\text{Opp}(A) = -\text{Dual}(A)$.

In $I(\mathbb{R})$ the only natural symmetry is

$$A = [a_1, a_2] \leftrightarrow -A = [-a_2, -a_1]$$

but in $I^*(\mathbb{R})$ the following ones stand out:

$$A = [a_1, a_2] \leftrightarrow \text{Id}(A) = [a_1, a_2]$$
$$\leftrightarrow -A = [-a_2, -a_1]$$
$$\leftrightarrow \text{Dual}(A) = [a_2, a_1]$$
$$\leftrightarrow \text{Opp}(A) = [-a_1, -a_2].$$

Properties of the difference:

1. $A \subseteq B, C \subseteq D \Rightarrow A - C \subseteq B - D$
2. $A - \text{Dual}(B) = A + \text{Opp}(B)$
3. The equation $A + X = B$ has the unique solution

$$X = B - \text{Dual}(A) = B + \text{Opp}(A)$$

Properties of multiplication by a real number:

1. If $\phi \in \{\text{Dual}, -, \text{Opp}, \text{Prop}\}$, then $\phi(rA) = r\phi(A)$
2. $r \bigvee_{i \in I} A(i) = \bigvee_{i \in I} rA(i)$ and $r \bigwedge_{i \in I} A(i) = \bigwedge_{i \in I} rA(i)$
3. $A \subseteq B \Rightarrow rA \subseteq rB$
4. $A \leq B \Rightarrow \begin{vmatrix} rA \leq rB \text{ si } r \geq 0 \\ rA \geq rB \text{ si } r < 0 \end{vmatrix}$
5. $r \underset{i \in I}{\text{Min}} A(i) = \begin{vmatrix} \underset{i \in I}{\text{Min}} rA(i) \text{ si } r \geq 0 \\ \underset{i \in I}{\text{Max}} rA(i) \text{ si } r < 0 \end{vmatrix}$

5.3 Arithmetic Operators

$$r \operatorname*{Max}_{i \in I} A(i) = \begin{cases} \operatorname*{Max}_{i \in I} rA(i) & \text{si } r \geq 0 \\ \operatorname*{Min}_{i \in I} rA(i) & \text{si } r < 0 \end{cases}$$

6. $r(A+B) = rA + rB$
7. $r(sA) = (rs)A$
8. If $rs \geq 0$, then $(r+s)A = rA + sA$

Properties of the product:

1. $A * B = B * A$
2. $A * (B * C) = (A * B) * C$
3. $r(A * B) = (rA) * B = A * (rB)$
4. $(-A) * (-B) = A * B$
5. $\text{Opp}(A) * \text{Opp}(B) = \text{Dual}(A * B)$
6. $\text{Dual}(A) * \text{Dual}(B) = \text{Dual}(A * B)$
7. $rA = [r, r] * A$
8. $rs \geq 0 \Rightarrow \begin{cases} rA \vee sA = [\min\{r,s\}, \max\{r,s\}] * A \\ rA \wedge sA = [\max\{r,s\}, \min\{r,s\}] * A \end{cases}$
9. $[1, 1] * A = A$
10. $(A \subseteq B, C \subseteq D) \Rightarrow A * C \subseteq B * D$
11. $([0,0] \leq A \leq B, [0,0] \leq C \leq D) \Rightarrow A * C \leq B * D$
12. A proper implies $A * (B + C) \subseteq A * B + A * C$
    A improper implies $A * (B + C) \supseteq A * B + A * C$
13. If $\frac{1}{A}$ and $\frac{1}{B}$ exist, then

$$A \subseteq B \Rightarrow \frac{1}{A} \subseteq \frac{1}{B}$$

$$A \leq B \Rightarrow \frac{1}{A} \geq \frac{1}{B}$$

$$\text{Dual}\left(\frac{1}{A}\right) = \frac{1}{\text{Dual}(A)}$$

Properties of the quotient:

1. $\dfrac{A}{B} = A * \dfrac{1}{B}$
2. $(A \subseteq B, C \subseteq D) \Rightarrow A/C \subseteq B/D$
3. $\text{Dual}\left(\dfrac{A}{B}\right) = \dfrac{\text{Dual}(A)}{\text{Dual}(B)}$
4. The equation $A * X = B$ (provided that $0 \notin \text{Prop}(A)$) has the unique solution

$$X = \frac{B}{\text{Dual}(A)}$$

5.3.2 Inner-Rounding and Computations

For any arithmetic operator ω, since $\text{Dual}(A \omega B) = \text{Dual}(A) \omega \text{Dual}(B)$, then

$$\text{Inn}(fR(A_1, \ldots, B_1, \ldots)) = \text{Dual}(\text{Out}(fR(\text{Dual}(A_1), \ldots, \text{Dual}(B_1), \ldots))).$$

This property allows obtaining the canonical inner and outer rounding of syntactic computations by only implementing the, for instance, outer-rounded arithmetic..

5.3.3 Sub-distributivity of the Operations * and +

Regional distributivity completes also on $I^*(\mathbb{R})$ its original structure on $I(\mathbb{R})$. We have seen in Property 12 of the product that

$$A \text{ proper} \Rightarrow A * (B + C) \subseteq A * B + A * C$$
$$A \text{ improper} \Rightarrow A * (B + C) \supseteq A * B + A * C.$$

A more general result is the following Sub-Distributive Law.

Lemma 5.3.6 *For $A, B, C \in I^*(\mathbb{R})$,*

$$\text{Impr}(A) * B + A * C \subseteq A * (B + C) \subseteq \text{Prop}(A) * B + A * C.$$

Proof The function $f_1 : \mathbb{R}^3 \to \mathbb{R}$ defined by $f_1(a, b, c) = a(b + c)$ is a modal syntactic function, uni-incident with JM-commutable operators, and so

$$f_1^*(A, B, C) \subseteq f_1 R(A, B, C) \subseteq f_1^{**}(A, B, C).$$

Let $f_2 : \mathbb{R}^4 \to \mathbb{R}$ be the function defined by $f_2(a_1, a_2, b, c) = a_1 b + a_2 c$ (of which the modal syntactic extension $f_2 R$ is tree-optimal). Two cases can be distinguished, depending on the modality of A.

First, if A is an improper interval, then

$$A * (B + C) = f_1 R(A, B, C) \subseteq f_1^{**}(A, B, C)$$
$$= \bigwedge_{a \in A'} f_1^{**}(a, B, C) = \bigwedge_{a \in A'} f_2^{**}(a, a, B, C)$$
$$\subseteq \bigwedge_{a_1 \in A'} \bigvee_{a_2 \in A'} f_2^{**}(a_1, a_2, B, C)$$
$$= f_2^{**}(A, \text{Prop}(A), B, C) = f_2 R(A, \text{Prop}(A), B, C)$$
$$= A * B + \text{Prop}(A) * C.$$

5.3 Arithmetic Operators

Second, if A is a proper interval, then

$$A * (B + C) = f_1 R(A, B, C) \subseteq f_1^{**}(A, B, C)$$
$$= \underset{(b,B)(c,C)}{\Omega \Omega} f_1^{**}(A, b, c) \subseteq \underset{(b,B)(c,C)}{\Omega \Omega} f_2^{**}(A, A, b, c)$$
$$= f_2^{**}(A, A, B, C) = f_2 R(A, A, B, C) = A * B + A * C.$$

Hence, for every $A, B, C \in I^*(\mathbb{R})$, we have

$$A * (B + C) \subseteq f_1^{**}(A, B, C) \subseteq A * B + \text{Prop}(A) * C$$

(the roles of B and C are, obviously, interchangeable).
By duality,

$$\text{Impr}(A) * B + A * C \subseteq f_1^{**}(A, B, C) \subseteq A * (B + C). \qquad \blacksquare$$

Example 5.3.1 For $A = [1, -1]$, $B = [3, 1]$ and $C = [-3, -1]$

$$A * (B + C) = [1, -1] * ([3, 1] + [-3, -1]) = [1, -1] * [0, 0] = [0, 0]$$
$$\text{Impr}(A) * B + A * C = [1, -1] * [3, 1] + [1, -1] * [-3, -1] = [4, -4]$$
$$\text{Prop}(A) * B + A * C = [-1, 1] * [3, 1] + [1, -1] * [-3, -1] = [0, 0]$$

and $[4, -4] \subseteq [0, 0] \subseteq [0, 0]$.

5.3.4 Metric Functions

Together with the basic relations and operations, there exist functions which provide $I^*(\mathbb{R})$ with a metric structure, thus making sense of concepts such as limit and convergence [89].

Definition 5.3.1 (Absolute value, mignitude, midpoint and width) If $X = [x_1, x_2]$ is a modal interval, the absolute value of X is

$$\text{abs}(X) = \max(|x_1|, |x_2|) = \text{abs}(\text{Dual}(X)),$$

the mignitude of X is

$$\text{mig}(X) = \min(|x_1|, |x_2|) = \text{mig}(\text{Dual}(X)),$$

the midpoint of X is

$$\text{mid}(X) = \frac{x_1 + x_2}{2} = \text{mid}(\text{Dual}(X)),$$

and the width (or span) of X is

$$\text{wid}(X) = |x_1 - x_2| = \text{wid}(\text{Dual}(X))$$

Properties of abs, mig, mid, and wid:

1. $A \subseteq B \Rightarrow \text{abs}(A) \leq \text{abs}(B)$ and $\text{mig}(A) \geq \text{mig}(B)$
2. $A \subseteq B \Rightarrow \text{mig}(A) \geq \text{mig}(B)$
3. if $0 \notin A'$, then $\text{abs}\left(\frac{1}{A}\right) = \frac{1}{\text{mig}(A)}$ and $\text{mig}\left(\frac{1}{A}\right) = \frac{1}{\text{abs}(A)}$

The absolute value is a norm in $(I^*(\mathbb{R}), +, *)$ because

1. $\text{abs}(A) \geq 0$ and $\text{abs}(A) = 0 \Leftrightarrow A = 0$
2. $\text{abs}(A + B) \leq \text{abs}(A) + \text{abs}(B)$.
3. $\text{abs}(\lambda A) = |\lambda|\text{abs}(A)$

The mignitude is a quasi-norm in $(I^*(\mathbb{R}), +, *)$ because

1. $\text{mig} A \geq 0$ and $\text{mig}(A) = 0 \Leftrightarrow 0 \notin A'$
2. $\text{mig}(A + B) \leq \text{mig}(A) + \text{mig}(B)$
3. $\text{mig}(\lambda A) = |\lambda|\text{mig}(A)$

Moreover,

1. $\text{mig}(A) \leq \text{abs}(A)$
2. $\text{mig}(A * B) = \text{mig}(A)\text{mig}(B)$
3. $\text{abs}(A * B) = \text{abs}(A)\text{abs}(B)$
4. $A \subseteq B \Rightarrow \text{wid}(A) \leq \text{wid}(B)$
5. $\text{wid}(A) = \text{abs}(A - A)$
6. $\text{wid}(\lambda A) = |\lambda|\text{wid}(A)$
7. $\text{wid}(A) = \text{abs}(A)(1 - \chi(A))$
8. $0 \in A \Rightarrow \text{abs}(A) \leq \text{wid}(A) \leq 2 \cdot \text{abs}(A)$
9. Any improper interval can be split into the addition of a proper interval with a symmetric improper one: if $X = [x_1, x_2]$ is an improper interval, then

$$X = [x_1, x_2] = [x_2, x_1] + |x_1 - x_2| * [1, -1] = \text{Prop}(X) + [\text{wid}(X), -\text{wid}(X)]$$

Let ρ be the Hausdorff distance between two compact sets A and B,

$$\rho(A, B) = \max \left\{ \max_{x \in A'} \min_{y \in B'} \rho(x, y), \max_{y \in B'} \min_{x \in A'} \rho(x, y) \right\}.$$

Particularizing this definition to intervals,

Definition 5.3.2 (Hausdorff distance) When A and B are modal intervals

$$\text{dist}(A, B) = \max \left\{ |\underline{a} - \underline{b}| , |\overline{a} - \overline{b}| \right\}$$

Properties of the distance:

1. $\text{dist}(A + C, B + C) = \text{dist}(A, B)$
2. $\text{dist}(A + C, B + D) \leq \text{dist}(A, B) + \text{dist}(C, D)$
3. $\text{dist}(A, 0) = \text{abs}(A)$
4. $\text{dist}(A, B) = \text{abs}(B - A)$ if $\text{wid}(A) \leq \text{wid}(B)$ else $\text{dist}(A, B) = \text{abs}(A - B)$
5. $\text{dist}(aB, aC) = |a| \cdot \text{dist}(B, C)$
6. $\text{dist}(A * B, A * C) \leq \text{abs}(A) \cdot \text{dist}(B, C)$
7. $\text{dist}(A * B, A * C) \geq \text{mig}(A) \cdot \text{dist}(B, C)$
8. $\text{dist}(A, B) \leq d \Leftrightarrow (\forall x \in A)(\exists y \in B) \text{dist}(x, y) \leq d$ and $(\forall y \in B)(\exists x \in A) \text{dist}(x, y) \leq d$

Definition 5.3.3 (Convergence)

$$\lim_{k \to \infty} A_k = A = \lim_{k \to \infty} \text{dist}(A_k, A) = 0$$

Properties of convergence

1. $\lim_{k \to \infty} A_k = A \Leftrightarrow (\lim_{k \to \infty} \underline{a}_k = \underline{a} \text{ y } \lim_{k \to \infty} \overline{a}_k = \overline{a})$
2. $(I(\mathbb{R}), \text{dist})$ is a complete metric space.
3. The arithmetic operations $+, -, *, /$ and the functions inf, sup, mag, mig, wid, and mid are continuous in $(I^*(\mathbb{R}), \text{dist})$.

5.4 Interval Arithmetic for the C++ Environment

The implementation of modal interval arithmetic is built by means of a library where the objects are intervals and the arithmetic operations with intervals must contain the exact results, therefore, it is necessary to control the truncation of the interval operations.

Moreover, the arithmetic must allow of controlling the floating-point exceptions such as division by zero, infinity, underflow, ..., to obtain an accurate and controlled result for any operation whose variables are those exceptional values.

The interval library **ivalDb** provides an easy way to make programs using Modal Intervals including the following features:

1. Basic operators (addition, difference, product, quotient, pow, etc.)
2. Trigonometric functions (sin, cos, tan, inverse functions)
3. Boolean operations

4. Proper and Improper Interval operations
5. Exception handling

5.4.1 About the Library

5.4.1.1 Object-Oriented

ivalDb is made in Borland C++, to exploit the potency of object oriented programming, handling every interval as an object fitted with many properties and functions.

This operating environment allows the following options:

1. To control the truncation of floating point operations without making calls to functions that act on the coprocessor computer, avoiding codifying the truncation by independent subroutines. This simplifies the task of creating the support library.
2. To define the intervals as objects, using the capability of C++ to define types and operator overloading.

5.4.1.2 Numeric Guarantee

ivalDb has a numeric guarantee thanks to the use of ©FDLIBM (Freely Distributable LIBM), a C-library developed by Sun Microsystems, Inc., for machines that support IEEE 754 floating-point arithmetic which assures, in the worst case, a ULP (Units-Bits of the Last Place) of error for all the given functions. Moreover, it assures multi-platform compatibility (PC, SUN...). FDLIBM provides a function which allows rounding a floating point number to $+\infty$ or to $-\infty$. Then, knowing that the maximal error that can be committed by the FDLIBM computations is one ULP, it is easy to implement a guaranteed interval function by adding a ULP to the upper bound of the solution interval and by subtracting a ULP to the lower bound of the solution interval, in order to get the outer rounding. Inner rounding is not implemented because it can be obtained by means of the process of dual computing, as in Theorem 2.2.2.

This method may be numerically conservative compared to other techniques used by other libraries to assure a numerical guarantee, but it is also more efficient in terms of time because it does not change the rounding mode of the computer.

5.4.1.3 Use in Different Environments

The library was adapted to allow programmers to use it on Linux and Windows. For Linux, programmers can use g++ to compile programs, but in Windows there are two choices, Borland C++ and Visual C++.

5.4.2 Available Functions and Operators

5.4.2.1 Creation of New Intervals

1. *Completely defined bound interval.* In this case the user must specify both lower and upper bounds to create the new interval.

       Input: Double,Double
       Output: ivalDb
       Syntax: `ivalDb A=ivalDb(a,b);`

2. *Point-wise interval.* The user must specify only one value, then it will be assigned to lower and upper bounds.

       Input: Double
       Output: ivalDb
       Syntax: `ivalDb A=ivalDb(a);`

3. *Using a previously defined interval.* The user creates the new interval using as parameter another interval. The new interval is exactly equal to the interval used to create it.

       Input: ivalDb
       Output: ivalDb
       Syntax: `ivalDb B=ivalDb(A);`

4. *Without parameters.* If no input parameter is specified, the new interval will be defined as $(-\infty, +\infty)$.

       Input: None
       Output: ivalDb
       Syntax: `ivalDb A;`

5.4.2.2 Access to Interval Bounds

1. The functions *GetInf, GetSup, GetMid* return the lower bound value, the upper bound value, and the mid value of an interval, respectively.

       Input: None
       Output: Double
       Syntax: `ivalDb A=(a,b);`
            `Double lb, ub, m;`
            `lb=A.GetInf();`
            `ub=A.GetSup();`
            `m=A.GetMid();`

2. The functions *SetInf* and *SetSup* establish the lower bound value and the upper bound value of an interval, respectively.

        Input: Double
        Output: None
        Syntax: `ivalDb A;`
                      `A.SetInf(a);`
                      `A.SetSup(b);`

3. *SetBounds* establishes both lower and upper bounds of an interval.

        Input: Double, Double
        Output: None
        Syntax: `ivalDb A;`
                      `A.SetBounds(a,b);`

5.4.2.3 Monary Operators

1. *Assignation operator*. Assigns one interval to another.

        Input: ivalDb
        Output: ivalDb
        Syntax: `ivalDb A,B=ivalDb(a,b);`
                      `A=B;`

2. *Negation operator*. Returns the opposite of an interval. The lower bound becomes the upper bound with opposite sign and the upper bound becomes the lower bound with opposite sign.

        Input: ivalDb
        Output: ivalDb
        Syntax: `ivalDb A,B=ivalDb(a,b);`
                      `A=-B;`

3. *Power operator*. Returns the result of raising to a power (with exponent an integer).

        Input: ivalDb, integer
        Output: ivalDb
        Syntax: `ivalDb A,B=ivalDb(a,b);`
                      `A=B^n;`

4. The *Prop*, *Impr* and *Du* are modal operators. If the interval is improper, the operator *Prop* converts it to a proper interval. If the interval is proper, the operator *Impr* converts it to an improper one. The operator *Du* returns the dual of the interval.

5.4 Interval Arithmetic for the C++ Environment

```
Input:   ivalDb
Output:  ivalDb
Syntax:  ivalDb A, B=(a,b);
         A=Prop(B);
         A=Impr(B);
         A=Du(B);
```

5.4.2.4 Binary Operators

1. The arithmetic operators *Addition*, *Difference*, *Product* and *Division* return the result of the sum, difference, product, and quotient, respectively, of two intervals. The rule used is described by Lemmas 5.3.1–5.3.5.

    ```
    Input:   ivalDb, ivalDb
    Output:  ivalDb
    Syntax:  ivalDb C, A=ivalDb(a1,a2),B=ivalDb(b1,b2);
             C=A+B;
             C=A-B;
             C=A*B;
             C=A/B;
    ```

2. *Meet* and *Join* operators. The meet of two intervals is the interval defined by the maximum value of the lower bounds and the minimum value of upper bounds, as lower and upper bounds, respectively. The join of two intervals is the interval defined by the minimum value of the lower bounds and the maximum value of upper bounds, as lower and upper bounds, respectively.

    ```
    Input:   ivalDb, ivalDb
    Output:  ivalDb
    Syntax:  ivalDb C, A=ivalDb(a1,a2),B=ivalDb(b1,b2);
             C=A&&B;
             C=A||B;
    ```

3. *Relational operators*. These operators allow of making a comparison between two intervals. The relations are greater than ($>$), greater than or equal to ($\geq$), less than ($<$), less than or equal to ($\leq$), and equality ($==$).

    ```
    Input:   ivalDb, ivalDb
    Output:  unsigned long
    Syntax:  ivalDb A=ivalDb(a1,a2),B=ivalDb(b1,b2);
             if (A>>B) cout<<"A is greater than B"<<endl;
             if (A>=B) cout<<"A is greater or equal than B"<<endl;
             if (A<<B) cout<<"A is least than B"<<endl;
             if (A<=B) cout<<"A is least or equal than B"<<endl;
             if (A==B) cout<<"A is equal than B"<<endl;
    ```

5.4.2.5 Interval Functions

1. The *abs* function returns the absolute value of an interval.

    Input: ivalDb
    Output: ivalDb
    Syntax: ivalDb A, B=ivalDb(a,b);
             A=abs(B);

2. The *sqr* function returns the square of an interval.

    Input: ivalDb
    Output: ivalDb
    Syntax: ivalDb A, B=ivalDb(a,b);
             A=sqr(B);

3. The *pow* function is the version of the power operator for a real exponent.

    Input: ivalDb, double
    Output: ivalDb
    Syntax: ivalDb A, B=ivalDb(a,b);
             A=pow(B,x);

4. The *sqrt* and *cbrt* functions return an interval that contain the square root and the cube root, respectively.

    Input: ivalDb
    Output: ivalDb
    Syntax: ivalDb A, B=ivalDb(a,b);
             A=sqrt(B);
             A=cbrt(B);

5. The *root* function returns the generic root from an interval.

    Input: ivalDb, int
    Output: ivalDb
    Syntax: ivalDb A, B=ivalDb(a,b);
             A=root(B,n);

6. The *exp* function calculates the exponential of an interval.

    Input: ivalDb
    Output: ivalDb
    Syntax: ivalDb A, B=ivalDb(a,b);
             A=exp(B);

7. The *log* function calculates the logarithm of an interval and *log10* calculates its base 10 logarithm.

5.4 Interval Arithmetic for the C++ Environment

```
Input:   ivalDb
Output:  ivalDb
Syntax:  ivalDb A, B=ivalDb(a,b);
         A=log10(B);
         A=log(B);
```

8. *trigonometric and hyperbolic functions*. The inputs for trigonometric functions are in radians.

- *sin*: Calculates the sine of an interval.
- *cos*: Calculates the cosine of an interval.
- *tan*: Calculates the tangent of an interval.
- *asin*: Calculates the arcsin of an interval.
- *acos*: Calculates the arccos of an interval.
- *atan*: Calculates the arctan of an interval.
- *sinh*: Calculates the hyperbolic sine of an interval.
- *cosh*: Calculates the hyperbolic cosine of an interval.
- *tanh*: Calculates the hyperbolic tangent of an interval.
- *asinh*: Calculates the inverse hyperbolic sine of an interval.
- *acosh*: Calculates the inverse hyperbolic cosine of an interval.
- *atanh*: Calculates the inverse hyperbolic tangent of an interval.

```
Input:   ivalDb
Output:  ivalDb
Syntax:  ivalDb A, B=ivalDb(a,b);
         A=sin(B);
         ......
         A=atanh(B);
```

5.4.2.6 Metric Interval Functions

1. The *Width* function returns the absolute value of the difference between the upper and lower bounds.

```
Input:   ivalDb
Output:  double
Syntax:  ivalDb A=ivalDb(a,b);
         double w;
         w=Width(A);
```

2. The *Centre* function returns the half of the sum of the upper and lower bounds.

```
Input:   ivalDb
Output:  double
Syntax:  ivalDb A=ivalDb(a,b);
         double c;
         c=Centre(A);
```

5.4.2.7 Boolean Interval Functions

1. The *IsProp* function returns true if the interval is proper, and otherwise returns false.

    Input: ivalDb
    Output: bool
    Syntax: ivalDb A=ivalDb(a,b);
             if (A.IsProp()) cout<<"Is proper"<<endl;
                else cout<<"Is improper"<<endl;

2. The *IsImpr* function returns true if the interval is improper, and otherwise returns false.

    Input: ivalDb
    Output: bool
    Syntax: ivalDb A=ivalDb(a,b);
             if (A.IsImpr()) cout<<"Is improper"<<endl;
                else cout<<"Is proper"<<endl;

3. The *IsInterval* function returns true if the lower bound is different than the upper bound.

    Input: ivalDb
    Output: bool
    Syntax: ivalDb A=ivalDb(a,b);
            if (A.IsInterval()) cout<<"Is an interval"<<endl;
               else cout<<"Is only a point"<<endl;

4. The *IsEmpty* function returns true if either the lower or upper bounds is NaN.

    Input: ivalDb
    Output: bool
    Syntax: ivalDb A=(NaN(),1.3);
             if (A.IsEmpty()) cout<<"Is an empty interval"<<endl;

5. The *IsIn* function returns true if the first interval is contained in the second interval.

    Input: ivalDb, ivalDb
    Output: bool
    Syntax: ivalDb A=ivalDb(a1,a2), B=ivalDb(b1,b2);
             if (IsIn(A,B)) cout<<"A is contained in B"<<endl;

6. The *IsOut* function returns true if the first interval is not contained in the second interval.

    Input: ivalDb, ivalDb
    Output: bool
    Syntax: ivalDb A=ivalDb(a1,a2), B=ivalDb(b1,b2);
             if (IsOut(A,B)) cout<<"A is not contained in B"<<endl;

5.4 Interval Arithmetic for the C++ Environment

7. The *IsIntersect* function returns true if the two intervals intersect.

   Input: ivalDb, ivalDb
   Output: bool
   Syntax: `ivalDb A=ivalDb(a1,a2), B=ivalDb(b1,b2);`
          `if (IsIntersect(A,B)) cout<<"A is intersecting B"<<endl;`

8. The *IsBig* function returns true if one interval is completely bigger than the other.

   Input: ivalDb, ivalDb
   Output: bool
   Syntax: `ivalDb A=ivalDb(a1,a2), B=ivalDb(b1,b2);`
          `if (A.IsBig(B)) cout<<"A is bigger than B"<<endl;`

9. The *IsSmall* function returns true if one interval is completely smaller than the other.

   Input: ivalDb, ivalDb
   Output: bool
   Syntax: `ivalDb A=ivalDb(a1,a2), B=ivalDb(b1,b2);`
          `if (A.IsSmall(B)) cout<<"A is smaller than B"<<endl;`

5.4.2.8 Auxiliary Interval Constants

These functions return useful intervals for use in some calculation.

- *PI*: Returns the interval version of π.
- *LN2*: Returns interval version of $\ln(2)$.
- *Zero*: Returns [0,0].
- *Infinity*: Returns $(-\infty, +\infty)$.

5.4.2.9 Not Member Functions

There are some functions that are defined outside **ivalDb** objects. They work only with double values.

- *IsNaN*: Returns true if the double value is NaN.
- *NaN*: Returns IEEE754 NaN value.
- *PInfinity*: Returns IEEE754 +infinity value.
- *MInfinity*: Returns IEEE754 −infinity value.
- *IsPInfinity*: Returns true if the double value is +infinity.
- *IsMInfinity*: Returns true if the double value is −infinity.
- *IsInfinity*: Returns true if the double value is either +infinity or −infinity.
- *AddULP*: Returns the next representable double-precision floating-point value following the double value entered in the direction of $+\infty$.
- *RestULP*: Returns the next representable double-precision floating-point value following the double value entered in the direction of $-\infty$.

Table 5.1 Exceptions for the addition

$x + y$	NaN	NOERROR	0	PInfinity	MInfinity
NaN	NaN	NaN	NaN	NaN	NaN
NOERROR	NaN	NOERROR PInfinity MInfinity	NOERROR	PInfinity	MInfinity
0	NaN	NOERROR	0	PInfinity	MInfinity
PInfinity	NaN	PInfinity	PInfinity	PInfinity	NaN
MInfinity	NaN	MInfinity	MInfinity	NaN	MInfinity

Table 5.2 Exceptions for the difference

$x - y$	NaN	NOERROR	0	PInfinity	MInfinity
NaN	NaN	NaN	NaN	NaN	NaN
NOERROR	NaN	NOERROR PInfinity MInfinity	NOERROR	MInfinity	PInfinity
0	NaN	NOERROR	0	MInfinity	PInfinity
PInfinity	NaN	PInfinity	PInfinity	NaN	PInfinity
MInfinity	NaN	MInfinity	MInfinity	MInfinity	NaN

```
Input:  double
Output: bool
Syntax: double A=NaN();
        if (IsNaN(A)) cout<<"A is NaN"<<endl;
        double B=PInfinity();
        if (IsInfinity(B)) cout<<"B is infinity"<<endl;
        double C=RestULP(B);
        if (!IsInfinity(C)) cout<<"is not infinity"<<endl;
```

5.4.2.10 Operations with the Exceptional Values

Tables 5.1–5.4 contain the results for the $+, -, *, /$ operators considering different values for the variables. The NOERROR value means any numerical value.

For the product of two numeric values NOERROR, we consider the maximum and the minimum values that are possible for double precision binary floating point numbers. These values are in the float.h library:

$$\text{DBL\_MIN} = 2.22507e-308$$
$$\text{DBL\_MAX} = 1.79769e+308$$

For example the result of the product $4.0e + 300 * 3.0e + 50$ is PInfinity because the truncation for the DBL_MAX is PInfinity and DBL_MIN is 0.

For example

5.4 Interval Arithmetic for the C++ Environment

Table 5.3 Exceptions for the product

x * y	NaN	NOERROR	0	PInfinity	MInfinity
NaN	NaN	NaN	NaN	NaN	NaN
NOERROR	NaN	NOERROR PInfinity MInfinity	0	PInfinity MInfinity	PInfinity MInfinity
0	NaN	0	0	NaN	NaN
PInfinity	NaN	PInfinity/MInfinity	NaN	PInfinity	MInfinity
MInfinity	NaN	MInfinity/PInfinity	NaN	MInfinity	PInfinity

Table 5.4 Exceptions for division

x/y	NaN	NOERROR	0	PInfinity	MInfinity
NaN	NaN	NaN	NaN	NaN	NaN
NOERROR	NaN	NOERROR PInfinity MInfinity 0	PInfinity MInfinity	0	0
0	NaN	0	NaN	0	0
PInfinity	NaN	PInfinity/MInfinity	PInfinity	NaN	NaN
MInfinity	NaN	PInfinity/MInfinity	MInfinity	NaN	NaN

- `ivalDb(PInfinity(),2.)+ivalDb(3.,MInfinity());` returns the interval [+inf,-inf].
- `ivalDb(1.,PInfinity())+ivalDb(3.,MInfinity());` returns [0,-NaN].
- `ivalDb(0.,1.)*ivalDb(PInfinity,3.);` returns [-NaN,-NaN].
- `ivalDb(1.,-1.)*ivalDb(-3.,MInfinity());` returns [-NaN, -NaN].
- `ivalDb(2.,1.)/ivalDb(4.,3.);` returns the interval [0.6666666, 0.25].
- `ivalDb(1.,2.)/ivalDb(PInfinity(),2.);` returns the interval [0.5,0].
- `ivalDb(1.,2.)/ivalDb(3.,-4.);` returns [-NaN,-NaN].

Chapter 6
Equations and Systems

6.1 Introduction

Similarly to the case of one interval equation $A * X = B$, it is possible to treat the general problem of finding solutions for a system of linear interval equations $\mathbf{A} * \mathbf{X} = \mathbf{B}$ and to obtain a semantics for them, compatible with the necessary rounding.

Several authors have studied the solutions of linear interval systems such as $\mathbf{A} * \mathbf{X} = \mathbf{B}$, where $\mathbf{A} = (A_{ij})$ is an interval (n,n)-matrix, $\mathbf{X} = (X_j)$ and $\mathbf{B} = (B_i)$ are interval $(n,1)$-matrices, distinguishing between a *formal solution*, i.e. intervals X which substituted in the system satisfy the equalities, and *interval enclosures of sets of solutions* for the different real-valued systems whose coefficients and right-hand sides are real numbers belonging to sets associated with the intervals A_{ij} and B_i. The main contributions to the problem of obtaining a formal solution can be found in [56, 57, 59, 81–83]. In this chapter, an approach to finding a formal solution (called for short simply a solution) of an interval linear system, when the coefficients and right-hand sides are modal intervals, will be treated, taking into account the double aspect: finding a solution and giving a logical meaning to this solution. An algorithm to obtain these solutions, as long as the algorithm converges, will be presented and sufficient conditions for convergence and non-convergence will be proved using the interval metric functions defined in Chap. 5.

6.2 Linear Equation

From the algebraic standpoint, in the system of modal intervals $I^*(\mathbb{R})$ the equation $A + X = B$ has a unique solution

$$X = B - \text{Dual}(A) = B + \text{Opp}(A)$$

and the equation $A * X = B$, with $0 \notin A'$, has a unique solution

$$X = B/\mathrm{Dual}(A).$$

Considering an arithmetic with rounding, the solutions are

$$X = \mathrm{Inn}(B - \mathrm{Dual}(A))$$

and

$$X = \mathrm{Inn}(B/\mathrm{Dual}(A)),$$

satisfying $A + X \subseteq B$ and $A * X \subseteq B$.

The semantic theorems provide a logical meaning to the solutions of both equations, whatever the modalities of A, B and X are, and compatible with rounding. If X is an inner-rounded solution with, for instance, A an improper interval, B a proper interval, and X a proper interval, then $A * X \subseteq B$ is equivalent to

$$(\forall x \in X')\, (\exists a \in A')\, (\exists b \in B')\, ax = b.$$

Example 6.2.1 The solution of

$$[-3, -7] * X = [2, 6]$$

is

$$X = [2, 6]/\mathrm{Dual}([-3, -7]) = [2, 6]/[-7, -3] = [-2, -2/7]$$

and the inner rounding $[-2, -0.286]$ of X satisfies

$$[-3, -7] * [-2, -0.286] \subseteq [2, 6],$$

which is equivalent to

$$(\forall x \in [-2, -0.286]')\, (\exists a \in [-7, -3]')\, (\exists b \in [2, 6]')\, ax = b.$$

6.3 Formal Solutions to a Linear System

Based on the well known Gauss algorithm to solve linear systems in $\mathbb{R}$, an interval algorithm to obtain solutions for an interval linear system, when the algorithm converges, is presented together with sufficient conditions of convergence and non-convergence. Let us consider an interval system of linear equations

6.3 Formal Solutions to a Linear System

$$\begin{cases} A_{11} * X_1 + \ldots + A_{1n} * X_n = B_1 \\ \ldots\ldots\ldots\ldots\ldots\ldots\ldots\ldots\ldots\ldots \\ A_{n1} * X_1 + \ldots + A_{nn} * X_n = B_n \end{cases}$$

concisely represented by

$$\mathbf{A} * \mathbf{X} = \mathbf{B}$$

together with the closely associated ones $\mathbf{A} * \mathbf{X} \subseteq \mathbf{B}$ and $\mathbf{A} * \mathbf{X} \supseteq \mathbf{B}$, with the interval coefficients matrix $\mathbf{A} = (A_{ij})_{i,j=1,\ldots,n}$, the interval right-hand matrix $\mathbf{B} = (B_i)_{i=1,\ldots,n}$, and the interval unknowns matrix $\mathbf{X} = (X_j)_{j=1,\ldots,n}$. Isolating any interval unknown X_i in the ith equation of the system of linear equations, the result is

$$X_i = \frac{B_i - \sum_{j \neq i} \mathrm{Dual}(A_{ij}) * \mathrm{Dual}(X_j)}{\mathrm{Dual}(A_{ii})},$$

supposing that $0 \notin A_{ii}$. This suggests the following definition.

Definition 6.3.1 (Jacobi interval operator) The *Jacobi interval operator* associated to the interval system $S : \mathbf{A} * \mathbf{X} = \mathbf{B}$, of n linear equations with n unknowns is the function from $I^*(\mathbb{R}^n)$ to $I^*(\mathbb{R}^n)$ such that the image of an interval $Y = (Y_1, \ldots, Y_n)$ is the interval $\mathfrak{J}_S(Y) = (\mathfrak{J}_S(Y_1), \ldots, \mathfrak{J}_S(Y_n))$ defined by

$$\mathfrak{J}_S(Y_i) = \frac{B_i - \sum_{j \neq i} \mathrm{Dual}(A_{ij}) * \mathrm{Dual}(Y_j)}{\mathrm{Dual}(A_{ii})}, \tag{6.1}$$

for $i = 1, \ldots, n$.

The Jacobi interval operator satisfies the following properties: Let $S : \mathbf{A} * \mathbf{X} = \mathbf{B}$ be a linear system,

1) Y is a solution of $\mathbf{A} * \mathbf{X} = \mathbf{B}$ is equivalent to $\mathfrak{J}_S(Y) = Y$.

    Proof If Y solution of $\mathbf{A} * \mathbf{X} = \mathbf{B}$, then for every $i = 1, \ldots, n$

$$\mathfrak{J}_S(Y_i) = \frac{B_i - \sum_{j \neq i} \mathrm{Dual}(A_{ij}) * \mathrm{Dual}(Y_j)}{\mathrm{Dual}(A_{ii})} = \frac{A_{ij} * Y_i}{\mathrm{Dual}(A_{ii})} = Y_i.$$

    Conversely if $\mathfrak{J}_S(Y) = Y$, then for each i-component

$$\frac{B_i - \sum_{j \neq i} \mathrm{Dual}(A_{ij}) * \mathrm{Dual}(Y_j)}{\mathrm{Dual}(A_{ii})} = Y_i$$

$$\Rightarrow A_{i1} * Y_1 + \ldots + A_{ii} * Y_i + \ldots + A_{1n} * X_n = B_i.$$

So
$$\mathbf{A} * \mathbf{Y} = \mathbf{B}.$$ ∎

2) If $\mathbf{Y}$ is a solution of $\mathbf{A} * \mathbf{X} \subseteq \mathbf{B}$, then $\mathfrak{J}_S(\mathbf{Y})$ is a solution of $\mathbf{A} * \mathbf{X} \supseteq \mathbf{B}$.

Proof For $i = 1, \ldots, n$, the ith equation yields

$$A_{i1} * \mathfrak{J}_S(Y_1) + \ldots + A_{ii} * \mathfrak{J}_S(Y_i) + \ldots + A_{in} * \mathfrak{J}_S(Y_n)$$

$$= A_{i1} * \frac{B_1 - \sum_{j \neq 1} \text{Dual}(A_{1j}) * \text{Dual}(Y_j)}{\text{Dual}(A_{11})} + \ldots$$

$$+ A_{ii} * \frac{B_i - \sum_{j \neq i} \text{Dual}(A_{ij}) * \text{Dual}(Y_j)}{\text{Dual}(A_{ii})} + \ldots$$

$$+ A_{in} * \frac{B_n - \sum_{j \neq n} \text{Dual}(A_{nj}) * \text{Dual}(Y_j)}{\text{Dual}(A_{nn})}$$

$$\supseteq A_{i1} * \frac{A_{11} * Y_1}{\text{Dual}(A_{11})} + \ldots$$

$$+ B_i - \text{Dual}(A_{i1}) * \text{Dual}(Y_1) - \ldots - \text{Dual}(A_{in}) * \text{Dual}(Y_n) + \ldots$$

$$+ A_{in} * \frac{A_{nn} * Y_n}{\text{Dual}(A_{nn})}$$

$$= B_i.$$ ∎

3) If $\mathbf{Y}$ is a solution of $\mathbf{A} * \mathbf{X} \subseteq \mathbf{B}$, then $\mathbf{Y} \subseteq \mathfrak{J}_S(\mathbf{Y})$.

Proof Deducing Y_i from the ith equality of $\mathbf{A} * \mathbf{Y} \subseteq \mathbf{B}$, $i = 1, \ldots, n$

$$Y_i \subseteq \frac{B_i - \sum_{j \neq i} \text{Dual}(A_{ij}) * \text{Dual}(Y_j)}{\text{Dual}(A_{ii})} = \mathfrak{J}_S(Y_i).$$ ∎

Conversely and with similar proofs, we get the following properties
2') If $\mathbf{Y}$ is a solution of $\mathbf{A} * \mathbf{X} \supseteq \mathbf{B}$, then $\mathfrak{J}_S(\mathbf{Y})$ is a solution of $\mathbf{A} * \mathbf{X} \subseteq \mathbf{B}$.
3') If $\mathbf{Y}$ is a solution of $\mathbf{A} * \mathbf{X} \supseteq \mathbf{B}$, then $\mathbf{Y} \supseteq \mathfrak{J}_S(\mathbf{Y})$.

Moreover,
4) $\mathfrak{J}_S$ is $\subseteq$-antitonic, i.e., if $\mathbf{Y} \subseteq \mathbf{Z}$, then $\mathfrak{J}_S(\mathbf{Y}) \supseteq \mathfrak{J}_S(\mathbf{Z})$.

Proof From $(\text{Dual}(Y_1), \ldots, \text{Dual}(Y_n)) \supseteq (\text{Dual}(Z_1), \ldots, \text{Dual}(Z_n))$ and the isotonicity of the arithmetic operations. ∎

6.3.1 Solving a Linear System

Let $(x_1, \ldots, x_n)$ be a real solution of the real-valued system Ax=b, where A=(a_{ij}) and b=(b_i) are real matrices (respectively, $(n \times n)$ and $(n \times 1)$) with a_{ij} any real

6.3 Formal Solutions to a Linear System

numbers belonging to the intervals A'_{ij} and b_i any real numbers belonging to the intervals B'_i. Let X^0 be the interval vector $([x_1, x_1], \ldots, [x_n, x_n])$. If $\text{Impr}(\mathbf{A})$ is the interval matrix formed by $\text{Impr}(A_{ij})$ and analogously for $\text{Prop}(\mathbf{B})$, then X^0 is a solution of the system

$$S_1 : \text{Impr}(\mathbf{A}) * \mathbf{X} \subseteq \text{Prop}(\mathbf{B}) \tag{6.2}$$

since

$$\text{Impr}(\mathbf{A}) * \mathbf{X}^{(0)} \subseteq \mathbf{A} * ([x_1, x_1] \ldots [x_n, x_n])^\top = \mathbf{b} \subseteq \text{Prop}(\mathbf{B}),$$

after the identifications $a_{ij} \leftrightarrow [a_{ij}, a_{ij}]$ and $b_i \leftrightarrow [b_i, b_i]$.

Using the Jacobi interval operator associated to this system (6.2) we can define

$$Y_i^0 = \mathfrak{J}_{S_1}(X_i^0) = \frac{\text{Prop}(B_i) - \sum_{j \neq i} \text{Dual}(\text{Impr}(A_{ij})) * \text{Dual}(X_j^0)}{\text{Dual}(\text{Impr}(A_{ii}))}$$

for $i = 1, \ldots, n$. The interval vector $Y^0 = (Y_1^0, \ldots, Y_n^0)$ is proper and, by the properties of the Jacobi interval operator, it is a solution of $\text{Impr}(\mathbf{A}) * \mathbf{X} \supseteq \text{Prop}(\mathbf{B})$. Also Y^0 is an initial solution for the system $S : \mathbf{A} * \mathbf{X} \supseteq \mathbf{B}$, since

$$\mathbf{A} * Y^{(0)} \supseteq \text{Impr}(\mathbf{A}) * Y^{(0)} \supseteq \text{Prop}(\mathbf{B}) \supseteq \mathbf{B}.$$

By means of the Jacobi interval operator associated to the system $S : \mathbf{A} * \mathbf{X} \supseteq \mathbf{B}$,

$$\mathfrak{J}_S(Y_i) = \frac{B_i - \sum_{j \neq i} \text{Dual}(A_{ij}) * \text{Dual}(Y_j)}{\text{Dual}(A_{ii})}$$

it is possible to get a sequence of vector intervals, $Y^{(0)}$, $Y^{(1)} = \mathfrak{J}_S(Y^{(0)})$, $Y^{(2)} = \mathfrak{J}_S(Y^{(1)}), \ldots$, which, by the properties of the Jacobi interval operator, satisfies

$$Y^{(0)} \supseteq Y^{(1)} \subseteq Y^{(2)} \supseteq Y^{(3)} \subseteq Y^{(4)} \supseteq Y^{(5)} \subseteq \ldots \tag{6.3}$$

and such that $Y^{(t)}$ is a solution of $\mathbf{A} * \mathbf{X} \supseteq \mathbf{B}$ if t is even and a solution of $\mathbf{A} * \mathbf{X} \subseteq \mathbf{B}$ if t is odd.

6.3.2 Algorithm with Rounding

The computational scheme of the interval vectors sequence of solutions for the associated systems $\mathbf{A} * \mathbf{X} \subseteq \mathbf{B}$ and $\mathbf{A} * \mathbf{X} \supseteq \mathbf{B}$ in an arithmetic with rounding will be the following:

1. To arrange equations and unknowns to achieve that $0 \notin A'_{ii}$, for every $i = 1, \ldots, n$.

2. To obtain the real solution $(x_1, \ldots, x_n)$ for, e.g., the real-valued system mid$(\mathbf{A})x = $ mid$(\mathbf{B})$ consisting of the middle points of the intervals A'_{ij} and B'_i and to build the interval vector $X^{(0)} = ([x_1, x_1], \ldots, [x_n, x_n])$.
3. To compute for $i = 1, \ldots, n$

$$X_i^{(1)} = \mathfrak{J}_S(X_i^{(0)}) = \text{Out} \left(\frac{\text{Prop}(B_i) - \sum_{j \neq i} \text{Dual}(\text{Impr}(A_{ij})) * \text{Dual}(X_j^{(0)})}{\text{Dual}(\text{Impr}(A_{ii}))} \right)$$

and to build the interval vector $Y^{(0)} = (X_1^{(1)}, \ldots, X_n^{(1)})$, which is the initial solution of $\mathbf{A} * \mathbf{X} \supseteq \mathbf{B}$.
4. To compute for $t = 1, 2, \ldots$

$$\mathfrak{J}_S(Y_i^{(t)}) = \text{Out} \left(\frac{B_i - \sum_{j \neq i} \text{Dual}(A_{ij}) * \text{Dual}(Y_j^{(t-1)})}{\text{Dual}(A_{ii})} \right) \quad (i = 1, \ldots, n),$$

if t is even, or

$$\mathfrak{J}_S(Y_i^{(t)}) = \text{Inn} \left(\frac{B_i - \sum_{j \neq i} \text{Dual}(A_{ij}) * \text{Dual}(Y_j^{(t-1)})}{\text{Dual}(A_{ii})} \right) \quad (i = 1, \ldots, n),$$

if t is odd, and to build the sequence

$$Y^{(0)} \supseteq Y^{(1)} \subseteq Y^{(2)} \supseteq Y^{(3)} \subseteq Y^{(4)} \supseteq Y^{(5)} \subseteq \ldots.$$

Example 6.3.1 For the system

$$\begin{cases} [5, 2] * X_1 + [-1, -2] * X_2 + [2, 1] * X_3 = [3, 2] \\ [1, 0] * X_1 + [4, 3] * X_2 + [1, 2] * X_3 = [1, 3] \\ [1, 0] * X_1 + [2, 3] * X_2 + [5, 3] * X_3 = [4, 3] \end{cases}$$

starting from the solution

$$(x_1, x_2, x_3) = (0.510135, 0.206081, 0.682432),$$

of the real system formed by the mid-points

$$\begin{cases} 3.5\,x_1 - 1.5\,x_1 + 1.5\,x_1 = 2.5 \\ 0.5\,x_1 + 3.5\,x_1 + 1.5\,x_1 = 2.0 \\ 0.5\,x_1 + 2.5\,x_1 + 4.0\,x_1 = 3.5 \end{cases}$$

and using the Jacobi interval operator, the sequence of solutions is

6.3 Formal Solutions to a Linear System

$X^{(0)} = ([0.510135, 0.510135], [0.206081, 0.206081], [0.682432, 0.682432])$,

$Y^{(0)} = X^{(1)} = ([0.168243, 1.36486], [-0.291667, 0.772523], [0.374324, 1.19595]$,

$Y^{(1)} = ([0.604775, 0.256194], [0.114358, 0.202703], [0.941351, 0.227477])$,

$Y^{(2)} = ([0.264, 1.00062], [-0.182042, 0.848348], [0.633302, 0.797297])$,

$Y^{(3)} = ([0.516349, 0.51033], [0.0256745, 0.468468], [0.856425, 0.151652])$,

$Y^{(4)} = ([0.351124, 0.949849], [-0.124258, 0.898899], [0.68646, 0.531532])$,

$Y^{(1)} = ([0.604775, 0.256194], [0.114358, 0.202703], [0.941351, 0.227477])$,

$Y^{(2)} = ([0.264, 1.00062], [-0.182042, 0.848348], [0.633302, 0.797297])$,

$Y^{(3)} = ([0.516349, 0.51033], [0.0256745, 0.468468], [0.856425, 0.151652])$,

$Y^{(4)} = ([0.351124, 0.949849], [-0.124258, 0.898899], [0.68646, 0.531532])$,

$Y^{(5)} = ([0.505196, 0.672105], [-0.012528, 0.645646], [0.80433, 0.101101])$,

$Y^{(10)} = ([0.461, 0.950438], [-0.090324, 0.970044], [0.732071, 0.157491])$,

$Y^{(20)} = ([0.494958, 0.957191], [-0.0841278, 0.996055], [0.747791, 0.0207396])$,

$Y^{(30)} = ([0.499337, 0.958182], [-0.0834366, 0.99948], [0.749711, 0.00273116])$,

$Y^{(40)} = ([0.499913, 0.958313], [-0.0833469, 0.999932], [0.749962, 0.000359694])$,

$Y^{(49)} = ([0.5, 0.958296], [-0.0833251, 0.999953], [0.750005, 1.35104e-05])$,

$Y^{(50)} = ([0.49998, 0.958331], [-0.0833351, 0.999991], [0.749995, 4.73658e-05])$.

The distances between consecutive terms are

$$\text{dist}(Y^{(1)}, Y^{(0)}) = (1.10867, 0.56982, 0.968468),$$
$$\text{dist}(Y^{(2)}, Y^{(1)}) = (0.744426, 0.645646, 0.56982),$$
$$\text{dist}(Y^{(3)}, Y^{(2)}) = (0.490289, 0.37988, 0.645646),$$
$$\text{dist}(Y^{(4)}, Y^{(3)}) = (0.439518, 0.43043, 0.37988),$$
$$\text{dist}(Y^{(5)}, Y^{(4)}) = (0.277744, 0.253253, 0.43043),$$
$$\text{dist}(Y^{(10)}, Y^{(9)}) = (0.115948, 0.127535, 0.112557),$$
$$\text{dist}(Y^{(20)}, Y^{(19)}) = (0.0150141, 0.0167947, 0.0148224),$$
$$\text{dist}(Y^{(30)}, Y^{(29)}) = (0.0019744, 0.00221163, 0.00195193),$$

$$\text{dist}(Y^{(40)}, Y^{(39)}) = (0.000259995, 0.000291288, 0.000257095),$$

$$\text{dist}(Y^{(50)}, Y^{(49)}) = (3.42131e-05, 3.83854e-05, 3.38554e-05),$$

$$\text{dist}(Y^{(60)}, Y^{(59)}) = (4.58956e-06, 5.06639e-06, 4.52995e-06).$$

Indeed

$$Y^{(0)} \supseteq Y^{(1)} \subseteq Y^{(2)} \supseteq Y^{(3)} \subseteq Y^{(4)} \supseteq Y^{(5)} \subseteq \ldots$$

and $Y^{(t)}$ is a solution of $\mathbf{A} * \mathbf{X} \supseteq \mathbf{B}$ if t is even and a solution of $\mathbf{A} * \mathbf{X} \subseteq \mathbf{B}$ if t is odd.

6.3.3 Sufficient Conditions for Convergence

Starting from any interval vector $Y^{(0)} = (Y_1^{(0)}, \ldots, Y_n^{(0)})$ let us build the sequence $(Y^{(0)}, Y^{(1)}, \ldots, Y^{(t)}, \ldots)$ with $Y^{(t)} = \mathfrak{J}_S(Y^{(t-1)})$. For the distance q between two consecutive terms of the sequence we can obtain, for every $i = 1, \ldots, n$

$$\text{dist}(Y_i^{(2t+1)}, Y_i^{(2t)})$$

$$= \text{dist}(\mathfrak{J}_S(Y_i^{(2t)}), \mathfrak{J}_S(Y_i^{(2t-1)}))$$

$$= \text{dist}\left(\frac{B_i - \sum_{j \neq i} \text{Dual}(A_{ij}) * \text{Dual}(Y_j^{(2t)})}{\text{Dual}(A_{ii})}, \right.$$

$$\left. \frac{B_i - \sum_{j \neq i} \text{Dual}(A_{ij}) * \text{Dual}(Y_j^{(2t-1)})}{\text{Dual}(A_{ii})}\right)$$

$$\leq \text{abs}\left(\frac{1}{\text{Dual}(A_{ii})}\right) \text{dist}\left(B_i - \sum_{j \neq i} \text{Dual}(A_{ij} * Y_j^{(2t)}),\right.$$

$$\left. B_i - \sum_{j \neq i} \text{Dual}(A_{ij} * Y_j^{(2t-1)})\right)$$

$$= \frac{1}{\text{mig}(A_{ii})} \text{wid}\left(\sum_{j \neq i} \text{Dual}(A_{ij} * Y_j^{(2t)}), \sum_{j \neq i} \text{Dual}(A_{ij} * Y_j^{(2t-1)})\right)$$

$$\leq \frac{1}{\text{mig}(A_{ii})} \sum_{j \neq i} \text{wid}\left(A_{ij} * Y_j^{(2t)}, A_{ij} * Y_j^{(2t-1)}\right)$$

6.3 Formal Solutions to a Linear System

$$= \frac{1}{\text{mig}(A_{ii})} \sum_{j \neq i} \text{abs}(A_{ij})\text{dist}(Y_j^{(2t)}, Y_j^{(2t-1)})$$

$$\leq \frac{\sum_{j \neq i} \text{abs}(\text{Prop}(A_{ij}))}{\text{mig}(\text{Prop}(A_{ii}))} \max_{j \in \{1,\ldots,n\}} \text{dist}(Y_j^{(2t)}, Y_j^{(2t-1)}).$$

Therefore, if Prop(**A**) is a strictly diagonally dominant interval matrix, then for every $i = 1, \ldots, n$

$$\frac{\sum_{j \neq i} \text{abs}(\text{Prop}(A_{ij}))}{\text{mig}(\text{Prop}(A_{ii}))} < \alpha < 1, \qquad (6.4)$$

and, for every i,

$$\text{dist}(Y_i^{(2t+1)}, Y_i^{(2t)}) \leq \alpha^t \max_{j \in \{1,\ldots,n\}} \text{dist}(Y_j^{(1)}, Y_j^{(0)}).$$

So, $(Y^{(0)}, Y^{(1)}, \ldots, Y^{(t)} \ldots)$ is a Cauchy sequence and it has a limit Y satisfying $Y = \mathfrak{J}_S(Y)$ and, due to property (1) of the Jacobi operator, this limit is a solution of the system $\mathbf{A} * \mathbf{X} = \mathbf{B}$. This solution is unique because if we started from any other interval vector $\mathbf{Z}^{(0)} = (Z_1^{(0)}, \ldots, Z_n^{(0)})$, and built the sequence $(\mathbf{Z}^{(0)}, \mathbf{Z}^{(1)}, \ldots, \mathbf{Z}^{(t)}, \ldots)$, with $\mathbf{Z}^{(t)} = \mathfrak{J}_S(\mathbf{Z}^{(t-1)})$, repeating the previous reasoning twice, we would obtain:

1. $(\mathbf{Z}^{(0)}, \mathbf{Z}^{(1)}, \ldots, \mathbf{Z}^{(t)}, \ldots)$ is convergent and
2. it is true that

$$\text{dist}(Y_i^{(t)}, Z_i^{(t)}) \leq \alpha^t \max_{j \in \{1,\ldots,n\}} \text{dist}(Y_j^{(0)}, Z_j^{(0)}).$$

So both sequences have the same limit. Therefore if the condition (6.4) is true for every $i = 1, \ldots, n$, then there exists a unique solution.

Moreover, if $Y^{(0)}$ is any initial solution of $\mathbf{A} * \mathbf{X} \supseteq \mathbf{B}$, then by means of the Jacobi interval operator associated with the system it is possible to get a sequence of vector intervals, $Y^{(0)}, Y^{(1)} = \mathfrak{J}_S(Y^{(0)}), Y^{(2)} = \mathfrak{J}_S(Y^{(1)}), \ldots$, satisfying

$$Y^{(0)} \supseteq Y^{(1)} \subseteq Y^{(2)} \supseteq Y^{(3)} \subseteq Y^{(4)} \supseteq Y^{(5)} \subseteq \ldots$$

and such that $Y^{(t)}$ is a solution of $\mathbf{A}*\mathbf{X} \supseteq \mathbf{B}$ if t is even and a solution of $\mathbf{A}*\mathbf{X} \subseteq \mathbf{B}$ if t is odd. This sequence converges to a limit Y, the same one for any initial solution, which is the unique solution of $\mathbf{A} * \mathbf{X} = \mathbf{B}$.

Thus for the system of Example 6.3.1, for which the condition (6.4) is true, the limit of $(Y^{(0)}, Y^{(1)}, Y^{(2)}, \ldots)$ exists and it is

$$Y = ([0.5, 0.985\overline{3}], [-0.08\overline{3}, 1], [0.75, 0]),$$

which is the unique solution of the system $\mathbf{A} * \mathbf{X} = \mathbf{B}$.

6.3.4 Sufficient Condition for Non-convergence

By the properties of the width d of an interval, for every $i = 1, \ldots, n$

$$\begin{aligned}
\operatorname{wid}(\mathfrak{J}_S(Y_i^{(t)})) &= \operatorname{wid}(\operatorname{Prop}(\mathfrak{J}_S(Y_i^{(t)}))) \\
&= \operatorname{wid}\left(\frac{\operatorname{Prop}(B_i) - \sum_{j \neq i} \operatorname{Prop}(A_{ij}) * \operatorname{Prop}(Y_j^{(t)})}{\operatorname{Prop}(A_{ii})}\right) \\
&\geq \operatorname{wid}\left(\frac{\operatorname{Prop}(B_i) - \sum_{j \neq i} [\underline{a_{ij}}, \overline{a_{ij}}] * \operatorname{Prop}(Y_j^{(t)})}{[\underline{a_{ii}}, \underline{a_{ii}}]}\right) \\
&= \operatorname{wid}\left(\frac{\operatorname{Prop}(B_i)}{\underline{a_{ii}}}\right) + d\left(\frac{\sum_{j \neq i} \overline{a_{ij}} * \operatorname{Prop}(Y_j^{(t)})}{\underline{a_{ii}}}\right) \\
&= \frac{\operatorname{wid}(\operatorname{Prop}(B_i))}{|\underline{a_{ii}}|} + \frac{\sum_{j \neq i} |\overline{a_{ij}}| * \operatorname{wid}(\operatorname{Prop}(Y_j^{(t)}))}{|\underline{a_{ii}}|}.
\end{aligned}$$

for $\operatorname{Prop}(A_{ij}) = [\underline{a_{ij}}, \overline{a_{ij}}]$. So,

$$\operatorname{wid}(\mathfrak{J}_S(Y_i^{(t)})) \geq \frac{\sum_{j \neq i} \operatorname{abs}(\operatorname{Prop}(A_{ij}))}{\operatorname{mig}(\operatorname{Prop}(A_{ii}))} \min_{j \in \{1,\ldots,n\}} \operatorname{wid}(Y_j^{(t)}).$$

Therefore, if for every $i = 1, \ldots, n$

$$\frac{\sum_{j \neq i} \operatorname{abs}(\operatorname{Prop}(A_{ij}))}{\operatorname{mig}(\operatorname{Prop}(A_{ii}))} > \alpha > 1,$$

then for any i,

$$\operatorname{wid}(Y_i^{(t)}) \geq \alpha^t \min_{j \in \{1,\ldots,n\}} \operatorname{wid}(Y_j^{(0)}).$$

and the width of the successive intervals of the sequence $Y^{(0)}$, $Y^{(1)} = \mathfrak{J}_S(Y^{(0)})$, $Y^{(2)} = \mathfrak{J}_S(Y^{(1)}), \ldots$, will increase without bound. So, the Jacobi operator will not provide a solution for systems of type $A * X = B$.

6.3.5 Solution in the Case of Non-convergence

Let us consider a linear system $A * X = B$ in the case of non-convergence of the Jacobi algorithm, i.e., where

6.3 Formal Solutions to a Linear System

$$\begin{cases} A_{11} * X_1 + \ldots + A_{1n} * X_n = B_1 \\ \ldots\ldots\ldots\ldots\ldots\ldots\ldots\ldots\ldots\ldots \\ A_{n1} * X_1 + \ldots + A_{nn} * X_n = B_n \end{cases}$$

To get a solution, the problem can be put in the form of solving an optimization problem in two different ways, using linear or non-linear techniques.

By means of a linear optimization scheme the problem becomes

$$\min \sum_{i=1,\ldots,n} (\text{Inf}(A_{i1} * X_1 + \ldots + A_{in} * X_n) - \text{Inf}(B_i)$$

$$+ \text{Sup}(A_{i1} * X_1 + \ldots + A_{in} * X_n) - \text{Sup}(B_i))$$

subject to the restrictions:

$$\text{Inf}(A_{11} * X_1 + \ldots + A_{1n} * X_n) \geq \text{Inf}(B_1)$$
$$\text{Sup}(A_{11} * X_1 + \ldots + A_{1n} * X_n) \leq \text{Sup}(B_1)$$
$$\ldots\ldots\ldots\ldots\ldots\ldots\ldots\ldots\ldots\ldots\ldots$$
$$\text{Inf}(A_{n1} * X_1 + \ldots + A_{nn} * X_n) \geq \text{Inf}(B_n)$$
$$\text{Sup}(A_{n1} * X_1 + \ldots + A_{nn} * X_n) \leq \text{Sup}(B_n).$$

Due to the definition of the interval product of modal intervals, the products involved in these expressions are not linear, but only piece-wise linear. Therefore, it is necessary to introduce 0–1 variables to convert the problem to a mixed integer one, solvable by standard techniques. The number of these 0–1 variables is four for each unknown, to take into account the different signs of the interval bounds, plus two for each coefficient of the system with bounds of different sign, to take into account that, in some cases, the bounds of the interval product are the minimum or maximum of the product of the bounds.

Example 6.3.2 For the system

$$\begin{cases} [5,7] * X_1 + [3.01,-3] * X_2 + [3.01,-3] * X_3 + [3.01,-3] * X_4 = [26,-43.11] \\ [3.01,-3] * X_1 + [5,7] * X_2 + [2.99,-3] * X_3 + [2.99,-3] * X_4 = [-5,-44.94] \\ [2.99,-3] * X_1 + [2.99,-3] * X_2 + [5,7] * X_3 + [3.01,-3] * X_4 = [-18,-48.98] \\ [3.01,-3] * X_1 + [3.01,-3] * X_2 + [2.99,-3] * X_3 + [5,7] * X_4 = [23,-45] \end{cases}$$

the Jacobi algorithm does not converge. To find the solution by means of a mixed integer programming procedure it is necessary to introduce forty 0–1 variables and the execution time is high for such a small system. But, with the a priori knowledge that the unknowns are negative, the number of 0–1 variables is zero and the solution (using the LINDO © linear and integer programming software, www.lindo.com),

$$X_1 = [-0.999998, -2.999997],$$
$$X_2 = [-4.999997, -3.000001],$$
$$X_3 = [-2.000006, -5.000002],$$
$$X_4 = [-6.000002, -1.000000],$$

is available after negligible execution time.

The number of 0–1 variables can be large even for small systems, dramatically increasing the execution time. This last disadvantage does not necessarily compel us to discard this approach, since in many problems the physical context determines the sign of the unknowns, so the number of 0–1 variables can be small even for a non-small system and this procedure can then be very useful. Obviously, the physical context implies something logically interpretable and the solution of a linear system has a logical meaning, as will be shown in Sect. 6.4.

The second approach is to use any standard procedure of non-linear optimization applied to the objective function

$$f_{obj} = \min \sum_{i=1,...,n} ((\text{Inf}(A_{i1} * X_1 + ... + A_{in} * X_n) - \text{Inf}(B_i))^2$$
$$+ (\text{Sup}(A_{i1} * X_1 + ... + A_{in} * X_n) - \text{Sup}(B_i))^2)$$

with no restrictions.

Example 6.3.3 For the system $\mathbf{A} * \mathbf{X} = \mathbf{B}$ with

$$\mathbf{A} = \begin{pmatrix} [4,6] & [-9,0] & [0,12] & [2,3] & [5,9] & [-23,-9] & [15,23] \\ [0,1] & [6,10] & [-1,1] & [-1,3] & [-5,1] & [1,15] & [-3,-1] \\ [0,3] & [-20,-9] & [12,77] & [-6,30] & [0,3] & [-18,1] & [0,1] \\ [-4,1] & [-1,1] & [-3,1] & [3,5] & [5,9] & [1,2] & [1,4] \\ [0,3] & [0,6] & [0,20] & [-1,5] & [8,14] & [-6,1] & [10,17] \\ [-7,-2] & [1,2] & [7,14] & [-3,1] & [0,2] & [3,5] & [-2,1] \\ [-1,5] & [-3,2] & [0,8] & [1,11] & [-5,10] & [2,7] & [6,82] \end{pmatrix}$$

$$\mathbf{X} = \begin{pmatrix} X_1 \\ X_2 \\ X_3 \\ X_4 \\ X_5 \\ X_6 \\ X_7 \end{pmatrix} \text{ and } \mathbf{B} = \begin{pmatrix} [-10,95] \\ [35,14] \\ [-6,2] \\ [30,7] \\ [4,95] \\ [-6,46] \\ [-2,65] \end{pmatrix},$$

the solution (using, for example the GRG2 ©non-linear optimization software, www.optimalmethods.com) is

$$X_1 = [-1.22474317578, .50542987670],$$
$$X_2 = [18.26444337097, -9.51750410301],$$
$$X_3 = [-.02818650587, 1.16075521933],$$
$$X_4 = [16.40769576636, -14.45553419850],$$
$$X_5 = [-1.34356527337, 3.98821848038],$$
$$X_6 = [-3.52893852104, 4.54345836822],$$
$$X_7 = [5.43086236811, -.67400838684],$$

after an execution time of less than 1 s. But if the right-hand side B_7 is replaced by the interval $[65, -2]$, the final value of f_{obj} is greater than 0 and the minimum obtained is not a solution of the system.

Example 6.3.4 For $n \times n$ systems with coefficients defined by the matrix (from [32])

$$\begin{pmatrix} 1 & 2 & 3 & \cdots & n-1 & n \\ 2 & 2 & 3 & \cdots & n-1 & n \\ 3 & 3 & 3 & \cdots & n-1 & n \\ \vdots & \vdots & \vdots & \ddots & \vdots & \vdots \\ n-1 & n-1 & n-1 & \cdots & n-1 & n \\ n & n & n & \cdots & n & n \end{pmatrix}$$

intervalized by $[a_{ij} - 0.1, a_{ij} + 0.1]$ if $i \neq j$, $[a_{ij} + 0.1, a_{ij} - 0.1]$ if $i = j$ and right-hand sides $B_i = [i + 10, i - 10]$, the final value $f_{obj} = 0$ is obtained in 8 s when $n = 30$ and 48 s when $n = 50$.

6.4 Logical Meaning of the Solution

All the previous results can be obtained within the frame of classical intervals completed with Kaucher's generalized arithmetic, but with no interpretation of the systems of equations and their solutions in terms of the real-valued system of which the coefficients, right-hand sides, and solutions belong to the sets of real numbers associated with the intervals involved in the interval system. As modal interval analysis provides a logical basis for the classical intervals, it is possible to interpret the system and its solution when they verify certain conditions, established precisely by the semantic theorems.

Let us suppose that every component of the solution $(X_1, \ldots, X_n)$ is a proper interval. In this case, by Theorem 4.4.3 (*-interpretability for multidimensional computations), if $A_{i_1 j_1}, \ldots, A_{i_p j_p}, B_{k_{q+1}}, \ldots, B_{k_n}$ are proper intervals,

$A_{i_{p+1}j_{p+1}}, \ldots, A_{i_n j_n}, B_{k_1}, \ldots, B_{k_q}$ are independent improper, then

$$\begin{cases} A_{11} * X_1 + \ldots + A_{1n} * X_n \subseteq B_1 \\ \cdots\cdots\cdots\cdots\cdots\cdots\cdots\cdots\cdots\cdots\cdots \\ A_{n1} * X_1 + \ldots + A_{nn} * X_n \subseteq B_n \end{cases}$$

means

$(\forall x_1 \in X'_1) \ldots (\forall x_n \in X'_n)$
$(\forall a_{i_1 j_1} \in A'_{i_1 j_1}) \ldots (\forall a_{i_p j_p} \in A'_{i_p j_p})(\forall b_{k_1} \in B'_{k_1}) \ldots (\forall b_{k_q} \in B'_{k_q})$
$(\exists a_{i_{p+1} j_{p+1}} \in A'_{i_{p+1} j_{p+1}}) \ldots (\exists a_{i_n j_n} \in A'_{i_n j_n})(\exists b_{k_q+1} \in B'_{k_q+1}) \ldots (\exists b_{k_n} \in B'_{k_n})$

$$\begin{pmatrix} a_{11} * x_1 + \ldots + a_{1n} * x_n = b_1 \\ \cdots\cdots\cdots\cdots\cdots\cdots\cdots\cdots\cdots \\ a_{n1} * x_1 + \ldots + a_{nn} * x_n = b_n \end{pmatrix}.$$

In a similar way, Theorem 4.4.4 (**-interpretability for multi-dimensional computations) provides a semantics for the solutions of the system $\mathbf{A} * \mathbf{X} \supseteq \mathbf{B}$.

6.4.1 Semantics in the General Case

Let us suppose that some components of the solution of $\mathbf{A} * \mathbf{X} \subseteq \mathbf{B}$ are improper, for example X_i; substituting

$$X_i = \text{Prop}(X_i) + [\text{wid}(X_i), -\text{wid}(X_i)]$$

in all the equations, the system will become

$$\begin{cases} A_{11} * X_1 + \ldots + A_{1i} * (\text{Prop}(X_i) + [\text{wid}(X_i), -\text{wid}(X_i)]) + \ldots + A_{1n} * X_n \subseteq B_1 \\ A_{21} * X_1 + \ldots + A_{2i} * (\text{Prop}(X_i) + [\text{wid}(X_i), -\text{wid}(X_i)]) + \ldots + A_{2n} * X_n \subseteq B_2 \\ \cdots\cdots\cdots\cdots\cdots\cdots\cdots\cdots\cdots\cdots\cdots\cdots\cdots\cdots\cdots\cdots\cdots\cdots\cdots\cdots\cdots \\ A_{n1} * X_1 + \ldots + A_{ni} * (\text{Prop}(X_i) + [\text{wid}(X_i), -\text{wid}(X_i)]) + \ldots + A_{nn} * X_n \subseteq B_n. \end{cases}$$

By Theorem 4.4.3, if $A_{i_1 j_1}, \ldots, A_{i_p j_p}, B_{k_q+1}, \ldots, B_{k_n}$ are proper intervals and $A_{i_{p+1}j_{p+1}}, \ldots, A_{i_n j_n}, B_{k_1}, \ldots, B_{k_q}$ are improper, then

$(\forall x_1 \in X'_1) \ldots (\forall x_i \in X'_i) \ldots (\forall x_n \in X'_n)$
$(\forall a_{i_1 j_1} \in A'_{i_1 j_1}) \ldots (\forall a_{i_p j_p} \in A'_{i_p j_p})(\forall b_{k_1} \in B'_{k_1}) \ldots (\forall b_{k_q} \in B'_{k_q})$
$(\exists y_{i1} \in [\text{wid}(X_i), -\text{wid}(X_i)]') \ldots (\exists y_{in} \in [\text{wid}(X_i), -\text{wid}(X_i)]')$
$(\exists a_{i_{p+1} j_{p+1}} \in A'_{i_{p+1} j_{p+1}}) \ldots (\exists a_{i_n j_n} \in A'_{i_n j_n})(\exists b_{k_q+1} \in B'_{k_q+1}) \ldots (\exists b_{k_n} \in B'_{k_n})$

$$\begin{pmatrix} a_{11} * x_1 + \ldots + a_{1i} * (x_i + y_{i1}) + \ldots + a_{1n} * x_n = b_1 \\ a_{21} * x_1 + \ldots + a_{2i} * (x_i + y_{i2}) + \ldots + a_{2n} * a_n = b_2 \\ \ldots\ldots\ldots\ldots\ldots\ldots\ldots\ldots\ldots\ldots\ldots\ldots\ldots\ldots\ldots\ldots\ldots\ldots\ldots\ldots \\ a_{n1} * x_1 + \ldots + a_{ni} * (x_i + y_{in}) + \ldots + a_{nn} * x_n = b_n \end{pmatrix}.$$

Remark 6.4.1 The variables $y_{i1}, y_{i2}, \ldots, y_{in}$ act as control variables for each value of x_i within the interval X_i'.

Example 6.4.1 Following a previous example, the interval vector

$$X = ([0.5, 0.958296], [-0.0833251, 0.999953], [0.750005, 1.35104e - 05])$$

is a solution for the system

$$\begin{cases} [5,2] * X_1 + [-1,-2] * X_2 + [2,1] * X_3 \subseteq [3,2] \\ [1,0] * X_1 + [4,3] * X_2 + [1,2] * X_3 \subseteq [1,3] \\ [1,0] * X_1 + [2,3] * X_2 + [5,3] * X_3 \subseteq [4,3]. \end{cases}$$

Since the first component $[0.5, 0.958296]$ and the second one $[-0.0833251, 0.999953]$ are proper intervals but the third one $[0.750005, 1.35104e - 05]$ is improper, the semantic is

$(\forall x_1 \in [0.5, 0.958296]')(\forall x_2 \in [-0.0833251, 0.999953]')$
$(\forall x_3 \in [1.35104e - 05, 0.750005]')(\forall a_{23} \in [1,2]')$
$(\forall a_{32} \in [2,3]')(\forall b_1 \in [2,3]')(\forall b_3 \in [3,4]')$
$(\exists y_{31} \in [-0.75, 0.75]')(\exists y_{32} \in [-0.75, 0.75]')(\exists y_{33} \in [-0.75, 0.75]')$
$(\exists a_{11} \in [2,5]')(\exists a_{12} \in [-2,-1]')(\exists a_{13} \in [1,2]')$
$(\exists a_{21} \in [0,1]')(\exists a_{22} \in [-3,4]')(\exists a_{31} \in [0,1]')(\exists a_{33} \in [3,5]')(\exists b_2 \in [1,3]')$

$$\begin{pmatrix} a_{11} * x_1 + a_{12} * x_2 + a_{13} * (x_3 + y_{31}) = b_1 \\ a_{21} * x_1 + a_{22} * x_2 + a_{23} * (x_3 + y_{32}) = b_2 \\ a_{31} * x_1 + a_{32} * x_2 + a_{33} * (x_3 + y_{33}) = b_3 \end{pmatrix}.$$

Using an analogous process, the semantics for the solutions of $\mathbf{A} * \mathbf{X} \supseteq \mathbf{B}$ can be obtained using the dual Theorem 4.4.4 (of **-interpretability).

6.5 System Solution Sets

The two-step procedure for obtaining a formal solution to a linear system and then the semantic meaning of the solution, can not be the most suitable way to handle systems which appear in applications, especially when the solution involves intervals of different modalities which means different quantifiers affecting to the unknowns or coefficients.

If $\mathbf{Ax} = \mathbf{b}$ is a real-valued system of linear equations, considering intervals A'_{ij} of variation for the coefficients and intervals B'_i of variation for the right-hand sides, a *solution set* can be defined: the set of solutions for the system with logical specifications for the selection of the coefficients

$$\Xi_{\alpha\beta} = \{x \in \mathbb{R}^n \mid (\forall a_{i_1 j_1} \in A'_{i_1 j_1}) \ldots (\forall a_{i_p j_p} \in A'_{i_p j_p})(\forall b_{k_1} \in B'_{k_1}) \ldots (\forall b_{k_q}, B'_{k_q})$$
$$(\exists a_{i_{p+1} j_{p+1}} \in A'_{i_{p+1} j_{p+1}}) \ldots (\exists a_{i_n j_n} \in A'_{i_n j_n})$$
$$(\exists b_{k_{q+1}} \in B'_{k_{q+1}}) \ldots (\exists b_{k_n} \in B'_{k_n}) \, \mathbf{Ax} = \mathbf{b}\}.$$

For example the so-called *tolerable solution set*

$$\Xi_{tol} = \{x \in \mathbb{R}^n \mid (\forall a_{11} \in A'_{11}) \ldots (\forall a_{nn} \in A'_{nn}) \, (\exists b_1 \in B'_1) \ldots (\exists b_n \in B'_n) \, \mathbf{Ax} = \mathbf{b}\}$$

and the *united solution set*

$$\Xi_{uni} = \{x \in \mathbb{R}^n \mid (\exists a_{11} \in A'_{11}) \ldots (\exists a_{nn} \in A'_{nn}) \, (\exists b_1 \in B'_1) \ldots (\exists b_n \in B'_n) \, \mathbf{Ax} = \mathbf{b}\}.$$

The characterization of these solution sets can be obtained by means of interval enclosures, i.e., inner and outer or other interval estimates. These subjects will be treated in Chap. 10.

Chapter 7
Twins and f^* Algorithm

7.1 Introduction

This chapter deals with the construction of an algorithm to obtain inner and outer approximations of the f^* extension of a continuous function f, in the case of non-monotony of f in the studied domain. One convenient approach, but not the only one, is to simultaneously work with both inner and outer approximations. This kind of interval representation, referred to as twins, have already been studied in the field of classical intervals [55, 64]. First of all, twins with modal intervals will be presented.

Summarizing some results seen in Chap. 2, in the construction of the classical and modal intervals, from a lattice viewpoint and disregarding their logic-semantic features, starting from the order structure $(\mathbb{R}, \leq)$, a classical interval of real bounds $\underline{a}, \overline{a} \in \mathbb{R}$ was defined by

$$A' = [\underline{a}, \overline{a}] = \{x \in \mathbb{R} \mid \underline{a} \leq x \leq \overline{a}\}$$

with the condition $\underline{a} \leq \overline{a}$. In the set of the classical intervals

$$I(\mathbb{R}) = \{[\underline{a}, \overline{a}] \mid \underline{a} \leq \overline{a}\}$$

two relations were defined: if $A = [\underline{a}, \overline{a}], B = [\underline{b}, \overline{b}] \in I(\mathbb{R})$

$$A' \leq B' \Leftrightarrow (\underline{a} \leq \underline{b}, \overline{a} \leq \overline{b})$$
$$A' \subseteq B' \Leftrightarrow (\underline{a} \geq \underline{b}, \overline{a} \leq \overline{b}).$$

Both are order relations. $(I(\mathbb{R}), \leq)$ is a lattice with the infimum and supremum, called respectively *minimum* and *maximum*, given by

$$\text{Min}(A', B') = \text{Inf}_{\leq}(A', B') = [\min(\underline{a}, \underline{b}), \min(\overline{a}, \overline{b})]$$

$$\text{Max}(A', B') = \text{Sup}_{\leq}(A', B') = [\max(\underline{a}, \underline{b}), \max(\overline{a}, \overline{b})],$$

and $(I(\mathbb{R}), \subseteq)$ is also a lattice with the infimum and supremum, called respectively *meet* and *join*, given by

$$A' \wedge B' = \text{Inf}_\subseteq(A', B') = [\max(\underline{a}, \underline{b}), \min(\overline{a}, \overline{b})]$$
$$A' \vee B' = \text{Sup}_\subseteq(A', B') = [\min(\underline{a}, \underline{b}), \max(\overline{a}, \overline{b})].$$

The operation $\text{Inf}_\subseteq(A', B')$, i.e. $A' \wedge B'$, is not defined in $I(\mathbb{R})$ when

$$\max(\underline{a}, \underline{b}) > \min(\overline{a}, \overline{b}),$$

because the result is not a classical interval. In order to close this operation it was necessary to extend the classical intervals to the modal intervals, by defining the set

$$I^*(\mathbb{R}) = \{[\underline{a}, \overline{a}] \mid \underline{a}, \overline{a} \in \mathbb{R}\}$$

with the operations $\text{Inf}_\leq$, $\text{Sup}_\leq$, $\text{Inf}_\subseteq$ and $\text{Sup}_\subseteq$ defined in the same formal way as in $I(\mathbb{R})$. From the natural inclusion $I(\mathbb{R}) \subseteq I^*(\mathbb{R})$ the classical intervals of $I(\mathbb{R})$ were called *proper* intervals and the intervals of $I^*(\mathbb{R}) - I(\mathbb{R})$ *improper* intervals.

In fact, to start from the order structure $(\mathbb{R}, \leq)$ is irrelevant because this construction can be done starting from any structure $(M, \preccurlyeq)$, where $\preccurlyeq$ is a partial order relation on a set M. The results are $(I^*(M), \leq)$ and $(I^*(M), \subseteq)$ which are lattices of the same kind as $(M, \preccurlyeq)$.

7.2 Twins

Now the interval structures $(I^*(\mathbb{R}), \leq)$ and $(I^*(\mathbb{R}), \subseteq)$ will be used as starting points to construct new interval elements, *twins*.

7.2.1 Twins Associated to the Relations $\leq$ and $\subseteq$

Starting from the structure $(I^*(\mathbb{R}), \leq)$, a $\leq$-*proper twin* of bounds $\underline{A}, \overline{A} \in I^*(\mathbb{R})$, with the condition $\underline{A} \leq \overline{A}$, is the set of modal intervals

$$\mathbb{A}_\leq = |[\underline{A}, \overline{A}]|_\leq = \{X \in I^*(\mathbb{R}) \mid \underline{A} \leq X \leq \overline{A}\}.$$

Figure 7.1 illustrates this concept.

In the set of $\leq$-proper twins

$$I(I^*(\mathbb{R}))_\leq = \{|[\underline{A}, \overline{A}]|_\leq \mid \underline{A} \leq \overline{A}\}$$

7.2 Twins

Fig. 7.1 ≤-proper twin

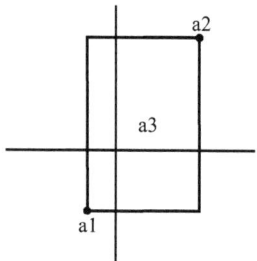

two relations are defined: if $\mathbb{A}_\leq = |[\underline{A}, \overline{A}]|_\leq, \mathbb{B}_\leq = |[\underline{B}, \overline{B}]|_\leq \in I(I^*(\mathbb{R}))$

$$\mathbb{A}_\leq \leq \mathbb{B}_\leq \Leftrightarrow (\underline{A} \leq \underline{B}, \overline{A} \leq \overline{B})$$
$$\mathbb{A}_\leq \subseteq \mathbb{B}_\leq \Leftrightarrow (\underline{A} \geq \underline{B}, \overline{A} \leq \overline{B}).$$

The infimum and supremum for these ≤-proper twins exist and they are given by

$$\text{Inf}_\leq(\mathbb{A}_\leq, \mathbb{B}_\leq) = |[\text{Inf}_\leq(\underline{A}, \underline{B}), \text{Inf}_\leq(\overline{A}, \overline{B})]| = |[\text{Min}(\underline{A}, \underline{B}), \text{Min}(\overline{A}, \overline{B})]|$$

$$\text{Sup}_\leq(\mathbb{A}_\leq, \mathbb{B}_\leq) = |[\text{Sup}_\leq(\underline{A}, \underline{B}), \text{Sup}_\leq(\overline{A}, \overline{B})]| = |[\text{Max}(\underline{A}, \underline{B}), \text{Max}(\overline{A}, \overline{B})]|,$$

and

$$\text{Inf}_\subseteq(\mathbb{A}_\leq, \mathbb{B}_\leq) = |[\text{Sup}_\leq(\underline{A}, \underline{B}), \text{Inf}_\leq(\overline{A}, \overline{B})]| = |[\text{Max}(\underline{A}, \underline{B}), \text{Min}(\overline{A}, \overline{B})]|$$

$$\text{Sup}_\subseteq(\mathbb{A}_\leq, \mathbb{B}_\leq) = |[\text{Inf}_\leq(\underline{A}, \underline{B}), \text{Sup}_\leq(\overline{A}, \overline{B})]| = |[\text{Min}(\underline{A}, \underline{B}), \text{Max}(\overline{A}, \overline{B})]|,$$

in accordance with the notations introduced in Definition (2.2.16).

The main properties of these twin relations are stated in the following lemma.

Lemma 7.2.1 (Properties of ≤ and ⊆ in $I(I^*(\mathbb{R}))_\leq$)

(1) Both ≤ and ⊆ are partial order relations in $I(I^*(\mathbb{R}))_\leq$.
(2) $\text{Inf}_\leq(\mathbb{A}_\leq, \mathbb{B}_\leq) \in I(I^*(\mathbb{R}))_\leq$.
(3) $\text{Sup}_\leq(\mathbb{A}_\leq, \mathbb{B}_\leq) \in I(I^*(\mathbb{R}))_\leq$.
(4) $\text{Sup}_\subseteq(\mathbb{A}_\leq, \mathbb{B}_\leq) \in I(I^*(\mathbb{R}))_\leq$.
(5) The structure $(I(I^*(\mathbb{R}))_\leq, \leq)$ is a lattice.

Proof

(1) Reflexivity, antisymmetry, and transitivity are obvious.
(2) If $\mathbb{A}_\leq = |[\underline{A}, \overline{A}]|_\leq, \mathbb{B}_\leq = |[\underline{B}, \overline{B}]|_\leq$ and $\underline{A} = [a_1, a_2]$, $\overline{A} = [a_3, a_4]$, $\underline{B} = [b_1, b_2]$, $\overline{B} = [b_3, b_4]$, then

$$a_1 \leq a_3, \; a_2 \leq a_4, \; b_1 \leq b_3, \; b_2 \leq b_4,$$

Fig. 7.2 ≤-proper twins

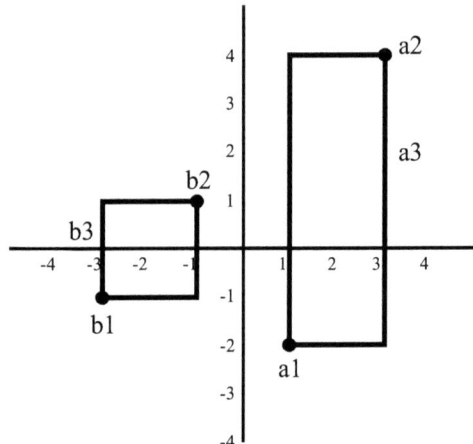

therefore

$$\text{Inf}_{\leq}(\underline{A}, \underline{B}) = [\inf(a_1, b_1), \inf(a_2, b_2)] \leq [\inf(a_3, b_3), \inf(a_4, b_4)] = \text{Inf}_{\leq}(\overline{A}, \overline{B}).$$

So $\text{Inf}_{\leq}(\mathbb{A}_{\leq}, \mathbb{B}_{\leq}) \in I(I^*(\mathbb{R}))_{\leq}$.

(3), (4) Similar reasoning as (2).

(5) By the previous properties and the definitions of infimum and supremum. ∎

Remark 7.2.1 It is not always true that $\text{Inf}_{\subseteq}(\mathbb{A}_{\leq}, \mathbb{B}_{\leq}) \in I(I^*(\mathbb{R}))_{\leq}$ for every $\mathbb{A}_{\leq}, \mathbb{B}_{\leq}$.

Example 7.2.1 The twins $\mathbb{A}_{\leq} = |[[1, -2], [3, 4]]|$ and $\mathbb{B}_{\leq} = |[[-3, -1], [2, 2]]|$ belong to $I(I^*(\mathbb{R}))_{\leq}$ because

$$\underline{A} = [1, -2] \leq [3, 4] = \overline{A},$$
$$\underline{B} = [-3, -1] \leq [2, 2] = \overline{B}.$$

They are represented in Fig. 7.2

It is true that $\mathbb{A}_{\leq} \not\leq \mathbb{B}_{\leq}$ and $\mathbb{A}_{\leq} \not\subseteq \mathbb{B}_{\leq}$. Infimum and supremum are

$$\text{Inf}_{\leq}(\mathbb{A}_{\leq}, \mathbb{B}_{\leq}) = |[[-3, -2], [-1, 1]]|$$
$$\text{Sup}_{\leq}(\mathbb{A}_{\leq}, \mathbb{B}_{\leq}) = |[[1, -1], [3, 4]]|$$
$$\text{Inf}_{\subseteq}(\mathbb{A}_{\leq}, \mathbb{B}_{\leq}) = |[[1, -1], [-1, 1]]|$$
$$\text{Sup}_{\subseteq}(\mathbb{A}_{\leq}, \mathbb{B}_{\leq}) = |[[-3, -2], [3, 4]]|,$$

represented in Fig. 7.3. Remark that $\text{Inf}_{\subseteq}(\mathbb{A}_{\leq}, \mathbb{B}_{\leq}) \notin I(I^*(\mathbb{R}))_{\leq}$.

7.2 Twins

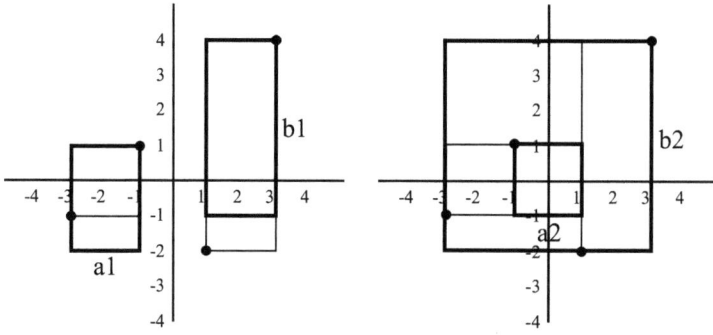

Fig. 7.3 $\leq$-infimum and supremum, $\subseteq$-infimum and supremum

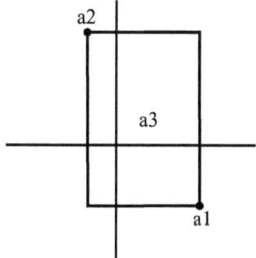

Fig. 7.4 $\subseteq$-proper twin

Analogously, starting from the structure $(I^*(\mathbb{R}), \subseteq)$, a $\subseteq$-*proper twin* of bounds $\underline{A}, \overline{A} \in I^*(\mathbb{R})$, with the condition $\underline{A} \subseteq \overline{A}$, can be also given by

$$\mathbb{A}_\subseteq = |[\underline{A}, \overline{A}]|_\subseteq = \{X \in I^*(\mathbb{R}) \mid \underline{A} \subseteq X \subseteq \overline{A}\}.$$

Figure 7.4 illustrates this definition.

In this set of $\subseteq$-proper twins

$$I(I^*(\mathbb{R}))_\subseteq = \{|[\underline{A}, \overline{A}]|_\subseteq \mid \underline{A} \subseteq \overline{A}\}$$

two relations are defined: if $\mathbb{A}_\subseteq = |[\underline{A}, \overline{A}]|_\subseteq, \mathbb{B}_\subseteq = |[\underline{B}, \overline{B}]|_\subseteq \in I(I^*(\mathbb{R}))$

$$\mathbb{A}_\subseteq \leq \mathbb{B}_\subseteq \Leftrightarrow (\underline{A} \subseteq \underline{B}, \overline{A} \subseteq \overline{B})$$
$$\mathbb{A}_\subseteq \subseteq \mathbb{B}_\subseteq \Leftrightarrow (\underline{A} \supseteq \underline{B}, \overline{A} \subseteq \overline{B}).$$

The infimum and supremum for these $\subseteq$-proper twins exist and they are given by

$$\text{Inf}_\leq(\mathbb{A}_\subseteq, \mathbb{B}_\subseteq) = |[\text{Inf}_\subseteq(\underline{A}, \underline{B}), \text{Inf}_\subseteq(\overline{A}, \overline{B})]| = |[\underline{A} \wedge \underline{B}, \overline{A} \wedge \overline{B}]|$$
$$\text{Sup}_\leq(\mathbb{A}_\subseteq, \mathbb{B}_\subseteq) = |[\text{Sup}_\subseteq(\underline{A}, \underline{B}), \text{Sup}_\subseteq(\overline{A}, \overline{B})]| = |[\underline{A} \vee \underline{B}, \overline{A} \vee \overline{B})|,$$

and

$$\text{Inf}_\subseteq(\mathbb{A}_\subseteq, \mathbb{B}_\subseteq) = |[\text{Sup}_\subseteq(\underline{A}, \underline{B}), \text{Inf}_\subseteq(\overline{A}, \overline{B})]| = |[\underline{A} \vee \underline{B}, \overline{A} \wedge \overline{B}]|$$
$$\text{Sup}_\subseteq(\mathbb{A}_\subseteq, \mathbb{B}_\subseteq) = |[\text{Inf}_\subseteq(\underline{A}, \underline{B}), \text{Sup}_\subseteq(\overline{A}, \overline{B})]| = |[\underline{A} \wedge \underline{B}, \overline{A} \vee \overline{B}]|,$$

in accordance with the notations introduced in Definition 2.2.15.

The main properties of these twin relations are stated in the following lemma.

Lemma 7.2.2 (Properties of $\leq$ and $\subseteq$ in $I(I^*(\mathbb{R}))_\subseteq$)

(1) Both $\leq$ and $\subseteq$ are partial order relations in $I(I^*(\mathbb{R}))_\subseteq$.
(2) $\text{Inf}_\leq(\mathbb{A}_\subseteq, \mathbb{B}_\subseteq) \in I(I^*(\mathbb{R}))_\subseteq$.
(3) $\text{Sup}_\leq(\mathbb{A}_\subseteq, \mathbb{B}_\subseteq) \in I(I^*(\mathbb{R}))_\subseteq$.
(4) $\text{Sup}_\subseteq(\mathbb{A}_\subseteq, \mathbb{B}_\subseteq) \in I(I^*(\mathbb{R}))_\subseteq$.
(5) The structure $(I(I^*(\mathbb{R}))_\subseteq, \leq)$ is a lattice.

Proof

(1) Reflexivity, antisymmetry, and transitivity are obvious.
(2) If $\mathbb{A}_\subseteq = |[\underline{A}, \overline{A}]|_\subseteq$, $\mathbb{B}_\subseteq = |[\underline{B}, \overline{B}]|_\subseteq$ and $\underline{A} = [a_1, a_2]$, $\overline{A} = [a_3, a_4]$, $\underline{B} = [b_1, b_2]$, $\overline{B} = [b_3, b_4]$, then

$$a_1 \geq a_3, \ a_2 \leq a_4, \ b_1 \geq b_3, \ b_2 \leq b_4,$$

therefore

$$\text{Inf}_\subseteq(\underline{A}, \underline{B}) = [\sup(a_1, b_1), \inf(a_2, b_2)] \subseteq [\sup(a_3, b_3), \inf(a_4, b_4)] = \text{Inf}_\subseteq(\overline{A}, \overline{B}).$$

So $\text{Inf}_\leq(\mathbb{A}_\subseteq, \mathbb{B}_\subseteq) \in I(I^*(\mathbb{R}))_\subseteq$.
(3), (4) Similar reasoning as (2).
(5) By the previous properties and the definitions of infimum and supremum. ■

Remark 7.2.2 It is not always true that $\text{Inf}_\subseteq(\mathbb{A}_\subseteq, \mathbb{B}_\subseteq) \in I(I^*(\mathbb{R}))_\subseteq$ for every $\mathbb{A}_\subseteq, \mathbb{B}_\subseteq$.

Example 7.2.2 The twins $\mathbb{A}_\subseteq = |[[3, -2], [-1, 4]]|$ and $\mathbb{B}_\subseteq = |[[1, -3], [-3, -1]]|$ belong to $I(I^*(\mathbb{R}))_\subseteq$ because

$$\underline{A} = [3, -2] \subseteq [-1, 4] = \overline{A},$$
$$\underline{B} = [1, -3] \subseteq [-3, -1] = \overline{B}.$$

They are represented in Fig. 7.5

7.2 Twins

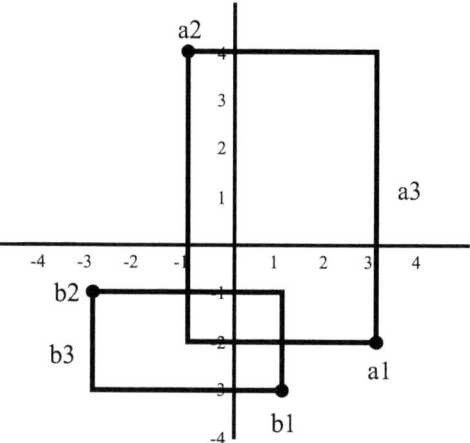

Fig. 7.5 $\subseteq$-proper twins

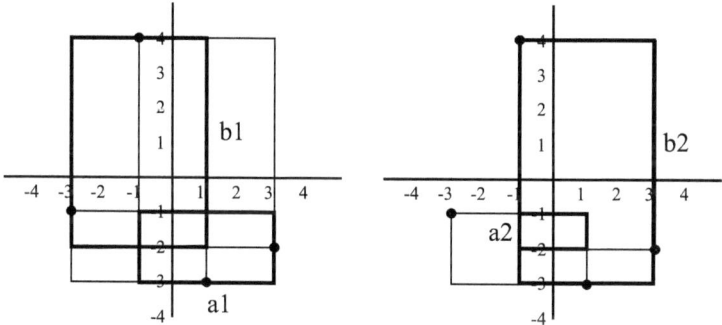

Fig. 7.6 $\leq$-infimum and supremum, $\subseteq$-infimum and supremum

It is true that $\mathbb{A}_\subseteq \not\leq \mathbb{B}_\leq$ and $\mathbb{A}_\leq \not\subseteq \mathbb{B}_\leq$. Infimum and supremum are

$$\text{Inf}_\leq(\mathbb{A}_\subseteq, \mathbb{B}_\subseteq) = |[[3,-3],[-1,-1]]|$$
$$\text{Sup}_\leq(\mathbb{A}_\subseteq, \mathbb{B}_\subseteq) = |[[1,-2],[-3,4]]|$$
$$\text{Inf}_\subseteq(\mathbb{A}_\subseteq, \mathbb{B}_\subseteq) = |[[1,-2],[-1,-1]]|$$
$$\text{Sup}_\subseteq(\mathbb{A}_\subseteq, \mathbb{B}_\subseteq) = |[[3,-3],[-1,4]]|,$$

represented in Fig. 7.6. Now $\text{Inf}_\subseteq(\mathbb{A}_\leq, \mathbb{B}_\leq) \in I(I^*(\mathbb{R}))_\subseteq$.

Fig. 7.7 Proper twin

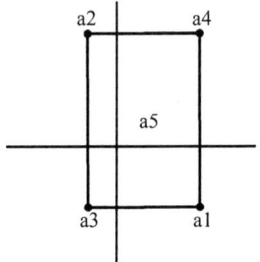

7.2.2 Proper Twins

Any twin of $I(I^*(\mathbb{R})_\leq)$ belongs to $I(I^*(\mathbb{R}))_\subseteq$ because

$$\mathbb{A}_\leq = |[\underline{A}, \overline{A}]|_\leq = \{X \in I^*(\mathbb{R}) \mid \underline{A} \leq X \leq \overline{A}\}$$
$$= \{X \in I^*(\mathbb{R}) \mid \underline{A} \wedge \overline{A} \subseteq X \subseteq \underline{A} \vee \overline{A}\} \in I(I^*(\mathbb{R}))_\subseteq$$

and any twin of $I(I^*(\mathbb{R})_\subseteq$ belongs to $I(I^*(\mathbb{R}))_\leq$ because

$$\mathbb{A}_\subseteq = |[\underline{A}, \overline{A}]|_\subseteq = \{X \in I^*(\mathbb{R}) \mid \underline{A} \subseteq X \subseteq \overline{A}\}$$
$$= \{X \in I^*(\mathbb{R}) \mid \text{Min } (\underline{A}, \overline{A}) \leq X \leq \text{Max } (\underline{A}, \overline{A})\} \in I(I^*(\mathbb{R}))_\leq.$$

So, any proper twin can be represented by a $\subseteq$-twin (or a $\leq$-twin), see Fig. 7.7.
Therefore, both sets of $\leq$-proper and $\subseteq$-proper twins are equal

$$I(I^*(\mathbb{R})_\leq = I(I^*(\mathbb{R}))_\subseteq,$$

and both $(I(I^*(\mathbb{R}))_\subseteq, \leq)$ and $(I(I^*(\mathbb{R}))_\subseteq, \subseteq)$ can be represented by the unique set $I(I^*(\mathbb{R}))$ of the *proper twins*.

Definition 7.2.1 (Proper twin) If $\underline{A}, \overline{A} \in I^*(\mathbb{R})$ such that $\underline{A} \leq \overline{A}$, the *proper twin* of bounds $\underline{A}$ and $\overline{A}$ is the set of modal intervals

$$\mathbb{A} = |[\underline{A}, \overline{A}]| = |[\underline{A}, \overline{A}]|_\leq = \{X \in I^*(\mathbb{R}) \mid \underline{A} \leq X \leq \overline{A}\};$$

if $\underline{A} \subseteq \overline{A}$, the *proper twin* of bounds $\underline{A}$ and $\overline{A}$ is the set of modal intervals

$$\mathbb{A} = |[\underline{A}, \overline{A}]| = |[\underline{A}, \overline{A}]|_\subseteq = \{X \in I^*(\mathbb{R}) \mid \underline{A} \subseteq X \subseteq \overline{A}\}.$$

In the set of proper twins $I(I^*(\mathbb{R}))$ two order relations $\leq$ and $\subseteq$ are defined as follows.

7.2 Twins

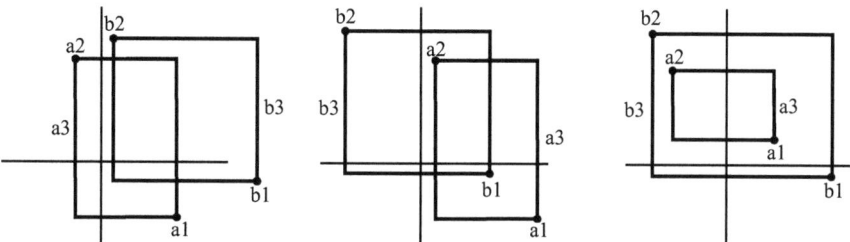

Fig. 7.8 $\leq$-twin, $\subseteq$-twin and $\sqsubseteq$-twin relations between proper twin

Definition 7.2.2 (Less or equal twin relation) If $\mathbb{A} = |[\underline{A}, \overline{A}]|, \mathbb{B} = |[\underline{B}, \overline{B}]| \in I(I^*(\mathbb{R}))$,

$$\mathbb{A} \leq \mathbb{B} \Leftrightarrow (\underline{A} \leq \underline{B}, \overline{A} \leq \overline{B}).$$

Definition 7.2.3 (Inclusion twin relation) If $\mathbb{A} = |[\underline{A}, \overline{A}]|, \mathbb{B} = |[\underline{B}, \overline{B}]| \in I(I^*(\mathbb{R}))$,

$$\mathbb{A} \subseteq \mathbb{B} \Leftrightarrow (\underline{A} \subseteq \underline{B}, \overline{A} \subseteq \overline{B}).$$

A third relation, denoted by $\sqsubseteq$, can be defined when both twins are considered just as set of modal intervals and it is set-inclusion.

Definition 7.2.4 (Set-theoretical inclusion twin relation) If $\mathbb{A} = |[\underline{A}, \overline{A}]|, \mathbb{B} = |[\underline{B}, \overline{B}]| \in I(I^*(\mathbb{R}))$,

$$\mathbb{A} \sqsubseteq \mathbb{B} \Leftrightarrow (\underline{A} \supseteq \underline{B}, \overline{A} \subseteq \overline{B}).$$

Figure 7.8 shows geometrical representations to illustrate these relations.

The infimum and supremum for these relations between proper twins exist and they are defined as follows.

Definition 7.2.5 (Infimum and supremum) If $\mathbb{A}, \mathbb{B} \in I(I^*(\mathbb{R}))$, then

$$\text{Inf}_{\leq}(\mathbb{A}, \mathbb{B}) = |[\text{Inf}_{\leq}(\underline{A}, \underline{B}), \text{Inf}_{\leq}(\overline{A}, \overline{B})]| = |[\text{Min}(\underline{A}, \underline{B}), \text{Min}(\overline{A}, \overline{B})]|,$$

$$\text{Sup}_{\leq}(\mathbb{A}, \mathbb{B}) = |[\text{Sup}_{\leq}(\underline{A}, \underline{B}), \text{Sup}_{\leq}(\overline{A}, \overline{B})]| = |[\text{Max}(\underline{A}, \underline{B}), \text{Max}(\overline{A}, \overline{B})]|,$$

$$\text{Inf}_{\subseteq}(\mathbb{A}, \mathbb{B}) = |[\text{Inf}_{\subseteq}(\underline{A}, \underline{B}), \text{Inf}_{\subseteq}(\overline{A}, \overline{B})]| = |[\underline{A} \wedge \underline{B}, \overline{A} \wedge \overline{B}]|,$$

$$\text{Sup}_{\subseteq}(\mathbb{A}, \mathbb{B}) = |[\text{Sup}_{\subseteq}(\underline{A}, \underline{B}), \text{Sup}_{\subseteq}(\overline{A}, \overline{B})]| = |[\underline{A} \vee \underline{B}, \overline{A} \vee \overline{B}]|,$$

$$\text{Inf}_{\sqsubseteq}(\mathbb{A}, \mathbb{B}) = |[\text{Sup}_{\subseteq}(\underline{A}, \underline{B}), \text{Inf}_{\subseteq}(\overline{A}, \overline{B})]| = |[\underline{A} \vee \underline{B}, \overline{A} \wedge \overline{B}]|,$$

$$\text{Sup}_{\sqsubseteq}(\mathbb{A}, \mathbb{B}) = |[\text{Inf}_{\subseteq}(\underline{A}, \underline{B}), \text{Sup}_{\subseteq}(\overline{A}, \overline{B})]| = |[\underline{A} \wedge \underline{B}, \overline{A} \vee \overline{B}]|.$$

Fig. 7.9 Meet and join with proper twins

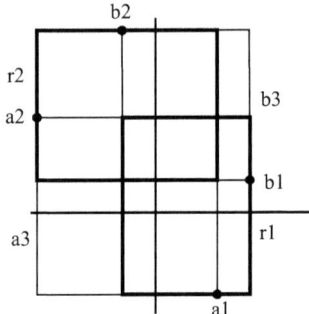

The main properties of these twin relations $\leq, \subseteq$ and $\sqsubseteq$ are stated in the following lemma.

Lemma 7.2.3 (Properties of $\leq, \subseteq$ and $\sqsubseteq$ in $I(I^*(\mathbb{R}))$)

(1) $\leq, \subseteq$ and $\sqsubseteq$ are partial order relations in $I(I^*(\mathbb{R}))$.
(2) $\text{Inf}_\leq(\mathbb{A}, \mathbb{B}) \in I(I^*(\mathbb{R}))$.
(3) $\text{Sup}_\leq(\mathbb{A}, \mathbb{B}) \in I(I^*(\mathbb{R}))$.
(4) $\text{Sup}_\subseteq(\mathbb{A}, \mathbb{B}) \in I(I^*(\mathbb{R}))$.
(5) $\text{Sup}_\sqsubseteq(\mathbb{A}, \mathbb{B}) \in I(I^*(\mathbb{R}))$.
(6) The structure $(I(I^*(\mathbb{R})), \leq)$ is a lattice.

Proof Properties of $\leq$ and $\subseteq$ are the Lemmas 7.2.1 and 7.2.2. Properties of $\sqsubseteq$ are analogous.

Remark 7.2.3 It is not always true that $\text{Inf}_\subseteq(\mathbb{A}, \mathbb{B}) \in I(I^*(\mathbb{R}))$ and $\text{Inf}_\sqsubseteq(\mathbb{A}, \mathbb{B}) \in I(I^*(\mathbb{R}))$ for every $\mathbb{A}, \mathbb{B}$.

Similarly to the interval case, $\text{Inf}_\subseteq(\mathbb{A}, \mathbb{B})$ and $\text{Sup}_\subseteq(\mathbb{A}, \mathbb{B})$ are called *meet* and *join*, respectively, and denoted by $\mathbb{A} \wedge \mathbb{B}$ and $\mathbb{A} \vee \mathbb{B}$. Figure 7.9 gives geometrical representations to illustrate these operations.

The infimum and supremum can be extended to a family of twins and they are defined as follows.

Definition 7.2.6 (Infimum and supremum) For a bounded family of proper twins $\mathbb{A}(I) = \{\mathbb{A}(i) = [\underline{A}(i), \overline{A}(i)] \in I(I^*(\mathbb{R})) \mid i \in I\}$ (I is the index's domain),

$$\text{Inf}_\leq(\mathbb{A}(i)) = |[\text{Inf}_\leq(\underline{A}(i)), \text{Inf}_\leq(\overline{A}(i))]| = |[\text{Min}(\underline{A}(i)), \text{Min}(\overline{A}(i))]|,$$

$$\text{Sup}_\leq(\mathbb{A}(i)) = |[\text{Sup}_\leq(\underline{A}(i)), \text{Sup}_\leq(\overline{A}(i))]| = |[\text{Max}(\underline{A}(i)), \text{Max}(\overline{A}(i))]|,$$

$$\text{Inf}_\subseteq(\mathbb{A}(i)) = |[\text{Inf}_\subseteq(\underline{A}(i)), \text{Inf}_\subseteq(\overline{A}(i))]| = |[\bigwedge_{i \in I} \underline{A}(i), \bigwedge_{i \in I} \overline{A}(i)]|,$$

$$\text{Sup}_\subseteq(\mathbb{A}(i)) = |[\text{Sup}_\subseteq(\underline{A}(i)), \text{Sup}_\subseteq(\overline{A}(i))]| = |[\bigvee_{i \in I} \underline{A}(i), \bigvee_{i \in I} \overline{A}(i)]|,$$

7.2 Twins

$$\operatorname{Inf}_{\sqsubseteq}(\mathbb{A}(i)) = |[\operatorname{Sup}_{\subseteq}(\underline{A}(i)), \operatorname{Inf}_{\subseteq}(\overline{A}(i))]| = |[\bigvee_{i \in I} \underline{A}(i), \bigwedge_{i \in I} \overline{A}(i)]|,$$

$$\operatorname{Sup}_{\sqsubseteq}(\mathbb{A}(i)) = |[\operatorname{Inf}_{\subseteq}(\underline{A}(i)), \operatorname{Sup}_{\subseteq}(\overline{A}(i))]| = |[\bigwedge_{i \in I} \underline{A}(i), \bigvee_{i \in I} \overline{A}(i)]|.$$

The twins, $\operatorname{Inf}_{\subseteq}(\mathbb{A}(i))$ and $\operatorname{Sup}_{\subseteq}(\mathbb{A}(i))$ are called the *meet* and *join* and denoted by

$$\bigwedge_{i \in I} \mathbb{A}(i) = \operatorname{Inf}_{\sqsubseteq}(\mathbb{A}(i)) = |[\operatorname{Inf}_{\subseteq}(\underline{A}(i)), \operatorname{Inf}_{\subseteq}(\overline{A}(i))]| = |[\bigwedge_{i \in I} \underline{A}(i), \bigwedge_{i \in I} \overline{A}(i)]|$$

$$\bigvee_{i \in I} \mathbb{A}(i) = \operatorname{Sup}_{\sqsubseteq}(\mathbb{A}(i)) = |[\operatorname{Sup}_{\subseteq}(\underline{A}(i)), \operatorname{Sup}_{\subseteq}(\overline{A}(i))]| = |[\bigvee_{i \in I} \underline{A}(i), \bigvee_{i \in I} \overline{A}(i)]|,$$

respectively.

7.2.3 The Set of Twins

In order to close the operations $\operatorname{Inf}_{\subseteq}(\mathbb{A}, \mathbb{B})$ and $\operatorname{Inf}_{\sqsubseteq}(\mathbb{A}, \mathbb{B})$ the set of proper twins

$$I(I^*(\mathbb{R})) = \{|[\underline{A}, \overline{A}]|_{\leq} \mid \underline{A} \leq \overline{A}\} = \{|[\underline{A}, \overline{A}]|_{\subseteq} \mid \underline{A} \subseteq \overline{A}\},$$

can be extended to the *set of twins*

$$I^*(I^*(\mathbb{R})) = \{|[\underline{A}, \overline{A}]| \mid \underline{A}, \overline{A} \in I^*(\mathbb{R})\}$$

and the structure $(I(I^*(\mathbb{R})), \leq, \subseteq, \sqsubseteq)$ to the structure $(I^*(I^*(\mathbb{R})), \leq, \subseteq, \sqsubseteq)$ with the same formal Definitions 7.2.1–7.2.6 for $\mathbb{A} \leq \mathbb{B}$, $\mathbb{A} \subseteq \mathbb{B}$, $\mathbb{A} \sqsubseteq \mathbb{B}$, infimum and supremum. Now it is true that the structures $(I^*(I^*(\mathbb{R})), \leq)$, $(I^*(I^*(\mathbb{R})), \subseteq)$ and $(I^*(I^*(\mathbb{R})), \sqsubseteq)$ are three lattices.

By analogy with the set of modal intervals $I^*(\mathbb{R})$, several kinds of twins can be distinguished. If $\mathbb{A} = |[\underline{A}, \overline{A}]|$,

- $\mathbb{A}$ is *proper* when $\underline{A} \subseteq \overline{A}$ or $\underline{A} \leq \overline{A}$.
- $\mathbb{A}$ is *proper-transposed* when $\underline{A} \supseteq \overline{A}$ or $\underline{A} \geq \overline{A}$.
- $\mathbb{A}$ is *punctual* when $\underline{A} = \overline{A}$.
- $\mathbb{A}$ is *crossed* when it is neither proper nor proper-transposed.
- $\mathbb{A}$ is *twin-interval* when it is proper or proper-transposed.
- $\mathbb{A}$ is *improper* when it is proper-transposed or crossed.

A proper twin $\mathbb{A} = |[\underline{A}, \overline{A}]| \in I(I^*(\mathbb{R}))$ can be identified with the set

$$\mathbb{A} = \{X \in I^*(\mathbb{R}) \mid \underline{A} \subseteq X \subseteq \overline{A}\}.$$

Any twin $\mathbb{A} = |[\underline{A}, \overline{A}]| \in I^*(I^*(\mathbb{R}))$ can be associated to a proper twin defined by

$$\operatorname{Twpr}(\mathbb{A}) = |[\underline{A} \wedge \overline{A}, \underline{A} \vee \overline{A}]|.$$

Obviously, if $\mathbb{A}$ is a proper twin, $\operatorname{Twpr}(\mathbb{A}) = \mathbb{A}$.

7.3 Symmetries

Following a similar development to the one of the modal intervals, lattice operators will lead to defining the main twin symmetries.

Definition 7.3.1 (Twin symmetries) If $\mathbb{A} = |[\underline{A}, \overline{A}]| = |[\underline{A}, \overline{A}]| \in I^*(I^*(\mathbb{R}))$,

- $-\mathbb{A} = |[-\overline{A}, -\underline{A}]|$.
- $\mathrm{Dual}(\mathbb{A}) = |[\mathrm{Dual}(\underline{A}), \mathrm{Dual}(\overline{A})]|$.
- $\mathrm{Transp}(\mathbb{A}) = |[\overline{A}, \underline{A}]|$.
- $\mathrm{Twal}(\mathbb{A}) = |[\mathrm{Dual}(\overline{A}), \mathrm{Dual}(\underline{A})]|$.
- $\mathrm{Opp}(\mathbb{A}) = |[-\mathrm{Dual}(\underline{A}), -\mathrm{Dual}(\overline{A})]|$.
- $\mathrm{Twopp}(\mathbb{A}) = |[-\mathrm{Dual}(\overline{A}), -\mathrm{Dual}(\underline{A})]|$.

This basic symmetries change and transform the order relations and lattice operators as the following lemmas state.

Lemma 7.3.1 (Elementary symmetries) If $\mathbb{A} = |[\underline{A}, \overline{A}]|$ and $\mathbb{B} = |[\underline{B}, \overline{B}]|$,

(1) $\mathbb{A} \le \mathbb{B} \Leftrightarrow -\mathbb{A} \ge -\mathbb{B}$.
(2) $\mathbb{A} \subseteq \mathbb{B} \Leftrightarrow \mathrm{Dual}(\mathbb{A}) \supseteq \mathrm{Dual}(\mathbb{B})$.
(3) $\mathbb{A} \sqsubseteq \mathbb{B} \Leftrightarrow \mathrm{Transp}(\mathbb{A}) \sqsupseteq \mathrm{Transp}(\mathbb{B})$.

Proof

(1) If $\mathbb{A} = |[\underline{A}, \overline{A}]|_\le$ and $\mathbb{B} = |[\underline{B}, \overline{B}]|_\le$,

$$\mathbb{A} \le \mathbb{B} \Leftrightarrow \underline{A} \le \underline{B}, \overline{A} \le \overline{B} \Leftrightarrow -\underline{B} \le -\underline{A}, -\overline{B} \le -\overline{A} \Leftrightarrow -\mathbb{B} \le -\mathbb{A}.$$

The same reasoning if $\mathbb{A} = |[\underline{A}, \overline{A}]|_\subseteq$ and $\mathbb{B} = |[\underline{B}, \overline{B}]|_\subseteq$.

(2) If, for example $\mathbb{A} = |[\underline{A}, \overline{A}]|_\subseteq$ and $\mathbb{B} = |[\underline{B}, \overline{B}]|_\subseteq$,

$$\mathbb{A} \subseteq \mathbb{B} \Leftrightarrow \underline{A} \subseteq \underline{B}, \overline{A} \subseteq \overline{B}$$
$$\Leftrightarrow \mathrm{Dual}(\underline{B}) \subseteq \mathrm{Dual}(\underline{A}), \mathrm{Dual}(\overline{B}) \subseteq \mathrm{Dual}(\overline{A})$$
$$\Leftrightarrow \mathrm{Dual}(\mathbb{B}) \subseteq \mathrm{Dual}(\mathbb{A}).$$

(3) If, for example $\mathbb{A} = |[\underline{A}, \overline{A}]|_\subseteq$ and $\mathbb{B} = |[\underline{B}, \overline{B}]|_\subseteq$,

$$\mathbb{A} \sqsubseteq \mathbb{B} \Leftrightarrow (\underline{A} \supseteq \underline{B}, \overline{A} \subseteq \overline{B}) \Leftrightarrow \mathrm{Transp}(\mathbb{B}) \sqsubseteq \mathrm{Transp}(\mathbb{A}). \qquad\blacksquare$$

Lemma 7.3.2 (Composed symmetries) If $\mathbb{A} = |[\underline{A}, \overline{A}]|$,

(1) $\mathrm{Twal}(\mathbb{A}) = \mathrm{Transp}(\mathrm{Dual}(\mathbb{A}))$.
(2) $\mathrm{Opp}(\mathbb{A}) = -\mathrm{Dual}(\mathbb{A})$.
(3) $\mathrm{Twopp}(\mathbb{A}) = -\mathrm{Twal}(\mathbb{A})$.

7.3 Symmetries

Proof

(1) If, for example $\mathbb{A} = ||[\underline{A}, \overline{A}]||_\subseteq$,

$$\text{Transp}(\text{Dual}(\mathbb{A})) = \text{Transp}(||[\text{Dual}(\underline{A}), \text{Dual}(\overline{A})]||_\subseteq)$$
$$= ||[\text{Dual}(\overline{A}), \text{Dual}(\underline{A})]||_\subseteq$$
$$= \text{Twal}(\mathbb{A}).$$

Similar reasonings for (2) and (3). ∎

Summarizing some of the previous concepts, from the lattice of modal intervals $(I^*(\mathbb{R}), \subseteq)$, a new lattice $(I^*(I^*(\mathbb{R})), \subseteq)$ can be built following the standard process. One element $\mathbb{A} \in I^*(I^*(\mathbb{R}))$, named a *twin*, is defined by

$$\mathbb{A} = ||[\underline{A}, \overline{A}]||$$

where $\underline{A} \in I^*(\mathbb{R})$ is the *lower bound* and $\overline{A} \in I^*(\mathbb{R})$ is the *upper bound* of $\mathbb{A}$, and

$$I^*(I^*(\mathbb{R})) = \{\mathbb{A} = ||[\underline{A}, \overline{A}]|| \mid \underline{A}, \overline{A} \in I^*(\mathbb{R})\}$$

is the set of twins over $I^*(\mathbb{R})$. If $\underline{A} \leq \overline{A}$ or $\underline{A} \subseteq \overline{A}$ the twin is called a *proper twin*, which can be identified with the set

$$\mathbb{A} = \{X \in I^*(\mathbb{R}) \mid \underline{A} \subseteq X \subseteq \overline{A}\} \quad \text{or} \quad \mathbb{A} = \{X \in I^*(\mathbb{R}) \mid \underline{A} \leq X \leq \overline{A}\}$$

with elements that are the modal intervals between both bounds $\underline{A}$ and $\overline{A}$.

The set-theoretical inclusion between twins $\mathbb{A} = ||[\underline{A}, \overline{A}]||$ and $\mathbb{B} = ||[\underline{B}, \overline{B}]||$ is defined by means of the interval inclusion between their bounds

$$\mathbb{A} \subseteq \mathbb{B} \Leftrightarrow (\underline{A} \supseteq \underline{B}, \overline{A} \subseteq \overline{B})$$

The lattice operations *meet* and *join* on $I^*(I^*(\mathbb{R}))$ for a bounded family of twins $\mathbb{A}(I) = \{\mathbb{A}(i) = [\underline{A}(i), \overline{A}(i)] \in I^*(I^*(\mathbb{R})) \mid i \in I\}$ (I is the index's domain) are defined by

$$\bigwedge_{i \in I} \mathbb{A}(i) = [\bigwedge_{i \in I} \underline{A}(i), \bigwedge_{i \in I} \overline{A}(i)]$$
$$\bigvee_{i \in I} \mathbb{A}(i) = [\bigvee_{i \in I} \underline{A}(i), \bigvee_{i \in I} \overline{A}(i)],$$

using $\mathbb{A} \wedge \mathbb{B}$ and $\mathbb{A} \vee \mathbb{B}$ for the corresponding case of two operands. These operators do not have the same set-theoretical meaning as in $I^*(\mathbb{R})$.

7.4 Semantic Extension of a Function to $I^*(I^*(\mathbb{R}))$

The *-semantic and **-semantic extensions of a continuous function f from $\mathbb{R}^k$ to $\mathbb{R}$ to a twin $\mathbf{X} = (\mathbb{X}_1, \ldots \mathbb{X}_k) = (|[\underline{X_1}, \overline{X_1}]|, \ldots, |[\underline{X_k}, \overline{X_k}]|) \in I^*(I^*(\mathbb{R}^k))$ are defined by

$$f^*(\mathbb{X}_1, \ldots \mathbb{X}_k) = |[f^*(\underline{X_1}, \ldots, \underline{X_k}), f^*(\overline{X_1}, \ldots, \overline{X_k})]|$$
$$f^{**}(\mathbb{X}_1, \ldots \mathbb{X}_k) = |[f^{**}(\underline{X_1}, \ldots, \underline{X_k}), f^{**}(\overline{X_1}, \ldots, \overline{X_k})]|.$$

In particular, if f is an arithmetic operator,

$$\mathbb{A} + \mathbb{B} = |[\underline{A} + \underline{B}, \overline{A} + \overline{B}]|$$
$$\mathbb{A} - \mathbb{B} = |[\underline{A} - \underline{B}, \overline{A} - \overline{B}]|$$
$$\mathbb{A} * \mathbb{B} = |[\underline{A} * \underline{B}, \overline{A} * \overline{B}]|$$
$$\mathbb{A}/\mathbb{B} = |[\underline{A}/\underline{B}, \overline{A}/\overline{B}]|.$$

An important property is the iso-monotonicity with respect the $\sqsubseteq$ inclusion-twin.

Theorem 7.4.1 *If f is a continuous function f from $\mathbb{R}^k$ to $\mathbb{R}$ and $\mathbf{X}, \mathbf{Y} \in I^*(I^*(\mathbb{R}^k))$, then*

$$\mathbf{X} \sqsubseteq \mathbf{Y} \Rightarrow (f^*(\mathbf{X}) \sqsubseteq f^*(\mathbf{X}), \; f^{**}(\mathbf{X}) \sqsubseteq f^{**}(\mathbf{X})).$$

Proof As $\mathbf{X} \sqsubseteq \mathbf{Y} \Leftrightarrow (\underline{\mathbf{Y}} \subseteq \underline{\mathbf{X}}, \overline{\mathbf{X}} \subseteq \overline{\mathbf{Y}})$, then

$$f^*(\mathbf{X}) = |[f^*(\underline{\mathbf{X}}), f^*(\overline{\mathbf{X}})]| \sqsubseteq |[f^*(\underline{\mathbf{Y}}), f^*(\overline{\mathbf{Y}})]| = f^*(\mathbf{Y}),$$
$$f^{**}(\mathbf{X}) = |[f^{**}(\underline{\mathbf{X}}), f^{**}(\overline{\mathbf{X}})]| \sqsubseteq |[f^{**}(\underline{\mathbf{Y}}), f^{**}(\overline{\mathbf{Y}})]| = f^{**}(\mathbf{Y}),$$

by the iso-monotonicity of the semantic extensions with respect to the modal inclusion. ∎

7.5 The f^* Algorithm

In Chap. 3 was shown two important *-Semantic Theorems 3.3.1 and 3.3.2, which state an equivalence between a first order predicate logical formula, involving equalities relating to a continuous real function, and a modal interval inclusion. This equivalence makes the computation of f^* a useful tool for solving many problems where these types of logical formulas are involved.

The difficulty of computing f^* for a modal interval X depends on the function f. Under some monotonicity conditions, when f is monotonic with respect to all its

7.5 The f* Algorithm

variables and their incidences and it has a tree-optimal syntactic extension fR, then it is possible to compute $f^*(X)$ using interval arithmetic. By the Theorem of Coercion 4.2.15,

$$f^*(X) = fR(XD) = f^{**}(X),$$

and the computation of f^* is reduced to simple modal interval arithmetic operations. When the function involved does not satisfy these conditions, those theorems can be partially applied in order to reduce the complexity of the problem: when f^* and f^{**} are different or when f is not monotonic with respect to all its variables and their incidences, it is not possible to approximate $f^*(X)$ using simple arithmetic computations, and only interpretability theorems can be applied. For example by Theorem 4.2.10,

$$f^*(X) \subseteq fR(XDt^*),$$

or by Theorem 4.2.12,

$$f^*(X) \subseteq fR(XDT^*),$$

and fR will provide only an approximation to f^*. But using any interpretable rational extension causes an overestimation of the interval evaluation, due to possible multiple occurrences of the variables.

On the other hand, any algorithm for approximating f^* must provide inner and outer estimates in order to guarantee a specific degree of approximation. Therefore, the set of twins is the convenient background to handle both approximations simultaneously. An algorithm based on branch-and-bound techniques, which allows obtaining a twin defined by inner and an outer approximations of f^*, is described in this section.

7.5.1 Approximate *-Semantic Extension

Let $X = (U, V)$ be a modal interval vector split into its proper (U) and improper (V) components. Let $\{U_1, \ldots, U_r\}$ be a partition of U and, for every $j = 1, \ldots, r$, let $\{V_{1_j}, \ldots, V_{s_j}\}$ be partitions of V for every j. Each interval $U_j \times V_{k_j}$ is called a *Cell*, each V_{*_j}-partition is called a *Strip*, and the U-partition is called the *Strips' List*.

Figure 7.10 shows a geometrical representation of an example of these partitions, when X has only one proper component and one improper component.

The algorithm we present is based on the following theorem.

Fig. 7.10 Partition, *Strips* and *Cells*

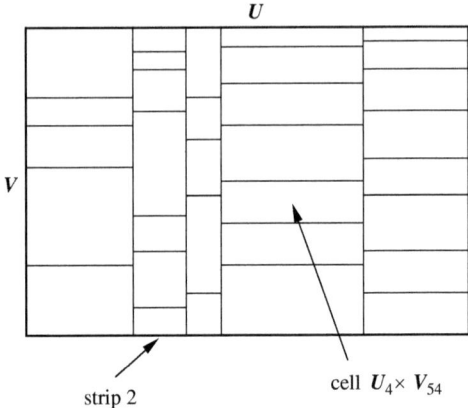

Theorem 7.5.1 (Twin inclusion for f^*) *Given an $\mathbb{R}^k$ to $\mathbb{R}$ real continuous function f, then*

$$f^*(X) \in \bigvee_{j \in \{1,\ldots,r\}} \bigwedge_{k_j \in \{1_j,\ldots,s_j\}} |[\text{Inn}(fR(\check{u}_j, V_{k_j})), \text{Out}(fR(U_j, \check{v}_{k_j}))]|$$

where $\check{u}_j$ is any fixed point of U'_j ($j = 1,\ldots,r$) and $\check{v}_{k_j}$ is any fixed point of V'_{k_j} ($k_j = 1_j,\ldots,s_j$) (for example the midpoints of the intervals, or their bounds).

Proof Starting from the definition of the interval *-semantic extension of f to X,

$$f^*(X) = \bigvee_{u \in U'} \bigwedge_{v \in V'} [f(u,v), f(u,v)] \quad (7.1)$$

$$= \bigvee_{j \in \{1,\ldots,r\}} \bigvee_{u_j \in U'_j} \bigwedge_{v \in V'} [f(u_j, v), f(u_j, v)] \quad (7.2)$$

$$= \bigvee_{j \in \{1,\ldots,r\}} \bigvee_{u_j \in U'_j} \bigwedge_{k_j \in \{1_j,\ldots,s_j\}} \bigwedge_{v_{k_j} \in V'_{k_j}} [f(u_j, v_{k_j}), f(u_j, v_{k_j})] \quad (7.3)$$

$$= \bigvee_{j \in \{1,\ldots,r\}} \bigvee_{u_j \in U'_j} \bigwedge_{k_j \in \{1_j,\ldots,s_j\}} \bigwedge_{v_{k_j} \in V'_{k_j}} f^*(u_j, v_{k_j}) \quad (7.4)$$

$$\in \bigvee_{j \in \{1,\ldots,r\}} \bigwedge_{k_j \in \{1_j,\ldots,s_j\}} |[f^*(\check{u}_j, V_{k_j}), f^*(U_j, \check{v}_{k_j})]| \quad (7.5)$$

$$\sqsubseteq \bigvee_{j \in \{1,\ldots,r\}} \bigwedge_{k_j \in \{1_j,\ldots,s_j\}} |[\text{Inn}(fR(\check{u}_j, V_{k_j})), \text{Out}(fR(U_j, \check{v}_{k_j}))]| \quad (7.6)$$

(7.1) is the definition of f^*.
(7.2) is true by the associativity of the join operator.
(7.3) is true by the associativity of the meet operator.
(7.4) is true because the point-wise interval $[f(u_j, v_{k_j}), f(u_j, v_{k_j})]$ is obviously equal to $f^*(u_j, v_{k_j})$.

7.5 The f^* Algorithm

(7.5) is true because of the inclusion $f^*(u_j, v_{k_j}) \supseteq f^*(u_j, V_{k_j})$ implies

$$\bigvee_{j \in \{1,...,r\}} \bigvee_{u_j \in U'_j} \bigwedge_{k_j \in \{1_j,...,s_j\}} \bigwedge_{v_{k_j} \in V'_{k_j}} f^*(u_j, v_{k_j})$$

$$\supseteq \bigvee_{j \in \{1,...,r\}} \bigvee_{u_j \in U'_j} \bigwedge_{k_j \in \{1_j,...,s_j\}} f^*(u_j, V_{k_j})$$

$$\supseteq \bigvee_{j \in \{1,...,r\}} \bigwedge_{k_j \in \{1_j,...,s_j\}} f^*(\check{u}_j, V_{k_j})$$

and, similarly, the inclusion $f^*(u_j, v_{k_j}) \subseteq f^*(U_j, v_{k_j})$ implies

$$\bigvee_{j \in \{1,...,r\}} \bigvee_{u_j \in U'_j} \bigwedge_{k_j \in \{1_j,...,s_j\}} \bigwedge_{v_{k_j} \in V'_{k_j}} f^*(u_j, v_{k_j})$$

$$\subseteq \bigvee_{j \in \{1,...,r\}} \bigwedge_{k_j \in \{1_j,...,s_j\}} \bigwedge_{v_{k_j} \in V'_{k_j}} f^*(U_j, v_{k_j})$$

$$\subseteq \bigvee_{j \in \{1,...,r\}} \bigwedge_{k_j \in \{1_j,...,s_j\}} f^*(U_j, \check{v}_{k_j}).$$

(7.6) is true because

$$f^*(\check{u}_j, V_{k_j}) = f^{**}(\check{u}_j, V_{k_j}) \supseteq \text{Inn}(fR(\check{u}_j, V_{k_j}))$$

and

$$f^*(U_j, \check{v}_{k_j}) \subseteq \text{Out}(fR(U_j, \check{v}_{k_j})),$$

(by Theorems 4.2.1 and 4.2.2).

Obviously, when the width of the concerned intervals tends to zero, the proper twin of (7.6) tends to the proper twin formed by the inner and outer rounding of f^* over X. ∎

The final inclusion stated by this theorem can be written as

Inner approximation: $\displaystyle\bigvee_{j \in \{1,...,r\}} \bigwedge_{k_j \in \{1_j,...,s_j\}} \text{Inn}(fR(\check{u}_j, V_{k_j})) \subseteq f^*(X)$ (7.7)

Outer approximation: $\displaystyle\bigvee_{j \in \{1,...,r\}} \bigwedge_{k_j \in \{1_j,...,s_j\}} \text{Out}(fR(U_j, \check{v}_{k_j})) \supseteq f^*(X)$ (7.8)

for any partition of X.

In the case when all the components of X are proper intervals, the inner and outer estimates are

Inner approximation: $\displaystyle\bigvee_{i \in \{1,...,r\}} \text{Inn}(fR(\check{x}_i)) \subseteq f^*(X)$ (7.9)

Outer approximation: $\displaystyle\bigvee_{i \in \{1,...,r\}} \text{Out}(fR(X_i)) \supseteq f^*(X)$ (7.10)

where $\{X_1, \ldots, X_r\}$ is any partition of X and $\check{x}_i$ is any fixed point of X_i ($i = 1, \ldots, r$).

Example 7.5.1 Let us consider the problem of verifying the logical statement

$$\forall (u \in [0, 6]') \exists (v \in [2, 8]') \exists (z \in [-4, 9]') f(u, v, z) = 0 \qquad (7.11)$$

where $f(u,v,z)$ is the continuous function

$$f(u, v, z) = u^2 + v^2 + uv - 20u - 20v + 100 - 10 \sin z.$$

By the *-Semantic Theorem 3.3.1, this is equivalent to the interval inclusion

$$f^*([0, 6], [8, 2], [9, -4]) \subseteq [0, 0] \qquad (7.12)$$

for which the inclusion of any outer approximation of $f^*([0, 6], [8, 2], [9, -4])$ in $[0, 0]$ is sufficient. As f is not totally monotonic for any of its variables, this approximation can be obtained using Theorem 4.2.10

$$\begin{aligned} f^*(U, V, Z) &\subseteq fR(U, Vt^*, Z) \\ &= [0, 6]^2 + [5, 5]^2 + [0, 6] * [5, 5] \\ &\quad -20 * [0, 6] - 20 * [5, 5] + 100 - 10 * \sin([9, -4]) \\ &= [-85, 81]. \end{aligned}$$

or by applying Theorem 4.2.8,

$$\begin{aligned} f^*(U, V, Z) &\subseteq fR(U, VT^*, Z) \\ &= [0, 6]^2 + [8, 2]^2 + [0, 6] * [2, 8] - 20 * [0, 6] \\ &\quad -20 * [2, 8] + 100 - 10 * \sin([9, -4]) \\ &= [-106, 138]. \end{aligned}$$

Notice that the results obtained are surely very over-estimated approximations of f^* and they do not satisfy (7.12). In order to get a better approximation to f^*, the f^* algorithm will be applied step-by-step.

Table 7.1 represents a possible bisection configuration over the variables' space where the columns represent the $\{U_1, U_2, U_3\}$ partition of $U = [0, 6]$, the rows are the corresponding $\{V_{1_j}, V_2, V_{3_j}\}$ partition of $V = [8, 2]$ for every U_j, and the interval Z has not been divided. Each cell contains the corresponding approximations. The last row contains the approximations which result from applying the meet operator over the cells of each columns and finally, the bottom-left

7.5 The f^* Algorithm

Table 7.1 Illustrative example of Theorem 7.5.1

(u, v)	$u = [0, 2]$	$u = [2, 4]$	$u = [4, 6]$
$v = [8, 6]$	Out = $[-21, 17]$	Out = $[-43, 3]$	Out = $[-57, -3]$
	Inn = $[43, -47]$	Inn = $[27, -67]$	Inn = $[19, -79]$
$v = [6, 4]$	Out = $[-5, 29]$	Out = $[-31, 11]$	Out = $[-49, 1]$
	Inn = $[17, 51]$	Inn = $[-11, 27]$	Inn = $[-31, 11]$
$v = [4, 2]$	Out = $[19, 49]$	Out = $[-11, 27]$	Out = $[-33, 13]$
	Inn = $[-9, 33]$	Inn = $[-33, 13]$	Inn = $[-49, 1]$
$[-33, 17]$	Out = $[19, 17]$	Out = $[-11, 3]$	Out = $[-33, -3]$
$[19, -47]$	Inn = $[43, -47]$	Inn = $[27, -67]$	Inn = $[19, -79]$

cell contains the approximations resulting from applying the join operator to the obtained approximations for each column. So approximations to f^*, by (7.7) and (7.8), are

$$\text{Inn} = [19, -47] \subseteq f^*([0, 6], [8, 2], [9, -4]) \subseteq \text{Out} = [-33, 17]$$

and the resulting outer approximation is better than the ones obtained previously $[-85, 81]$. However, it is still not good enough to decide for the truth value of the corresponding logical formula and further bisections are required.

7.5.2 Basic Algorithm

In accordance with (7.7) and (7.8), the necessary steps for the implementation of an algorithm which computes an inner and an outer approximation of f^* is shown in Algorithm 1. In order to simplify the algorithm presentation, the following notation and concepts are introduced.

- Inn(*Cell*): inner approximation of a *Cell*.
- Out(*Cell*): outer approximation of a *Cell*.
- Inn(*Strip*): inner approximation of a *Strip*.
- Out(*Strip*): outer approximation of a *Strip*.
- wid(U): Function returning the width of the widest component of a U partition.
- *Tolerance*(Inn, Out): Function returning the distance between the inner and the outer approximation

$$\textit{Tolerance}(\text{Inn}, \text{Out}) = \max(|\text{Inf}(\text{Out}) - \text{Inf}(\text{Inn})|, |\text{Sup}(\text{Out}) - \text{Sup}(\text{Inn})|),$$

where Inf and Sup are the left and right bounds of the corresponding approximations.
- Enqueue: The result of adding an element to a list.
- Dequeue: The result of extracting an element from a list.

Algorithm 1 f^* algorithm

Require: Continuous function $f(u, v)$ with its associated vectors of proper and improper intervals (U, V). Desired tolerance φ for the output.
Ensure: Inn and Out approximations of $f^*(U, V)$.
1: Create a *Cell* = (U, V) and compute its inner and outer approximations. Create a *Strip* containing *Cell*. Compute the *Strip* approximations and insert *Strip* into the *StripSet*. Compute Inn and Out approximations.
2: **while** *Tolerance*(Inn, Out) $> \varphi$ **do**
3:    Select the first *Strip* from the *StripSet* and the first *Cell* from the *Strip*.
4:    **if** wid(V) > wid(U) from *Cell* **then**
5:       Bisect V by the widest component, dequeue *Cell*, compute the approximations of the resulting cells and enqueue them to the *Strip*. Compute inner and outer approximations of *Strip*.
6:    **else**
7:       Bisect U by the widest component and create two new strips, dequeue *Strip*. Compute the cells' approximations of the resulting strips and the approximations of the strips. Add the resulting strips to the *StripSet*.
8:    **end if**
9:    Compute global inner and outer approximations.
10: **end while**
11: **return** Inn and Out.

- Compute inner and outer approximation of *Cell*, that is

$$\text{Inn}(Cell) = \text{Inn}(fR(\check{u}, V)),$$
$$\text{Out}(Cell) = \text{Out}(fR(U, \check{v})).$$

- Compute inner and outer approximations of *Strip*, that is

$$\text{Inn}(Strip) = \bigwedge_{\{Cell \text{ in } Strip\}} \text{Inn}(Cell),$$
$$\text{Out}(Strip) = \bigwedge_{\{Cell \text{ in } Strip\}} \text{Out}(Cell).$$

- Compute global inner and outer approximations, that is

$$\text{Inn} = \bigvee_{\{Strip \text{ in } StripSet\}} \text{Inn}(Strip),$$
$$\text{Out} = \bigvee_{\{Strip \text{ in } StripSet\}} \text{Out}(Strip).$$

7.5.2.1 Bounding Criteria

As with any branch-and-bound algorithm, bounding criterions are desired in order to avoid a combinatorial blow-up. The following non-bisection criteria can be used to avoid useless bisections:

7.5 The f^* Algorithm

Table 7.2 Basic algorithm results

Inn	$[98, -67]$
Out	$[-12, 18]$
Tolerance(Inn, Out)	110
Number of bisections	$\simeq 2,000,000$

- a *Cell* is not bisected when Inn(*Cell*) $\supseteq$ Out(*Strip*), because no division of any improper component V will improve the approximation.

Similarly,

- a *Strip* is not bisected when Out(*Strip*) $\subseteq$ Inn, because no division through any proper component U will improve the approximation. Moreover, this *Strip* can be eliminated from the *StripSet*.

7.5.2.2 Stopping Criteria

Apart from the tolerance criterion, the algorithm stops when the width of all cell dimensions is smaller than a fixed precision (wid(X) $\leq \epsilon$). Moreover, when a cell dimension reaches ϵ, this dimension is no longer bisected.

When the f^* algorithm is used for proving first-order logic formulas, an extra stopping criteria can be applied, which stops the algorithm when the corresponding logical formula is satisfied or not. Notice that it is not necessary to achieve the specified *Tolerance*(Inn, Out) to prove the satisfaction of a logical formula. Therefore, the two following conditions can be introduced inside the **while** loop,

- **if** Out $\subseteq [0, 0]$ **then break**,
- **else if** Inn $\not\subseteq [0, 0]$ **then break**.

Example 7.5.2 Following the Example 7.5.1, by using the proposed basic algorithm and after 600 s on a Pentium IV M, the following approximation of f^* is in Table 7.2.

Despite the high computation time, the stopping criteria have not been fulfilled and it has not been possible to prove the satisfaction of the logical formula expressed by (7.11). It seems obvious that the proposed algorithm is far from being useful. The next section introduces a set of improvements which can drastically reduce the computational effort.

7.5.3 Improvements

In order to make the f^* algorithm suitable for practical application, a set of strategies have been introduced. Basically, these strategies try to reduce as much as possible the number of bisections and to obtain better local approximations of

Table 7.3 Selection strategy improvement

Inn	[41.848, −9.4659]
Out	[0.2163, −5.0535]
Tolerance(Inn, Out)	4.4124
Number of bisections	2, 109

the resulting partitions. Some of these improvements are simple algorithmic tricks and others are based on results of MIA.

Example 7.5.1 will continue being used to quantify the improvements caused by the different proposed strategies with respect to the basic algorithm. The comparison criteria can be the computation time and the required number of bisections.

7.5.3.1 Selection Strategy

Instead of using a FIFO strategy (First In, First Out) for selecting a *Strip* and a *Cell* from its respective containers, a more efficient strategy is proposed. It consists of selecting the *Strip* and the *Cell* with the biggest *Tolerance*(*Inn*, *Out*), and which approximations match at least one of its bounds with one of the bounds of the global approximation (Inn or Out). Using this selection strategy, a faster and more uniform convergence to the f^* value is achieved.

Example 7.5.3 Using the new selection strategy, the following result for Example 7.5.1, shown in Table 7.3, has been obtained in 4 s.

It can be observed that the reduction of computation time is drastic. Moreover, it has been proven that the logical formula stated in (7.11) is satisfied because Out $\subseteq [0, 0]$ is true.

7.5.3.2 Monotonicity Study

A set of additional criteria, based on the study of the monotonicity of the objective function f has been derived. By computing the partial derivatives of the function with regard to each variable and each of their incidences, it is possible to determined if f is monotonic with respect each variable and their incidences in order to improve the inner and outer approximations to f^*.

Theorem 7.5.2 (Monotonic twin inclusion for f^*) *If the $\mathbb{R}^k$ to $\mathbb{R}$ real continuous function f is monotonic with respect some variables, then $f^*(X)$ belongs to the proper twin*

$$\bigvee_{j \in \{1,\ldots,r\}} \bigwedge_{k_j \in \{1_j,\ldots,s_j\}} |[\text{Inn}(fR(\check{u}_j D, V_{k_j} D), \text{Out}(fR(U_j D, V_{k_j} Dt^*))]|,$$

where the D and t^*-transformations are applied to the indicated intervals and $\check{u}_j D$ means that every U_j-component is reduced to a point $\check{u}_j$ of its domain, except for

7.5 The f^* Algorithm

the intervals corresponding to variables for which f is totally monotonic, which undergo the D-transformation.

Proof For the right interval bound, taking into account that

$$f^*(X) = f^*(U, V) = \bigvee_{j \in \{1,...,r\}} \bigvee_{u_j \in U'_j} \bigwedge_{v \in V'} f^*(u_j, v)$$

$$= \bigvee_{j \in \{1,...,r\}} \bigvee_{u_j \in U'_j} f^*(u_j, V) = \bigvee_{j \in \{1,...,r\}} f^*(U_j, V), \qquad (7.13)$$

if $\{V_{1j}, \ldots, V_{s_j}\}$ is a partition of V corresponding to the jth strip,

$$f^*(U_j, V) \subseteq \left\{ \begin{array}{c} f^*(U_j, V_{1j}) \\ \cdots\cdots\cdots \\ f^*(U_j, V_{s_j}) \end{array} \right\} \Rightarrow f^*(U_j, V) \subseteq \bigwedge_{k_j \in \{1j,...,s_j\}} f^*(U_j, V_{k_j}).$$

So, from Theorem 4.2.10

$$f^*(X) \subseteq \bigvee_{j \in \{1,...,r\}} \bigwedge_{k_j \in \{1j,...,s_j\}} fR(U_j D, V_{k_j} Dt^*). \qquad (7.14)$$

For the left interval bound, let $U_j = (Y_j, Z_j)$ be totally monotonic for a subset Z_j of their components, uni- or multi-incident, and $\check{y}_j$ any point of Y_j. Taking into account (7.13),

$$f^*(X) = f^*(U, V) = \bigvee_{j \in \{1,...,r\}} f^*(Y_j, Z_j, V)$$

$$\supseteq \bigvee_{j \in \{1,...,r\}} f^*(\check{y}_j, Z_j, V)$$

$$= \bigvee_{j \in \{1,...,r\}} f^{**}(\check{y}_j, Z_j, V)$$

// from Theorem 3.4.6 and Remark 3.4.4

$$= \bigvee_{j \in \{1,...,r\}} \bigwedge_{k_j \in \{1j,...,s_j\}} \bigwedge_{v_{k_j} \in V'_{k_j}} \bigvee_{z_j \in Z'_j} f(\check{y}_j, z_j, v_{k_j})$$

$$= \bigvee_{j \in \{1,...,r\}} \bigwedge_{k_j \in \{1j,...,s_j\}} f^{**}(\check{y}_j, Z_j, V_{k_j})$$

$$\supseteq \bigvee_{j \in \{1,...,r\}} \bigwedge_{k_j \in \{1j,...,s_j\}} fR(\check{y}_j, Z_j D, V_{k_j} D)$$

//from Theorem 4.2.13

This chain of equalities and inclusions can be summarized as

$$f^*(X) \supseteq \bigvee_{j \in \{1,\ldots,r\}} \bigwedge_{k_j \in \{1_j,\ldots,s_j\}} fR(\check{u}_j D, V_{k_j} D), \qquad (7.15)$$

where $\check{u}_j D$ means that every U_j-component must be reduced to a point of its domain, except for the intervals corresponding to variables, uni-incident or multi-incident, for which f is totally monotonic, which undergo the D-transformation. ∎

So, from this Theorem, inner and outer approximation (7.13) of *Cell*, when there exist some monotonicities, can be written

$$\text{Inn}(Cell) = \text{Inn}(fR(\check{u}D, VD)) \qquad (7.16)$$

and

$$\text{Out}(Cell) = \text{Out}(fR(UD, VDt^*)). \qquad (7.17)$$

In this situation, it is possible to apply the following non-division criteria for a *Cell* or a *Strip*:

- If the function is totally monotonic with regard to an improper component of a *Cell*, then do not bisect the *Cell* through this improper component.
- If the function is totally monotonic with regard to a proper component along the *Strip*, then do not bisect the *Strip* through this proper component.

7.5.3.3 Optimality Study

If f is optimal in some strip it is possible to take advantage of this in order to accelerate the convergence. Let us suppose that in a strip j the function f is optimal, for which it is sufficient to be tree-optimal and uni-incident (see Theorem 4.2.5), tree-optimal and totally monotonic with respect all multi-incident variables (see Theorem 4.2.15), or uni-modal and totally monotonic with respect all multi-incident variables (see Theorem 4.3.1). From (7.13),

$$f^*(X) = \bigvee_{j \in \{1,\ldots,r\}} f^*(U_j, V),$$

taking into account the result of Theorem 4.2.15,

$$f^*(X) = \bigvee_{j \in \{1,\ldots,r\}} fR(U_j D, VD).$$

So, inner and outer approximations to $f^*(X)$, in the strip where f is optimal, are

$$\text{Inn}(Strip) = \text{Inn}(fR(UD, VD)) \qquad (7.18)$$

7.5 The f^* Algorithm

Table 7.4 Tree-optimality study improvement

Inn	$[43.062, -6.0]$
Out	$[16.250, -6.0]$
Tolerance(Inn, Out)	27
Number of bisections	7

and

$$\text{Out}(Strip) = \text{Out}(fR(U\,D, V\,D)). \qquad (7.19)$$

Moreover, it is possible to apply the following non-division criterion:

- If the function is optimal in a *Strip*, then do not bisect the *Strip*.

Example 7.5.4 Considering the previous criteria based on the monotonicity and optimality study of the function, in 0.2 s the results of Table 7.4 were obtained.

As the studied rational function is tree-optimal in all its domain, the uni-incident improper interval variable Z is not bisected anymore. Therefore, the bisection is carried out over two variables (u, v) instead of over three, which significantly reduces the computational effort.

The last approximations to f^* were good enough to prove (7.11), but not to know its interval value. Using the algorithm, while eliminating the extra stopping criteria specific for validating logical formulas, the result is

$$[19.00009, -6.000000] \subseteq f^*([0, 6], [8, 2], [9 - 4]) \subseteq [18.99982, -6.000000]$$

with a computation time of 2.7 s.

7.5.4 Termination, Soundness and Completeness

Concerning the algorithm *termination*, for a given fixed precision ϵ representing the maximal bisected width dimension, the f^* algorithm finishes in less than N iterations, where $N = \prod_{i=1}^{n} \frac{\text{wid}(X_i)}{\epsilon}$ and n is the number of variables. For a finite precision (e.g. 16 bits), the algorithm necessarily finishes because it can not produce more that N_0^k, where N_0 is the total number of representable floating point number by the machine. For example for 16 bit precision, $N_0 = 2^{16}$.

The f^* algorithm can be considered *sound* because it provides an inner approximation of the *-semantic extension, i.e., all the points of the inner approximations belongs to the solution.

The f^* algorithm is *complete* because it also provides an outer approximation of the *-semantic extension, which guarantees that all the solution points are included in the provided approximation.

Chapter 8
Marks

8.1 Introduction

Working on any digital scale, either a computation scale or a reading/writing measurement scale, digital values must be considered as intrinsically inexact. For example, consider an electrical circuit where a voltage measured with a voltmeter is 11.3 V and a resistance of 50 Ω is measured with an ohmmeter. These values are obviously associated to their measurement devices, which have their corresponding errors. A priori, one can think that these measurements and errors could be represented by intervals, but these values need to be represented in a digital scale and they could be considered valid or not in accordance to a certain tolerance. Therefore, in this case, intervals are not enough to handle this difficult problem.

For each reading and for each computation, the obtained values have to be associated to an area in which the real numbers are indiscernible and so any one of them could be considered as a representative of the measurement or the computation result. This chapter presents an approach to work around the inaccuracy associated to any measurement or computation process with physical quantities by means of a new interval tool: *marks*, which define intervals in which it is not possible to make any distinction between its elements, i.e., indiscernibility intervals.

Marks are designed to warn about bad computations caused by problems such as aggregation in rounding, faulty conversions between integers and real numbers, errors in the finite element analysis, some of which have provoked human casualties or billions of dollars in losses [18, 28, 40]. Also marks can overcome some deficiencies existing in the rounded solutions of systems in $I^*(\mathbb{R})$ when the unknowns appear dualized in some occurrences.

Equality and inequality relations will be defined on these marks, as well as the extensions of the operators on $\mathbb{R}$ to the corresponding operators with marks, with emphasis on the arithmetic operators, and the extension of any continuous function to the corresponding function with marks. Special attention will be paid to the logical meaning of the result of any mark operator or function.

8.2 Marks

Any numerical system is the idealization of a computing system associated to the concepts of quantity and order. The natural number system forms the most elementary numerical system holding both concepts, integer numbers are an idealization of the sum and difference, the rational numbers system is an extension of the integers by means of the generalization of the quotient, real numbers come from the concept of convergence for sequences of rational numbers to exact values even when they can not be a rational number. The real numbers system $\mathbb{R}$ is the idealization of geometrical measurements accessible or not with the set of rational numbers.

These consecutive levels are not independent but the result of a construction, where the basic operations preserve their formal properties. Generally speaking the construction of a new system is motivated by shortcomings of the previous system, which could be stated in terms of the same system yet with no solution within it, which compels the creation of a wider system which includes the old one.

8.2.1 Real Line, Digital Line and Interval Analysis

In spite of the flexibility of real numbers $\mathbb{R}$ for dealing with geometrical models, their use possesses shortcomings since they form a continuous system, which can not be overcome inside the very system itself. One of them is digital encoding. Any digital encoding system is finite and its elements represent real numbers but, conceptually, it can not be identified with a subset of real numbers because in any effective computation most of the truncations to a finite number of digits involve the loss of the original value and different ways for equivalent computations yield different results.

In classical error analysis, a real value is represented by a unique digital value and the usual analytical techniques provide an order of magnitude for the gap between both values, the ideal real value, defined by the geometry of the problem, and the effective value obtained by means of an algorithm or measurement. This digital representation is, in principle, a more or less deterministic selection between the digital left-rounded or right-rounded value of the real value defined by the ideal computation or measurement.

Interval analysis arises, see Chap. 1, from the convenience of preserving all the information inherent in a computation or measurement, keeping both approximations for the ideal value: the left-bound and the right-bound. This feature leads to a step-by-step control of the numerical error.

The modal interval approximation to these problems leads to an autonomous theoretical system which allows, in a systematic way, the semantic link between the digital items and their analytical referents on the real line.

Fig. 8.1 Physical system

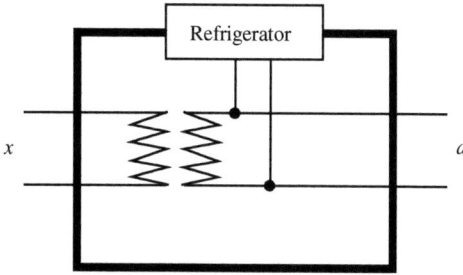

8.2.2 From the Set-Theoretical Interval System to Modal Intervals

The immediately higher level above the real system $\mathbb{R}$ is occupied by the classical or set-theoretical interval system $I(\mathbb{R})$. In $I(\mathbb{R})$, each real number is represented by the interval defined by their digital left-bound and right-bound approximations. To handle both bounds is compulsory because with only one-sided bounds it is not possible to obtain a bound for the result.

$I(\mathbb{R})$ looks like the correct extension of $\mathbb{R}$ because it contains the point intervals, which is a system isomorphic to $\mathbb{R}$, but the classical intervals system contain some limitations which were commented on in Chap. 1. Some of them can be easily overcome, for example the non-existence of an opposite element in the addition, but others are essential, for example the non-distributivity of the product with regard to addition, or semantic ambiguity, making $I(\mathbb{R})$ clearly limited.

The system of modal intervals $I^*(\mathbb{R})$ is a latticed and semantic completion of $I(\mathbb{R})$ of which structural and semantic limitations are drastically reduced.

8.2.3 Deficiencies in the System $I^*(\mathbb{R})$

The previous comment emphasized the necessity of extending a numerical system when the problems which can be stated within it do not have an answer within the system, such as the problem of truncations in the solution of some equations, or systems of equations: it is not possible to solve this problem within $I^*(\mathbb{R})$ because the effective modality of a modal interval, defined by its associated quantifier, can not be the same as its real modality. So the outer truncation rule can not be a universal rule in interval computations with linear operations.

Example 8.2.1 Let us consider the physical system of a transformer with a refrigerator. Let x be the input power, a be the output power, y be the power consumption of the refrigerator, b be the heat balance in the system, σ be the fraction of input power converted in heat by the transformer, and ρ be the fraction of refrigerator power transformed into heat.

Part of x is converted into heat (σx). The rest $(1 - \sigma)x$ of this power is transformed in electric energy $(a + y)$. Part of this electric power is consumed by the refrigerator y which has some losses so it also produces heat ρy. Therefore the refrigerator takes a heat equal to $(1 - \rho)y$ and brings it outside of the system. Then the mathematical model is

$$\begin{cases} (1-\sigma)x - y = a \\ \sigma x - (1-2\rho)y = b, \end{cases}$$

where the first equation corresponds to the electric balance, and the second is the thermal balance equation. The problem is to design the refrigerator, i.e., to calculate intervals X and Y for given A and B in such a way that

$$\forall (a \in A') \, \forall (y \in Y') \, \exists (x \in X') \, a = (1-\sigma)x - y$$

for the electrical balance and

$$\forall (x \in X') \, \exists (b \in B') \, \exists (y \in Y') \, b = \sigma x - (1-2\rho)y$$

for the thermal balance. By the *-semantic theorem, it is equivalent to solve the following interval linear system

$$\begin{cases} (1-\sigma) * X - \text{Dual}(Y) \subseteq A \\ \sigma * \text{Dual}(X) - (1-2\rho) * Y \subseteq B \end{cases} \quad (8.1)$$

with A, X and Y improper intervals and B proper. For $\sigma = 1/9$, $\rho = 1/9$, $A = [16, 8] \, W$ and $B = [\frac{1}{5}, \frac{2}{5}] \, W$ let us consider the system

$$\begin{cases} \frac{8}{9}[\underline{x}, \overline{x}] - [\overline{y}, \underline{y}] = [16, 8] \\ \frac{1}{9}[\overline{x}, \underline{x}] - \frac{7}{9}[\underline{y}, \overline{y}] = [\frac{1}{5}, \frac{2}{5}] \end{cases}$$

which can be split into two real systems

$$\begin{cases} \frac{8}{9}\underline{x} - \overline{y} = 16 \\ \frac{1}{9}\underline{x} - \frac{7}{9}\underline{y} = \frac{2}{5} \end{cases} \quad \text{and} \quad \begin{cases} \frac{8}{9}\overline{x} - \underline{y} = 8 \\ \frac{1}{9}\overline{x} - \frac{7}{9}\overline{y} = \frac{1}{5} \end{cases}$$

The solutions are

$$\underline{x} = \frac{4878}{235} \qquad \underline{y} = \frac{576}{235} \qquad \overline{x} = \frac{2439}{235} \qquad \overline{y} = \frac{288}{235},$$

therefore, with an exact arithmetic, the solution of the interval system is

$$X = [\frac{4878}{235}, \frac{2439}{235}] = [20.7574\ldots, 10.3787\ldots]$$

$$Y = [\frac{576}{235}, \frac{288}{235}] = [2.451\ldots, 1.225\ldots]$$

which is the optimal solution of the interval system (8.1) since

$$\begin{cases} \frac{8}{9}X - \text{Dual}(Y) = [16, 8] \\ \frac{1}{9}\text{Dual}(X) - \frac{7}{9}Y = [\frac{1}{5}, \frac{2}{5}]. \end{cases}$$

But taking into account truncations, the result could be

$$\tilde{X} = \text{Out}(X) = [20.75, 10.38] \supseteq X \quad , \quad \tilde{Y} = \text{Out}(Y) = [2.45, 1.23] \supseteq Y,$$

using outer rounding, but

$$\begin{cases} \frac{8}{9}\tilde{X} - \text{Dual}(\tilde{Y}) = [15.99\overline{4}, 7.99\overline{6}] \not\subseteq [16, 8] \\ \frac{1}{9}\text{Dual}(\tilde{X}) - \frac{7}{9}\tilde{Y} = [0.19\overline{6}, 0.4] \not\subseteq [\frac{1}{5}, \frac{2}{5}] \end{cases}$$

and the equations did not verify the required semantics. The same with inner rounding for X and Y or even merged rounding. So, a problem theoretically solvable has no computational solution.

This problem has a particular feature which transcends its linear context: in an interval linear system, an interval can appear with different modalities in different equations and the rules for truncation are inconsistent with the effective computations.

A new system, built to solve this kind of truncation problems, and called intervals of marks, will be developed in two chapters. This chapter contains the general theory of marks, as items which represent numerical readings or computations, and Chap. 9 will contain the system of intervals of marks.

8.2.4 The System of Marks

Two goals can be reached with the system of marks. A mark has to represent, in a consistent way, the punctual information provided by a digital scale and the system of marks will have to have an inner structure reflecting the losses of information inherent to any computational scale, without paying attention to the effective truncations which must be performed along the computations.

With regard to the first goal, we will provide an entity for marks which will deal with any elementary numerical information, deciding whether the quality of the information in a computation process make it admissible. Marks will guarantee the correctness of a computation, avoiding situations such as, for example the result of the computation of [75]

$$f(x, y) = 333.75y^6 + x^2(11x^2y^2 - y^6 - 121y^4 - 2) + 5.5y^8 + \frac{x}{2y}$$

for $x = 77617$ and $y = 33096$, with an extended precision of 34 digital digit arithmetic, which is the wrong value: $f = 1.172603940053\ldots$ when the correct value is $f = -0.827396059946\ldots$.

The second goal comes from the truncation problem in systems of equations, previously commented on. This aim compels defining intervals of marks as modal intervals with mark bounds, their relations, and their operations.

To achieve both goals, it will be necessary to build the marks as atomic items of information on a digital scale. The indiscernibility of any process of measurement or computation will be translated to an identification of the mark with an improper interval, but its nature is not an interval but a "spot" on the real line. Parameters such as the "relative technical tolerance" or "relative technical granularity" will be defined to represent the indetermination which the "spot" carries.

An interval defines a margin of indetermination such that the values which it contains are different and distinguishable whereas a mark has a margin of indetermination, as well, but there does not exist the possibility of distinguishing between the values which it contains. The reference to a real number contained in a mark will always be as a "reading of the mark", and different readings can never be comparable.

Relations of equality and inequality between marks together with operators for marks will be introduced, with a detailed study of the arithmetic operators and the rational functions of marks.

The set of marks will be defined together with relations inherited from the set of real numbers, since a mark essentially is a numerical reading, and relations of equalities and inequalities will be legitimate.

Computations performed using marks will reflect the gradual loss of information due to numerical errors and truncations and they will compel taking decisions either on the acceptance of the results or about the convenience of using more precision to get the required validity. So the concept of an operator for marks will be introduced, with a detailed study of the arithmetic operators. Next, functions for marks will be built from the operators from which they are built up, together with the semantic interpretations for the results.

8.3 Marks and Associated Intervals

The values of any working digital scale will always be expressed in floating point notation. $\mathbb{D}_n$ will denote the set of the digital numbers with $n + 1$ digits in the mantissa, of the form $a = a_0.a_1 \cdots a_n \times b^m$, where b is the basis of the scale (usually $b = 2$ or $b = 10$) and n the number of fractional digits. Always $a_0 \neq 0$ if $a \neq 0$.

8.3 Marks and Associated Intervals

The maximum relative separation between two points of the scale $\mathbb{D}_n$, different from the exceptional values, is b^{-n}, called the digital granularity.

If x is a real number expressed in floating point notation and $di(x)$ its representation on a digital scale $\mathbb{D}_n$, we always require $di(x) \in x * [1 - \frac{b^{-n}}{2}, 1 + \frac{b^{-n}}{2}]$. This imposition is compatible with the rounding arithmetic, in which also holds $x \in di(x) * [1 - \frac{b^{-n}}{2}, 1 + \frac{b^{-n}}{2}]$.

8.3.1 Mark on a Digital Scale

To speak about a value on a digital scale of any measurement device is, in fact, to make reference to a group of real numbers which have this value as visible referent in the scale. This group of real numbers is an interval which can be represented by means of its *center*, i.e., the value read off the digital scale and its *granularity*, i.e., the amplitude, considered in relative terms.

Since all the points of this interval have the same center of reference, it makes no sense to consider them individually, because they are indiscernible from the scale viewpoint. For this reason, these intervals will be considered as indiscernibility intervals and called *marks*.

Although the center is precisely determined from its reading on the scale, a *tolerance* must be fixed by the observer so that it guarantees the indiscernibility of the points belonging to the mark. In the process of setting a tolerance, several factors can intervene, from the accuracy of measurement devices, up to voluntarily imposed factors, in order to make indiscernible values inside a certain amplitude. The comparison between the tolerance and the granularity will provide validity to the mark.

There exist two other constituent elements of the mark: the *number of digits* used to represent a reading in the scale, which is equal to the precision of the device, and the *basis* of the numerical system in which the readings are displayed.

The bounds of the marks will not be, in general, elements of the digital scale, which discourages the traditional interval treatment, based on the bound elements infimum and supremum, since it would force possible corrections (truncations or rounding) on some of the interval's bounds with a loss of information and, consequently, the read value on the digital scale can be lost as reference.

Definition 8.3.1 (Mark) A *mark* on a digital scale $\mathbb{D}_n$ is an object denoted by $\langle c, t, g, n, b \rangle$ where c is the *center* of the mark, t is the *tolerance*, g is the *granularity*, n and b are, respectively, the *number of digits* and the *basis* of the digital scale.

Center, tolerance, granularity, and scale are the *attributes* of the mark. The center is a number $c \in DI_n$ representing the approximate value, when the mark represents a real constant, or the reading in a display, when the quantity comes from a measurement device.

The tolerance is a digital number $t \in]0, 1[$ which expresses the relative maximum separation among the points of the mark which the observer will consider as indiscernible from the center. The tolerance has to indicate the relative width of the intervals of indiscernibility which can be taken, depending on the phenomenon under study or the accuracy that might be required of the results.

The granularity is a digital number $g \in [0, 1[$ which reflects the inaccuracy of the measure. It is equal to b^{-n}, the digital granularity, when the mark represents to a real constant, or more generally, coincides with the relative error associated to the reading, when the quantity comes from a measurement device. It is reasonable to identify initially the granularity with the relative error of the reading, but one is also able to include other aspects related to the phenomenon under study, for instance, instabilities.

Granularity and tolerance must satisfy the *minimum condition of significance*:

$$b^{-n} \leq g < t < 1. \tag{8.2}$$

This condition prevents accepting, for example measures of millimeters with a ruler graduated in centimeters. Generally speaking, the tolerance must be greater than the sum of all the inaccuracy factors: the greater the inaccuracies, the greater the tolerance must be.

The set of marks with tolerance t, number of digits n and scale basis b, will be denoted by

$$\mathbb{M}(t, n, b) = \{\mathfrak{m} = \langle c, t, g, n, b \rangle \mid c \in DI_n, t \in]0, 1[, b^{-n} \leq g < t\}.$$

The numbers t, n and b are the *type* of the mark. For a given type, the mark is defined by center and granularity; in this case the mark will be denoted by $\langle c, g \rangle$.

To specify the value of b will be often irrelevant. In this case the set of marks will be denoted by $\mathbb{M}(t, n)$, neglecting b.

Example 8.3.1 The representation of the reading in a voltmeter of 12.23 V with a relative error of 1 % as a mark in a decimal scale DI_5, imposing a tolerance of 0.05, is

$$\langle 1.22300e1, 5.00000e - 2, 1.00000e - 2, 5, 10 \rangle,$$

or simply

$$\langle 1.223e1, 5.0e - 2, 1.0e - 2, 5, 10 \rangle$$

since $1.223e1, 5.0e - 2, 1.0e - 2 \in DI_5$. The representation of the number π as a mark of that scale DI_5, supposing the same tolerance, is

$$\langle 3.14159e0, 5.0e - 2, 1.0e - 5, 5, 10 \rangle.$$

8.3 Marks and Associated Intervals

8.3.2 Imprecision and Validity of a Mark

The minimum condition of significance $b^{-n} \le g < t < 1$, necessary to any mark $\mathfrak{m} = \langle c, t, g, n, b \rangle$, leads to the definition of the following parameters:

- The *imprecision index* is the value $\frac{g}{t}$. It satisfies $\frac{g}{t} < 1$, since otherwise the minimum condition of significance is not satisfied.
- The *validity index* is the complement to 1 of the imprecision index, that is to say, $1 - \frac{g}{t}$. The more accurate the mark, the smaller the $\frac{g}{t}$.

The minimum condition of significance $b^{-n} \le g < t < 1$ can be generalized in the form $b^{-n} \le g < \alpha t < 1$, where $\alpha \in]0, 1]$ is a parameter. When $g < \alpha t$, the granularity g will be called *compatible* with αt.

8.3.3 Associated Intervals to a Mark

Let $\mathfrak{m} = \langle c, g \rangle \in \mathbb{M}(t, n)$ be a mark, then the

- *Associated interval to* $\mathfrak{m}$, $Iv(\mathfrak{m}) = c * [1 + t, 1 - t]$.
- *Indiscernibility margin of* $\mathfrak{m}$, $Ind(\mathfrak{m}) = c * [1 - t, 1 + t]$.
- *External shadow of* $\mathfrak{m}$, $Exsh(\mathfrak{m}) = c * [1 + t, 1 - t] * [1 - g, 1 + g]$.
- *Internal shadow of* $\mathfrak{m}$, $Insh(\mathfrak{m}) = c * [1 + t, 1 - t] * [1 + g, 1 - g]$.

All these modal intervals can be positive or negative, but 0 never belongs to the interval domains.

The points of $Ind(\mathfrak{m})$ are called *readings* of $\mathfrak{m}$. Sometimes it can be necessary to consider the indiscernibility margin of a mark when the granularity g is compatible with αt. In this case $Ind_\alpha(\mathfrak{m}) = c * [1 - \alpha t, 1 + \alpha t]$.

The condition $g < t$ guarantees that the interval $Exsh(\mathfrak{m})$ is improper. Obviously, $Insh(\mathfrak{m})$ is always improper.

The *associated mark to the external shadow* of $\mathfrak{m} = \langle c, t, g, n, b \rangle$ is defined by

$$ExshM(\mathfrak{m}) = \langle c, t - g(1 + t), b^{-n}, n, b \rangle \in \mathbb{M}(t - g(1 + t), n, b).$$

The validity condition for this mark $ExshM(\mathfrak{m})$ is

$$b^{-n} < t - g(1 + t)$$

and its compatibility with αt is

$$b^{-n} < \alpha(t - g(1 + t)).$$

This inequality forces

$$g < \frac{t - \frac{b^{-n}}{\alpha}}{1+t}, \qquad (8.3)$$

which must prevail over $g < t$. The value

$$\tilde{t} = t - g(1+t)$$

is called the *effective tolerance* of the mark $\mathfrak{m} = \langle c, t, g, n, b \rangle$.

Example 8.3.2 The mark $\mathfrak{m} = \langle 3.2e0, 1.0e-2, 1.0e-3, 6, 10 \rangle$ has imprecision index $\frac{g}{t} = 0.001/0.01 = 0.1 < 1$, which means that the mark is valid and its validity index is $1 - \frac{g}{t} = 1 - 0.1 = 0.9$. The minimum condition of significance is $10^{-6} < 0.001 < 0.01 < 1$ which is true, and the generalized minimum condition of significance for the parameter $\alpha = 0.8$ takes the form $10^{-6} < 0.001 < 0.8 \cdot 0.01 < 1$, which is also true. So the granularity $g = 0.001$ is *compatible* with $\alpha t = 0.008$ and the effective tolerance is $\tilde{t} = t - g(1+t) = 0.00899$. Their associated intervals are

$$Iv(\mathfrak{m}) = c * [1+t, 1-t] = [3.232, 3.168],$$
$$Ind(\mathfrak{m}) = c * [1-t, 1+t] = [3.168, 3.232],$$
$$Exsh(\mathfrak{m}) = c * [1+t, 1-t] * [1-g, 1+g] = [3.228768, 3.171168],$$
$$Insh(\mathfrak{m}) = c * [1+t, 1-t] * [1+g, 1-g] = [3.235232, 3.164832].$$

The mark associated to the external shadow of $\mathfrak{m}$ is

$$ExshM(\mathfrak{m}) = \langle 3.2e0, 8.99e-3, 1.0e-5, 6, 10 \rangle$$

and its validity condition is

$$10^{-6} < 0.00899$$

and for $\alpha = 0.9$

$$10^{-6} < 0.9 * 0.00899$$

which forces

$$g = 0.001 < \frac{t - \frac{b^{-n}}{\alpha}}{1+t} = \frac{0.01 - 10^{-6}/0.9}{1 + 0.01} = 0.0098899\ldots$$

which is true.

8.4 Relations in the Set of the Marks

A mark is associated to a digital center and an indiscernibility radius. These two constituents will determine the equality and inequality relations. The relations among marks are going to be a structural copy of the relations among real numbers. As there does not exist any inclusion relationship between real numbers, consequently there will not exist any inclusion relationship between marks.

Two marks will be *comparable*, when they have the same tolerance and, to avoid basis transformation operations, they also have the same basis. It will not be necessary that the marks to be compared are of the same type, since the number of digits doesn't raise difficulties for those comparisons.

8.4.1 Equality Relations

Given two marks $m_1 = \langle c_1, g_1 \rangle$ and $m_2 = \langle c_2, g_2 \rangle$,

- m_1 is *materially equal* to m_2,

$$m_1 = m_2 \Leftrightarrow c_1 = c_2.$$

- m_1 is *weakly equal* to m_2 with respect to the parameter $\alpha \in]0,1]$,

$$m_1 \approx_\alpha m_2 \Leftrightarrow (c_2 \in Ind_\alpha(m_1) \quad \underline{\text{or}} \quad c_1 \in Ind_\alpha(m_2))$$

provided both granularities g_1 and g_2 are compatible with αt.

The main properties of the weak equality relation are

1. If m_1 and m_2 are comparable, $\alpha, \beta \in]0,1]$, $\alpha \leq \beta$ and g_1 and g_2 compatible with αt, then

$$m_1 \approx_\alpha m_2 \Rightarrow m_1 \approx_\beta m_2.$$

2. If m_1, m_2 and m_3 are comparable, $\alpha \in]0,1]$ and g_1, g_2, g_3 compatible with αt, then

$$(m_1 \approx_\alpha m_2, \ m_2 = m_3) \Rightarrow m_1 \approx_\alpha m_3.$$

3. If m_1, m_2 and m_3 are comparable, $\alpha, \beta \in]0,1]$, $\alpha + \beta < 1$ and g_1, g_2 and g_3 compatible with αt or with βt, respectively, then

$$(m_1 \approx_\alpha m_2, \ m_2 \approx_\beta m_3) \Rightarrow m_1 \approx_{\alpha+\beta} m_3.$$

This property is called the $(\alpha + \beta)$-*transitivity of the weak equality*.

8.4.2 Inequality Relations

Given two comparable marks $m_1 = \langle c_1, g_1 \rangle$ and $m_2 = \langle c_2, g_2 \rangle$,

- m_1 is *materially less than or equal to* m_2,

$$m_1 \leq m_2 \Leftrightarrow c_1 \leq c_2.$$

- m_1 is *weakly less than or equal to* m_2 with respect to the parameter $\alpha \in]0, 1]$,

$$m_1 \preceq_\alpha m_2 \Leftrightarrow (m_1 \leq m_2 \quad \text{or} \quad m_1 \approx_\alpha m_2)$$

if the granularities g_1 and g_2 are compatible with αt.
- m_1 is *materially greater than or equal to* m_2,

$$m_1 \geq m_2 \Leftrightarrow m_2 \leq m_1.$$

- m_1 is *weakly greater than or equal to* m_2 with respect to the parameter $\alpha \in]0, 1]$,

$$m_1 \succeq_\alpha m_2 \Leftrightarrow m_2 \preceq_\alpha m_1$$

The main properties of these inequality relations are:

1. $m_1 \leq m_2 \Leftrightarrow Iv(m_1) \leq Iv(m_2)$.
2. If m_1, m_2 and m_3 are comparable, $\alpha \in]0, 1]$ and g_1, g_2 and g_3 are compatible with αt, then

$$(m_1 \approx_\alpha m_2, m_2 \leq m_3) \Rightarrow m_1 \preceq_\alpha m_3.$$

3. If the comparable marks m_1 and m_2 are materially unequal, they will be weakly equal with respect to any parameter $\alpha \in]0, 1]$ if $g_1, g_2 < \alpha t$, i.e.,

$$m_1 \leq m_2 \Rightarrow (\max\{g_1, g_2\} < \alpha t \Rightarrow m_1 \preceq_\alpha m_2).$$

4. Given two comparable marks m_1 and m_2 if $\alpha \in]0, 1]$ and g_1 and g_2 are compatible with αt, then

$$(m_1 \preceq_\alpha m_2, m_2 \preceq_\alpha m_1) \Rightarrow m_1 \approx_\alpha m_2.$$

This property is called the *anti-symmetry of the weak inequality*.
5. Given three comparable marks m_1, m_2 and m_3, if $\alpha, \beta \in]0, 1]$, $\alpha + \beta < 1$ and g_1, g_2 and g_3 are compatible with αt or with βt, respectively, then

$$(m_1 \preceq_\alpha m_2, m_2 \preceq_\beta m_3) \Rightarrow m_1 \preceq_{\alpha+\beta} m_3.$$

This property is called the *$(\alpha + \beta)$-transitivity of weak inequality*.

8.4.3 Strict Inequality Relations

Given two comparable marks $m_1 = \langle c_1, g_1 \rangle$ and $m_2 = \langle c_2, g_2 \rangle$,

- m_1 is *materially less than* m_2,

$$m_1 < m_2 \Leftrightarrow Iv(m_1) < Iv(m_2).$$

- m_1 is *weakly less than* m_2 with respect to a parameter $\alpha \in]0, 1]$

$$m_1 \prec_\alpha m_2 \Leftrightarrow (\neg(m_1 \approx_\alpha m_2), c_1 < c_2)$$

if the granularities g_1 and g_2 are compatible with αt.

Let m_1 and m_2 be comparable marks, if $\alpha \in]0, 1]$ is such that g_1 and g_2 are compatible with αt, then

$$m_1 \preceq_\alpha m_2 \Leftrightarrow (m_1 \prec_\alpha m_2 \quad \text{or} \quad m_1 \approx_\alpha m_2).$$

Example 8.4.1 For the marks $m = \langle 3.2e0, 1.0e-2, 1.0e-3, 5, 10 \rangle$, $n = \langle 2.9e0, 1.0e-2, 1.0e-3, 5, 10 \rangle$ and $p = \langle 3.18e0, 1.0e-2, 1.0e-3, 5, 10 \rangle$ and a parameter $\alpha = 1$, the indiscernibility margins are

$$Ind_\alpha(m) = [3.168, 3.232],$$
$$Ind_\alpha(n) = [2.871, 2.929],$$
$$Ind_\alpha(p) = [3.1482, 3.2118],$$

therefore the followings relations are true:

m is weakly equal to p, $m \approx_\alpha p$.
m is materially greater than or equal to n, $m \geq n$.
m is weakly greater than or equal to n, $m \succeq_\alpha n$.
m is weakly greater than n, $m \succ_\alpha n$.

8.5 Mark Operators

Before defining any operation between marks and taking into account that the initial marks can come from direct or indirect readings in some scale, it is necessary to represent them in the computational scale in order to get suitable computation items. Therefore a read value $m_1 = \langle c_1, t_1, g_1, n_1, b \rangle \in \mathbb{M}(t_1, n_1)$ represented in a computation scale $\mathbb{D}_n$ becomes the mark

$$m = \langle c, t, g, n, b \rangle,$$

where c is the representation of the read value c_1 in $\mathbb{D}_n$ ($c = di(c_1)$), t is the lower truncation of t_1 in $\mathbb{D}_n$ ($t =\downarrow t_1$) and g is the nearest value of g_1 in $\mathbb{D}_n$ ($g \simeq g_1, g < t$).

The loss of information, intrinsic to any computation process, must be reflected in the granularity, not in the tolerance, and always, in order to maintain the homogeneity of the system, the result of any computation will have the same tolerance as the data. The granularity will embody any indetermination and, so, it will increase in each computation step. Nevertheless it will not have a main role but is an indicator of a loss of validity in the computation process, when increasing g leads to invalidating a result for which the imprecision index is greater than or equal to 1.

Any mark comes from a previous computation with marks or from a measurement. In this case, to represent this mark in the computation scale, it is necessary to increase its granularity in b^{-n}, if the computation scale is $\mathbb{D}_n$. From now on, this previous step is obviated and marks will be considered marks of the digital scale of the computation.

Operations between marks will be defined for marks of the same type and the result is also of the same type as the data. In this way, the tolerance will be constant along any computation, but the granularity will increase, reflecting the step-by-step loss of information, which constitutes the deviation of the computed value from the exact value. Along the computational process the granularity will go on increasing, following some propagation approach, that will be set by the phenomenon being modelled. The computation of the granularity will be necessary to check the validity of the resulting mark and the imprecision index in the resulting mark will depend on the approach used to propagate the granularity. The phenomenon under examination will determine this approach, with the only demand being that the resulting granularity must never be smaller than the granularities of the data. Accepting this norm, different approaches will be considered.

8.5.1 Mark Operators over $I^*(\mathbb{R})$

Let $f : \mathbb{R}^2 \to \mathbb{R}$ be a continuous function and $\mathfrak{m}_1 = \langle c_1, g_1 \rangle \in \mathbb{M}(t, n)$, $\mathfrak{m}_2 = \langle c_2, g_2 \rangle \in \mathbb{M}(t, n)$. A *mark operator over* $I^*(\mathbb{R})$ associated to f is any function

$$f_{\text{MI}} : \mathbb{M}(t, n) \times \mathbb{M}(t, n) \to I^*(\mathbb{R})$$

such that

$$f_{\text{MI}}(\mathfrak{m}_1, \mathfrak{m}_2) = f(x_1, x_2) * [1 + t, 1 - t],$$

where $(x_1, x_2) \in Ind(\mathfrak{m}_1) \times Ind(\mathfrak{m}_2)$.

The most important mark operators over $I^*(\mathbb{R})$ are:

8.5 Mark Operators

- $f_{\text{MI}}(m_1, m_2)$ is *admissible* for $(x_1, x_2) \in Ind(m_1) \times Ind(m_2)$ if

$$f^*(Iv(m_1), Iv(m_2)) \subseteq f(x_1, x_2) * [1+t, 1-t].$$

- $f_{\text{MI}}(m_1, m_2)$ is *centered* when

$$f^*(Iv(m_1), Iv(m_2)) \subseteq f(c_1, c_2) * [1+t, 1-t],$$

where c_1 and c_2 are the respective centers of the marks m_1 and m_2 and f^* the modal *-semantic extension of f.

Remark 8.5.1 If $f_{\text{MI}}(m_1, m_2)$ is centered, then it is admissible for (c_1, c_2).

Example 8.5.1 The mark operator associated to the real continuous function $f : \mathbb{R}_+^2 \to \mathbb{R}$ defined by $f(x, y) = (x + y)^2$ is a centered operator because

$$\begin{aligned}
f^*(Iv(m_1), Iv(m_2)) &= ([c_1(1+t), c_1(1-t)] + [c_2(1+t), c_2(1-t)])^2 \\
&= [(c_1 + c_2)^2(1+t)^2, (c_1 + c_2)^2(1-t)^2] \\
&\subseteq [(c_1 + c_2)^2(1+t), (c_1 + c_2)^2(1-t)] \\
&= f(c_1, c_2) * [1+t, 1-t],
\end{aligned}$$

where the first equality comes from the optimality of the syntactic extension of f in the domain $\mathbb{R}_+^2$. But the mark operator associated to the real continuous function $f : \mathbb{R}_+^2 \to \mathbb{R}$ defined by $f(x, y) = \sqrt{x+y}$ is not a centered operator because

$$\begin{aligned}
f^*(Iv(m_1), Iv(m_2)) &= \sqrt{([c_1(1+t), c_1(1-t)] + [c_2(1+t), c_2(1-t)])} \\
&= [\sqrt{c_1 + c_2}\sqrt{1+t}, \sqrt{c_1 + c_2}\sqrt{1-t}] \\
&\not\subseteq [\sqrt{c_1 + c_2}(1+t), \sqrt{c_1 + c_2}(1-t)] \\
&= f(c_1, c_2) * [1+t, 1-t],
\end{aligned}$$

where the first equality is also due to the optimality of the syntactic extension of f in the domain $\mathbb{R}_+^2$.

Operators which satisfy these inclusions allow of defining the mark operators of which the result is a mark of the same type as the arguments.

8.5.2 Mark Operators over $\mathbb{M}(t, n)$

Definition 8.5.1 (Mark operator) Let $f : \mathbb{R}^2 \to \mathbb{R}$ be a continuous function such that $f_{\text{MI}}(m_1, m_2)$ is centered. A *mark operator over* $\mathbb{M}(t, n)$ associated to f is a function

$$f_{\mathbb{M}(t,n)} : \mathbb{M}(t,n) \times \mathbb{M}(t,n) \longrightarrow \mathbb{M}(t,n)$$

defined by

$$f_{\mathbb{M}(t,n)}(\mathfrak{m}_1, \mathfrak{m}_2) = \langle di(f(c_1, c_2)), g_z \rangle \in \mathbb{M}(t,n)$$

where

1. $di(f(c_1, c_2))$ is the digital computation of the function f at (c_1, c_2) on the scale $\mathbb{D}_n$, supposing a minimum relative displacement of $di(f(c_1, c_2))$ with respect to the exact value $f(c_1, c_2)$, i.e., less than or equal to $\frac{b^{-n}}{2}$.
2. g_z is the granularity of the result, which it will be necessary to determine specifically for each operator. It is $g_z = \gamma_{1,2}$, if $di(f(c_1, c_2)) = f(c_1, c_2)$, or $g_z = \gamma_{1,2} + b^{-n}$, if the relative displacement of the computation $di(f(c_1, c_2))$ with respect to the exact value $f(c_1, c_2)$ is less than or equal to the digital granularity b^{-n}. The term $\gamma_{1,2}$ is the *main term of the granularity* which will depend on the interval operator associated to the function f, and one always has $\gamma_{1,2} \geq \max\{g_1, g_2\}$, i.e., $\gamma_{1,2}$ must be greater than or equal to the granularity of the data.

Remark 8.5.2 The expression $g_z = \gamma_{1,2} + b^{-n}$, is based on the equality

$$1 + g_z = (1 + \gamma_{1,2}) \cdot (1 + \frac{b^{-n}}{2}) \leq 1 + \gamma_{1,2} + b^{-n}.$$

This equality allows of taking b^{-n} as the secondary term of the granularity coming from the digital shift of the computation.

Granularity is essentially a fuzzy component and it can be defined in different ways, always with the condition that $\gamma_{1,2} \geq \max\{g_1, g_2\}$. Different approaches will reflect different aims of the computation:

1. The *minimal* or *semantic* approach, where the evolution of the granularity corresponds to the maximum projection of the granularities of the data: $\gamma_{1,2}$ is the smallest number satisfying

$$f(c_1, c_2) * [1 + \gamma_{1,2}, 1 - \gamma_{1,2}] \subseteq f^*(c_1 * [1 + g_1, 1 - g_1], c_2)$$

and

$$f(c_1, c_2) * [1 + \gamma_{1,2}, 1 - \gamma_{1,2}] \subseteq f^*(c_1, c_2 * [1 + g_2, 1 - g_2]).$$

Thus, the resulting granularity gives the biggest projection in the result of the granularity of each datum separately. This approach reflects situations of the "chain" type, when the lack of precision of the process comes from the biggest lack of precision of the data.

2. The *maximal* or *metric* approach, where the evolution of the granularity reflects the computation of errors: $\gamma_{1,2}$ is the smallest number satisfying

$$f(c_1, c_2) * [1 + \gamma_{1,2}, 1 - \gamma_{1,2}] \subseteq f^*(c_1 * [1 + g_1, 1 - g_1], c_2 * [1 + g_2, 1 - g_2]).$$

The maximal approach is an interval treatment of error propagation. In this approach, the granularity is a relative error and its propagation follows the rules of the propagation of errors. This approach is used when the objective is to secure that the relative error is smaller than a given tolerance.

3. Other approaches can also be considered, for example a statistical approach, considering the granularities as statistically distributed somehow.

Remark 8.5.3 The equality $f_{\mathbb{MI}}(\mathfrak{m}_1, \mathfrak{m}_2) = Iv(f_{\mathbb{M}(t,\infty)}(\mathfrak{m}_1, \mathfrak{m}_2))$ is, obviously, true.

Remark 8.5.4 In the definition of mark operator on $\mathbb{M}(t, n)$ the condition that $f_{\mathbb{MI}}(\mathfrak{m}_1, \mathfrak{m}_2)$ is centered is sufficient for the admissibility of the mark operator.

The following study of the main mark operators will be based on the minimal approach, but it can be adapted to the other approaches with only a few obvious variations.

8.6 Max and Min Operators

Before dealing with the arithmetic operators, let us begin with the election operators, which do not provoke any digital shift for the centers of the marks.

8.6.1 Maximum

Suppose $\mathfrak{m}_1 = \langle c_1, g_1 \rangle, \mathfrak{m}_2 = \langle c_2, g_2 \rangle \in \mathbb{M}(t, n)$.

Definition 8.6.1 (Maximum) The mark operator over $I^*(\mathbb{R})$ associated to the maximum is

$$\max(\mathfrak{m}_1, \mathfrak{m}_2) = \max(c_1, c_2) * [1 + t, 1 - t] \in I^*(\mathbb{R})$$

Lemma 8.6.1 *Putting* $Z = \max(c_1, c_2) * [1 + t, 1 - t] \in I^*(\mathbb{R})$, *then*

$$\text{Max}(Iv(\mathfrak{m}_1), Iv(\mathfrak{m}_2)) = Z,$$

that is to say, the maximum is a $\mathbb{MI}$-centered operator.

Proof

1. If $m_1 \geq 0$ and $m_2 \geq 0$, then

$$\text{Max }(Iv(m_1), Iv(m_2))$$
$$= [\max(c_1(1+t), c_2(1+t)), \max(c_1(1-t), c_2(1-t))] = Z$$

2. If $m_1 \leq 0$ and $m_2 \leq 0$, then

$$\text{Max }(Iv(m_1), Iv(m_2))$$
$$= [\max(c_1(1-t), c_2(1-t)), \max(c_1(1+t), c_2(1+t))] = Z$$

3. If $m_1 \geq 0$ and $m_2 \leq 0$, then

$$\text{Max }(Iv(m_1), Iv(m_2))$$
$$= [\max(c_1(1+t), c_2(1-t)), \max(c_1(1-t), c_2(1+t))] = Z$$

because $c_1 \geq c_2$ and $Z = [c_1(1+t), c_1(1-t)]$. ∎

Lemma 8.6.2 *The main term of the granularity of the maximum in the minimal approach, i.e.,*

$$\max(c_1, c_2) * [1 + \gamma_{1,2}, 1 - \gamma_{1,2}] \subseteq \text{Max }(c_1 * [1 + g_1, 1 - g_1], c_2)$$

and

$$\max(c_1, c_2) * [1 + \gamma_{1,2}, 1 - \gamma_{1,2}] \subseteq \text{Max }(c_1, c_2 * [1 + g_2, 1 - g_2]).$$

is $\gamma_{1,2} = \max(g_1, g_2)$.

Proof Let us suppose, for example $\max(c_1, c_2) = c_1$.

1. If $c_1 \geq 0$ and $c_2 \geq 0$, then

$$\max(c_1, c_2) * [1 + \gamma_{1,2}, 1 - \gamma_{1,2}] = [c_1(1 + \gamma_{1,2}), c_1(1 - \gamma_{1,2})]$$
$$\text{Max }(c_1 * [1 + g_1, 1 - g_1], c_2) = [c_1(1 + g_1), \max(c_1(1 - g_1), c_2)]$$
$$\text{Max }(c_1, (c_2 * [1 + g_2, 1 - g_2])) = [\max(c_1, c_2(1 + g_2)), c_1].$$

As $\gamma_{1,2} \geq \max(g_1, g_2)$, then

$$c_1(1 + \gamma_{1,2}) \geq c_1(1 + g_1)$$
$$c_1(1 - \gamma_{1,2}) \leq \max(c_1(1 - g_1), c_2)$$

8.6 Max and Min Operators

$$c_1(1 + \gamma_{1,2}) \geq \max(c_1, c_2(1+g_2))$$
$$c_1(1 + \gamma_{1,2}) \leq c_1,$$

therefore $\gamma_{1,2} = \max(g_1, g_2)$.
2. If $c_1 \leq 0$ and $c_2 < 0$, the result is obvious. ∎
3. If $c_1 < 0$ and $c_2 \geq 0$ or $c_2 < 0$, a similar line of reasoning yields the same conclusion. ∎

Theorem 8.6.1 (Computation algorithm for maximum operator) *The maximum over* $\mathbb{M}(t,n)$ *of two marks* $m_1 = \langle c_1, g_1 \rangle, m_2 = \langle c_2, g_2 \rangle \in \mathbb{M}(t,n)$ *is*

$$\max(m_1, m_2) = \langle c_z, t, g_z, n, b \rangle, \tag{8.4}$$

where

- *the center is the maximum of the centers:* $c_z = \max(c_1, c_2)$,
- *the tolerance* t, *the basis* b *and the number of digits* n *are the same as for the data,*
- *the granularity is* $g_z = \max(g_1, g_2)$

Proof From the previous two propositions. ∎

8.6.2 Minimum

The same reasoning, with the obvious adaptations, leads to the following results. Let $m_1 = \langle c_1, g_1 \rangle, m_2 = \langle c_2, g_2 \rangle \in \mathbb{M}(t,n)$.

Definition 8.6.2 (Minimum) The mark operator over $I^*(\mathbb{R})$ associated to the minimum is

$$\min(m_1, m_2) = \min(c_1 c_2) * [1+t, 1-t]$$

Lemma 8.6.3 *Putting* $Z = \min(c_1 c_2) * [1+t, 1-t] \in I^*(\mathbb{R})$, *then*

$$\text{Min } (Iv(m_1), Iv(m_2)) = Z,$$

that is to say, the minimum is a $\mathbb{MI}$-*centered operator.*

Theorem 8.6.2 (Computation algorithm for minimum operator) *The minimum over* $\mathbb{M}(t,n)$ *of two marks* $m_1 = \langle c_1, g_1 \rangle, m_2 = \langle c_2, g_2 \rangle \in \mathbb{M}(t,n)$ *is*

$$\min(m_1, m_2) = \langle c_z, t, g_z, n, b \rangle, \tag{8.5}$$

where

- the center is the minimum of the centers: $c_z = \min(c_1, c_2)$,
- the tolerance t, the basis b and the number of digits n are the same as those of the data,
- the granularity is $g_z = \max(g_1, g_2)$.

8.7 Arithmetic Operators

The real arithmetic operators, product, quotient, sum, and subtraction, will be extended to mark centered operators, i.e., for all them the inclusion

$$f^*(Iv(m_1), Iv(m_2)) \subseteq f(c_1, c_2) * [1+t, 1-t]$$

is fulfilled. The main term of the granularity will be determined according to the chosen approach, the minimal one. In the following arguments, the optimality of the arithmetic operators in every domain will be taken into account, so their *-extensions are equal to their syntactic extensions.

8.7.1 Product Operator

Suppose $m_1 = \langle c_1, g_1 \rangle$, $m_2 = \langle c_2, g_2 \rangle \in \mathbb{M}(t, n)$.

Definition 8.7.1 (Product of marks) The mark operator over $I^*(\mathbb{R})$ associated to the product is

$$m_1 * m_2 = (c_1 c_2) * [1+t, 1-t].$$

Lemma 8.7.1 *Putting $Z = (c_1 c_2) * [1+t, 1-t] \in I^*(\mathbb{R})$, then*

$$Iv(m_1) * Iv(m_2) \subseteq Z,$$

that is to say, multiplication is a $\mathbb{MI}$-centered operator.

Proof The inclusion

$$\begin{aligned} Iv(m_1) * Iv(m_2) &= (c_1 * [1+t, 1-t]) * (c_2 * [1+t, 1-t]) \\ &= (c_1 c_2) * [1+t, 1-t] * [1+t, 1-t] \\ &\subseteq (c_1 c_2) * [1+t, 1-t] * [1, 1]. \end{aligned}$$

implies that

$$Iv(m_1) * Iv(m_2) \subseteq (c_1 c_2) * [1+t, 1-t]. \qquad \blacksquare$$

8.7 Arithmetic Operators

Lemma 8.7.2 *The main term of the granularity of multiplication in the minimal approach*

$$(c_1 c_2) * [1 + \gamma_{1,2}, 1 - \gamma_{1,2}] \subseteq (c_1 * [1 + g_1, 1 - g_1]) * c_2$$

and

$$(c_1 c_2) * [1 + \gamma_{1,2}, 1 - \gamma_{1,2}] \subseteq c_1 * (c_2 * [1 + g_2, 1 - g_2])$$

is $\gamma_{1,2} = \max(g_1, g_2)$.

Proof From the commutativity of the interval product,

$$(c_1 * [1 + g_1, 1 - g_1]) * c_2 = (c_1 c_2) * [1 + g_1, 1 - g_1]$$

and

$$c_1 * (c_2 * [1 + g_2, 1 - g_2]) = (c_1 c_2) * [1 + g_2, 1 - g_2].$$

Therefore

$$\left.\begin{array}{l}[1 + \gamma_{1,2}, 1 - \gamma_{1,2}] \subseteq [1 + g_1, 1 - g_1] \\ [1 + \gamma_{1,2}, 1 - \gamma_{1,2}] \subseteq [1 + g_2, 1 - g_2]\end{array}\right\} \Rightarrow \gamma_{1,2} = \max\{g_1, g_2\}. \qquad \blacksquare$$

Theorem 8.7.1 (Computation algorithm for product operator) *The product over* $\mathbb{M}(t, n)$ *of two marks* $\mathrm{m}_1 = \langle c_1, g_1 \rangle, \mathrm{m}_2 = \langle c_2, g_2 \rangle \in \mathbb{M}(t, n)$ *is*

$$\mathrm{m}_1 * \mathrm{m}_2 = \langle c_z, t, g_z, n, b \rangle, \qquad (8.6)$$

where

- *the center is the digital product of the centers:* $c_z = di(c_1 c_2)$,
- *the tolerance* t, *the basis* b *and the number of digits* n *are the same as for the data,*
- *the granularity is* $g_z = \max(g_1, g_2) + b^{-n}$

Proof From the previous two propositions. $\blacksquare$

8.7.2 Quotient Operator

Suppose $\mathrm{m}_1 = \langle c_1, g_1 \rangle, \mathrm{m}_2 = \langle c_2, g_2 \rangle \in \mathbb{M}(t, n)$ with $c_2 \neq 0$.

Definition 8.7.2 (Quotient of marks) The mark operator over $I^*(\mathbb{R})$ associated to the quotient is

$$\mathrm{m}_1 / \mathrm{m}_2 = \frac{c_1}{c_2} * [1 + t, 1 - t].$$

Lemma 8.7.3 *Putting* $Z = \dfrac{c_1}{c_2} * [1+t, 1-t] \in I^*(\mathbb{R})$, *then*

$$\dfrac{Iv(m_1)}{Iv(m_2)} \subseteq Z,$$

that is to say, division is a $\mathbb{MI}$*-centered operator.*

Proof From the definition of indiscernibility margin, $0 \notin Ind(m_2)$. Therefore, for $(c_1 \geq 0, c_2 > 0)$ or $(c_1 < 0, c_2 < 0)$

$$\dfrac{Iv(m_1)}{Iv(m_2)} = \left[\dfrac{c_1}{c_2}\dfrac{1+t}{1-t}, \dfrac{c_1}{c_2}\dfrac{1-t}{1+t}\right] \subseteq \left[\dfrac{c_1}{c_2}(1+t), \dfrac{c_1}{c_2}(1-t)\right] = Z$$

for $(c_1 \geq 0, c_2 < 0)$ or $(c_1 < 0, c_2 > 0)$

$$\dfrac{Iv(m_1)}{Iv(m_2)} = \left[\dfrac{c_1}{c_2}\dfrac{1-t}{1+t}, \dfrac{c_1}{c_2}\dfrac{1+t}{1-t}\right] \subseteq \left[\dfrac{c_1}{c_2}(1-t), \dfrac{c_1}{c_2}(1+t)\right] = Z. \quad \blacksquare$$

Lemma 8.7.4 *The main term of the granularity of the quotient in the minimal approach*

$$\dfrac{c_1}{c_2} * [1+\gamma_{1,2}, 1-\gamma_{1,2}] \subseteq \dfrac{c_1 * [1+g_1, 1-g_1]}{c_2}$$

and

$$\dfrac{c_1}{c_2} * [1+\gamma_{1,2}, 1-\gamma_{1,2}] \subseteq \dfrac{c_1}{c_2 * [1+g_2, 1-g_2]}$$

is $\gamma_{1,2} = \max\{g_1, \dfrac{g_2}{1-g_2}\}$.

Proof From the first inclusion,

$$\dfrac{c_1}{c_2} * [1+\gamma_{1,2}, 1-\gamma_{1,2}] \subseteq \dfrac{c_1 * [1+g_1, 1-g_1]}{c_2} \quad \Rightarrow \quad \gamma_{1,2} \geq g_1.$$

The second inclusion is equivalent to

$$\dfrac{c_1}{c_2} * [1+\gamma_{1,2}, 1-\gamma_{1,2}] \subseteq \dfrac{c_1}{c_2} * \left[\dfrac{1}{1-g_2}, \dfrac{1}{1+g_2}\right],$$

and therefore

$$1+\gamma_{1,2} \geq \dfrac{1}{1-g_2} \quad \text{and} \quad 1-\gamma_{1,2} \leq \dfrac{1}{1+g_2}$$

8.7 Arithmetic Operators

or, equivalently, $\gamma_{1,2} \geq \frac{g_2}{1-g_2}$. So

$$\gamma_{1,2} = \max\{g_1, \frac{g_2}{1-g_2}\}.$$ ∎

Theorem 8.7.2 (Computation algorithm for the quotient operator) *The quotient over* $\mathbb{M}(t, n)$ *of two marks* $\mathrm{m}_1 = \langle c_1, g_1 \rangle, \mathrm{m}_2 = \langle c_2, g_2 \rangle \in \mathbb{M}(t, n)$, *with* $c_2 \neq 0$, *is*

$$\frac{\mathrm{m}_1}{\mathrm{m}_2} = \langle c_z, t, g_z, n, b \rangle, \tag{8.7}$$

where

- *the center is the digital quotient of the centers:* $c_z = di(\frac{c_1}{c_2})$,
- *the tolerance* t, *the basis* b *and the number of digits* n *are the same as for the data,*
- *the granularity is* $g_z = \max\{g_1, \frac{g_2}{1-g_2}\} + b^{-n}$.

Proof Obvious from the previous two propositions. ∎

8.7.3 Sum of Operands Having the Same Sign

Suppose $\mathrm{m}_1 = \langle c_1, g_1 \rangle, \mathrm{m}_2 = \langle c_2, g_2 \rangle \in \mathbb{M}(t, n)$ with $sgn(c_1) = sgn(c_2)$.

Definition 8.7.3 (Sum of marks having the same sign) The mark operator over $I^*(\mathbb{R})$ associated to a sum of marks having the same sign is

$$\mathrm{m}_1 + \mathrm{m}_2 = (c_1 + c_2) * [1 + t, 1 - t].$$

Lemma 8.7.5 *Defining* $Z = (c_1 + c_2) * [1 + t, 1 - t] \in I^*(\mathbb{R})$, *then*

$$Iv(\mathrm{m}_1) + Iv(\mathrm{m}_2) \subseteq Z$$

that is to say, the sum is a $\mathbb{MI}$-*centered operator.*

Proof The possible cases are:

1. $c_1 \geq 0$ and $c_2 \geq 0$

$$\begin{aligned} Iv(\mathrm{m}_1) + Iv(\mathrm{m}_2) &= [c_1(1+t), c_1(1-t)] + [c_2(1+t), c_2(1-t)] \\ &= [(c_1+c_2)(1+t), (c_1+c_2)(1-t)] = Z. \end{aligned}$$

2. $c_1 < 0$ and $c_2 < 0$

$$\begin{aligned} Iv(\mathrm{m}_1) + Iv(\mathrm{m}_2) &= [c_1(1-t), c_1(1+t)] + [c_2(1-t), c_2(1+t)] \\ &= [(c_1+c_2)(1-t), (c_1+c_2)(1+t)] = Z. \end{aligned}$$ ∎

Lemma 8.7.6 *The main term of the granularity of the sum in the minimal approach*

$$(c_1 + c_2) * [1 + \gamma_{1,2}, 1 - \gamma_{1,2}] \subseteq (c_1 * [1 + g_1, 1 - g_1]) + [c_2, c_2]$$

and

$$(c_1 + c_2) * [1 + \gamma_{1,2}, 1 - \gamma_{1,2}] \subseteq [c_1, c_1] + (c_2 * [1 + g_2, 1 - g_2])$$

is $\gamma_{1,2} = \max(g_1, g_2)$.

Proof As

$$c_1 * [1 + g_1, 1 - g_1] + [c_2, c_2] = [c_1 + c_2, c_1 + c_2] + [1 + c_1 g_1, 1 - c_1 g_1]$$
$$= (c_1 + c_2) * [1 + \tfrac{c_1}{c_1+c_2} g_1, 1 - \tfrac{c_1}{c_1+c_2} g_1].$$
$$[c_1, c_1] + c_2 * [1 + g_2, 1 - g_2] = [c_1 + c_2, c_1 + c_2] + [1 + c_2 g_2, 1 - c_2 g_2]$$
$$= (c_1 + c_2) * [1 + \tfrac{c_2}{c_1+c_2} g_2, 1 - \tfrac{c_2}{c_1+c_2} g_2].$$

It is necessary that $\gamma_{12} \geq \tfrac{c_1}{c_1+c_2} g_1$ and $\gamma_{12} \geq \tfrac{c_2}{c_1+c_2} g_2$. But c_1 and c_2 have the same sign, then $\tfrac{c_1}{c_1+c_2} g_1 \leq g_1$ and $\tfrac{c_2}{c_1+c_2} g_2 \leq g_2$. So it is enough to take $\gamma_{12} = \max\{g_1, g_2\}$. ∎

Notice that in this proof, the distributive property of the interval product with respect to the sum has been used, keeping in mind that $[c_1, c_1]$ and $[c_2, c_2]$ are pointwise intervals.

Theorem 8.7.3 (Computation algorithm for the sum operator) *The sum over $\mathbb{M}(t, n)$ of two marks $m_1 = \langle c_1, g_1 \rangle$, $m_2 = \langle c_2, g_2 \rangle \in \mathbb{M}(t, n)$ with $\text{sgn}(c_1) = \text{sgn}(c_2)$ is*

$$m_1 + m_2 = \langle c_z, t, g_z, n, b \rangle \tag{8.8}$$

where

- *the center is the digital sum of the centers:* $c_z = di(c_1 + c_2)$,
- *the tolerance t, the basis b and the number of digits n are the same as for the data,*
- *the granularity is* $g_z = \max(g_1, g_2) + b^{-n}$.

Proof Obvious from the previous propositions. ∎

8.7.4 Sum with Operators of Different Signs (Subtraction)

Let $m_1 = \langle c_1, g_1 \rangle$, $m_2 = \langle c_2, g_2 \rangle \in \mathbb{M}(t, n)$ with $\text{sgn}(c_1) \neq \text{sgn}(c_2)$ and $|c_1| \neq |c_2|$.

8.7 Arithmetic Operators

Definition 8.7.4 (Sum of marks of different signs) The mark operator over $I^*(\mathbb{R})$ associated to the sum of different signs is

$$m_1 + m_2 = (c_1 + c_2) * [1 + t, 1 - t].$$

Lemma 8.7.7 *Putting* $Z = (c_1 + c_2) * [1 + t, 1 - t] \in I^*(\mathbb{R})$, *then*

$$Iv(m_1) + Iv(m_2) \subseteq Z,$$

that is to say, the sum is a $\mathbb{MI}$-centered operator.

Proof Without any loss of generality, we may suppose $c_1 > 0$ and $c_2 < 0$. So

$$Iv(m_1) + Iv(m_2) = [c_1(1 + t) + c_2(1 - t), c_1(1 - t) + c_2(1 + t)].$$

and the different possible cases are $|c_1| > |c_2|$ and $|c_1| < |c_2|$.

1. If $|c_1| > |c_2|$, then

$$Z = (c_1 + c_2) * [1 + t, 1 - t] = [(c_1 + c_2)(1 + t), (c_1 + c_2)(1 - t)].$$

For the required inclusion it is necessary that

$$c_1(1 + t) + c_2(1 - t) \geq (c_1 + c_2)(1 + t)$$

and

$$c_1(1 - t) + c_2(1 + t) \leq (c_1 + c_2)(1 - t),$$

i.e.,

$$-c_2 t \geq c_2 t \quad \text{and} \quad c_2 t \leq -c_2 t$$

which holds because $c_2 < 0$.

2. If $|c_1| < |c_2|$, then

$$Z = (c_1 + c_2) * [1 + t, 1 - t] = [(c_1 + c_2)(1 - t), (c_1 + c_2)(1 + t)].$$

For the required inclusion it is necessary that

$$c_1(1 + t) + c_2(1 - t) \geq (c_1 + c_2)(1 - t)$$

and

$$c_1(1 - t) + c_2(1 + t) \leq (c_1 + c_2)(1 + t),$$

i.e.,
$$c_1 t \geq -c_1 t \quad \text{and} \quad -c_1 t \leq c_1 t$$

which holds because $c_1 > 0$. ∎

Lemma 8.7.8 *The main term of the granularity of the sum in the minimal approach*

$$(c_1 + c_2) * [1 + \gamma_{1,2}, 1 - \gamma_{1,2}] \subseteq (c_1 * [1 + g_1, 1 - g_1]) + [c_2, c_2]$$

and

$$(c_1 + c_2) * [1 + \gamma_{1,2}, 1 - \gamma_{1,2}] \subseteq [c_1, c_1] + (c_2 * [1 + g_2, 1 - g_2])$$

is $\gamma_{1,2} = \max\{\left|\frac{|c_1|}{|c_1|-|c_2|}\right| g_1, \left|\frac{|c_2|}{|c_1|-|c_2|}\right| g_2, g_1, g_2\}$

Proof There are two possible cases,

1) $c_1 > 0$ and $c_2 < 0$

$$\begin{aligned}
c_1 * [1 + g_1, 1 - g_1] + [c_2, c_2] &= [c_1(1 + g_1) + c_2, c_1(1 - g_1) + c_2] \\
&= [(c_1 + c_2)\left(1 + \frac{c_1}{c_1+c_2} g_1\right), \\
&\quad (c_1 + c_2)\left(1 - \frac{c_1}{c_1+c_2} g_1\right)] \\
&= (c_1 + c_2) * [1 + \left|\frac{|c_1|}{|c_1|-|c_2|}\right| g_1, 1 \\
&\quad - \left|\frac{|c_1|}{|c_1|-|c_2|}\right| g_1] \\
[c_1, c_1] + c_2 * [1 + g_2, 1 - g_2] &= [c_1 + c_2(1 - g_2), c_1 + c_2(1 + g_2)] \\
&= [(c_1 + c_2)\left(1 - \frac{c_2}{c_1+c_2} g_2\right), \\
&\quad (c_1 + c_2)\left(1 + \frac{c_2}{c_1+c_2} g_2\right)] \\
&= (c_1 + c_2) * [1 + \left|\frac{|c_2|}{|c_1|-|c_2|}\right| g_2, 1 \\
&\quad - \left|\frac{|c_2|}{|c_1|-|c_2|}\right| g_2]
\end{aligned}$$

and by these inequalities,

$$\gamma_{1,2} \geq \left|\frac{|c_1|}{|c_1| - |c_2|}\right| g_1 \quad \text{and} \quad \gamma_{1,2} \geq \left|\frac{|c_2|}{|c_1| - |c_2|}\right| g_2.$$

2) $c_1 < 0$ and $c_2 > 0$ is analogous to (1).

From both results,

$$\gamma_{1,2} = \max\{\left|\frac{|c_1|}{|c_1| - |c_2|}\right| g_1, \left|\frac{|c_2|}{|c_1| - |c_2|}\right| g_2, g_1, g_2\}.$$ ∎

8.7 Arithmetic Operators

Theorem 8.7.4 (Computation algorithm for the sum operator) *The sum over $\mathbb{M}(t,n)$ of two marks $\mathfrak{m}_1 = \langle c_1, g_1 \rangle, \mathfrak{m}_2 = \langle c_2, g_2 \rangle \in \mathbb{M}(t,n)$ with $\mathrm{sgn}(c_1) \neq \mathrm{sgn}(c_2)$ and $|c_1| \neq |c_2|$ is*

$$\mathfrak{m}_1 + \mathfrak{m}_2 = \langle c_z, t, g_z, n, b \rangle, \tag{8.9}$$

where

- *the center is the digital sum of the centers: $c_z = di(c_1 + c_2)$,*
- *the tolerance t, the basis b and the number of digits n are the same as for the data,*
- *the granularity is*

$$g_z = \max\left\{ g_1, g_2, \left|\frac{|c_1|}{|c_1|-|c_2|}\right| g_1, \left|\frac{|c_2|}{|c_1|-|c_2|}\right| g_2 \right\} + b^{-n}.$$

Proof Obvious from the previous propositions. ∎

Remark 8.7.1 If $|c_1| = |c_2|$, the sum of the marks of opposite sign is not defined. Anyway, there exists the possibility of declaring, by fiat, that

$$\mathfrak{m} + (-\mathfrak{m}) - \langle 0, g \rangle \in \mathbb{M}(t,n) \tag{8.10}$$

but this is just a definition and not a computation.

It is important to realize that computations with marks do not provide the true final values, but marks only warn about the quality of results, whether or not they are reliable. This fact is illustrated in the following example.

Example 8.7.1 For $\mathfrak{m} = \langle 3.2e0, 1.0e-2, 1.0e-3, 4, 10 \rangle$ and $\mathfrak{n} = \langle 2.9e0, 1.0e-2, 1.0e-3, 4, 10 \rangle$, marks on DI_4,

$$\mathfrak{m} + \mathfrak{n} = \langle 6.1000e0, 1.0e-2, 1.1e-3, 4, 10 \rangle$$
$$\mathfrak{m} * \mathfrak{n} = \langle 9.2800e0, 1.0e-2, 1.1e-3, 4, 10 \rangle$$
$$\mathfrak{m}/\mathfrak{n} = \langle 1.1035e0, 1.0e-2, 1.101e-3, 4, 10 \rangle$$

but

$$\mathfrak{m} - \mathfrak{n} = \langle 0.3000e0, 1.0e-2, 1.0766e-2, 4, 10 \rangle$$

which is an invalid mark since $\frac{g}{t} = 1.0766 > 1$. Nevertheless with a tolerance $t = 0.05$ this mark is valid.

Using a scale DI_5 with five decimal digits, i.e., for $\mathfrak{m} = \langle 3.2e0, 1.0e-2, 1.0e-3, 5, 10 \rangle$ and $\mathfrak{n} = \langle 2.9e0, 1.0e-2, 1.0e-3, 5, 10 \rangle$ the results are

$$m + n = \langle 6.10000e0, 1.0e-2, 1.01e-3, 5, 10 \rangle$$
$$m * n = \langle 9.28000e0, 1.0e-2, 1.01e-3, 5, 10 \rangle$$
$$m/n = \langle 1.10345e0, 1.0e-2, 1.011e-3, 5, 10 \rangle$$

but

$$m - n = \langle 0.30000e0, 1.0e-2, 1.06767e-2, 5, 10 \rangle$$

which will be a valid or invalid mark depending on the tolerance.

Computing with marks warns and prevents against any undesired subtractive cancellation, as the following examples show.

Example 8.7.2 Coming back to the problem, commented on in Sect. 8.2.4, about the computation of

$$f(x, y) = 333.75y^6 + x^2(11x^2y^2 - y^6 - 121y^4 - 2) + 5.5y^8 + \frac{x}{2y}.$$

In the scale DI_{15}, the computation of the extension of f to the marks

$$\mathfrak{x} = \langle 77617.0, 0.01, 10^{-15}, 15, 10 \rangle \text{ and } \mathfrak{y} = \langle 33096.0, 0.01, 10^{-15}, 15, 10 \rangle$$

yields a mark with center equal to $1.1726039400531e0$ and a huge granularity, greater than 1, and thus greater than the tolerance $t = 0.01$, or any other, i.e., this is an invalid mark.

Example 8.7.3 Let us consider the evaluation of the function $f(x) = (1-\cos x)/x^2$ for $x = 1.2e - 05$ rounded to 10 significant digits [39]. As

$$\cos x = 0.9999999999 \quad \text{and} \quad 1 - \cos x = 0.0000000001,$$

the result is

$$(1 - \cos x)/x^2 = 0.6944..,$$

which is wrong since $0 \leq f(x) \leq 0.5$ for all $x \neq 0$. The cause is the subtraction $1 - \cos x$ which is exact but has only one significant digit and it is the same size as the error in the computation of $\cos x$. The computation with marks yields, as a result for the difference, $1 - \cos x$, a mark with a big granularity, greater than 1, thus it is greater than any valid tolerance and, consequently, an invalid mark.

The next examples illustrate the importance of granularity and the tolerance, depending on the problem. In technical cases the granularity can be defined as the relative error of the corresponding measures. In other cases, the data are not measurements but real values of a coordinate axis and the granularity could

8.7 Arithmetic Operators

be defined as the error in the digital scale. But this can not be made into a fixed rule, and granularity can be strongly conditioned by the problem under consideration.

Example 8.7.4 Coming back to Example 4.5.1 of Chap. 4, for an ideal lens with the distance g between an object and the lens, a focal length of f, and distance b between the lens and the image of the object, all of which are related by

$$g = \frac{1}{\frac{1}{f} - \frac{1}{b}},$$

suppose now f is a mark with center 20 and granularity 0.05, equal to the relative error, and b is a mark with center 25 and granularity $1/25 = 0.04$. For the distance between the object and the lens the result is the mark

$$g = <1.000000e + 02, 3.571429e - 01>$$

which, by (8.3), is a valid mark for a tolerance of $t = 0.6$, but not for a tolerance of $t = 0.3$.

Example 8.7.5 Given the function [47]

$$Œ(y, z) = 108 - (815 - 1500/z)/y$$

and initial values $x_0 = 4$, $x_1 = 4.25$, let $x_{n+1} = Œ(x_n, x_{n-1})$ be for $n = 1, 2, 3, \ldots$. The sequence x_n converges to the limit L ($L = 5$) which can be approximated by computing x_n until x_{N-1} differs negligibly from x_N; then this x_N approximates L. So, the task is to compute x_N for some moderately big integer N, say $N = 80$. It seems that all floating-point hardware, all Randomized Arithmetic, and most implementations of Significance Arithmetic, give $L = 100$. With a FORTRAN program carrying 64 bits and a MATLAB program carrying 53 bits on an Intel 302 (i386/387 IBM PC clone).

n	True x_n	FORTRAN x_n	MATLAB x_n
...	...	...	...
12	4.9956558915066	4.9956595420973	4.9674550955522
...	...	...	...
74	4.9999999999999	100	100
75	4.9999999999999	100	100
...	...	...	...

So, different calculations produce the same wrong result $x_{80} \approx 100$. Solving the problem with marks instead of real numbers, this result is labelled as possible false.

Considering the initial marks

$$x_0 = < 4.0, 1.0e - 15 > \qquad x_1 = < 4.25, 1.0e - 15 >$$

with tolerance 0.01, number of digits $n = 15$, and basis $b = 10$ of the digital scale, the results are

n	x_n	
0	$< 4.0, 1.0e - 15 >$	
1	$< 4.25, 1.0e - 15 >$	
2	$< 4.470588e + 00, 4.762632e - 13 >$	
3	$< 4.644737e + 00, 1.062112e - 11 >$	
4	$< 4.770538e + 00, 2.298526e - 10 >$	
5	$< 4.855701e + 00, 4.882527e - 09 >$	
6	$< 4.910847e + 00, 1.024947e - 07 >$	
7	$< 4.945537e + 00, 2.135771e - 06 >$	
8	$< 4.966962e + 00, 4.430382e - 05 >$	
9	$< 4.980042e + 00, 9.165344e - 04 >$	
10	$< 4.987909e + 00, 1.894596e - 02 >$	
11	$< 4.991363e + 00, 3.985458e - 01 >$	invalid mark
12	$< 4.967455e + 00, 1.374410e + 01 >$	invalid mark
...	...	

For $n = 11$, the mark is invalid because its granularity is greater than the tolerance. Taking a greater tolerance, $t = 0.5$, this mark would become a valid mark. The result corresponding to step $n = 12$ is again an invalid mark because its granularity is, not only greater than the tolerance, but it is greater than 1 and no tolerance could make this mark valid. Due to the recursive nature of the computation, all marks from x_{11} represent possibly wrong results. For example the center of the mark computed for $n = 13$ is 4.429690 and the center of the mark computed for $n = 14$ is -7.817237.

Example 8.7.6 The graph of the function [47]

$$Spike(x) = 1 + x^2 + \log(|1 + 3(1 - x)|)/80.$$

presents a "spike" at $x = 4/3$, i.e., $Spike(4/3) = -\infty$. Drawing this function taking a set of x-values, for example $x_n = 1/2 + n/669$, for $n = 1, 2, 3, \ldots, 1003$, the result is a continuous and smooth curve in the interval $[1/2 + 1/669, 3/2]$, in which the spike is not detected despite the fact that $4/3 \in [1/2 + 1/669, 3/2]$. Computations with marks can solve this problem. Considering these x-values as the centers of the marks, with tolerance $1.0e - 2$, number of digits $n = 15$, and basis $b = 10$ of the digital scale, and different values for the granularity, the results for the x-values near 4/3 are in the following tables.

8.7 Arithmetic Operators

For a granularity of $1.0e-3$, the union of all the set-intervals defined by the mark centers and granularity, that is,

$$\bigcup_{(n,\{1,2,\ldots,1003\})} (x_n * [1-g, 1+g]),$$

contains the represented segment of the real line and the mark $< 1.332586e + 00, 1.0e - 3 >$, considered as a "brush-stroke" of x-values, contains the spike-point $x = 4/3$. The results are:

Granularity = $1.0e - 3$		
x	$Spike(x)$	
...	...	
$< 1.314649e + 00, 1.0e - 3 >$	$< 2.692283e + 00, 7.036000e - 02 >$	
$< 1.316143e + 00, 1.0e - 3 >$	$< 2.695173e + 00, 7.656522e - 02 >$	
$< 1.317638e + 00, 1.0e - 3 >$	$< 2.697973e + 00, 8.395238e - 02 >$	
$< 1.319133e + 00, 1.0e - 3 >$	$< 2.700663e + 00, 9.289474e - 02 >$	invalid mark
$< 1.320628e + 00, 1.0e - 3 >$	$< 2.703219e + 00, 1.039412e - 01 >$	invalid mark
...	...	
$< 1.332586e + 00, 1.0e - 3 >$	$< 2.699531e + 00, 1.783000e + 00 >$	invalid mark
$< 1.334081e + 00, 1.0e - 3 >$	$< 2.703517e + 00, 1.785000e + 00 >$	invalid mark
...	...	
$< 1.346039e + 00, 1.0e - 3 >$	$< 2.770982e + 00, 1.059412e - 01 >$	invalid mark
$< 1.347534e + 00, 1.0e - 3 >$	$< 2.776398e + 00, 9.489474e - 02 >$	invalid mark
$< 1.349028e + 00, 1.0e - 3 >$	$< 2.781680e + 00, 8.595238e - 02 >$	
$< 1.350523e + 00, 1.0e - 3 >$	$< 2.786853e + 00, 7.856522e - 02 >$	
$< 1.352018e + 00, 1.0e - 3 >$	$< 2.791934e + 00, 7.236000e - 02 >$	
$< 1.353513e + 00, 1.0e - 3 >$	$< 2.796941e + 00, 6.707407e - 02 >$	
...	...	

There exists a group of invalid marks, between $< 1.319133e + 00, 1.0e - 3 >$ and $< 1.347534e + 00, 1.0e - 3 >$ of which the granularities are greater than the chosen tolerance. This fact indicates that the results of the evaluation of $Spike(x)$ are possibly wrong.

Note as well that by taking a greater tolerance, some of these marks would become valid. But the tolerance can not be greater than 1, therefore, there does not exist any tolerance which validates the mark $< 2.699531e + 00, 1.783000e + 00 >$ because its granularity is greater than 1.

For a granularity of $1.0e - 4$ the union of all the marks does not cover the real line and the mark $< 1.332586e + 00, 1.0e - 4 >$ does not contain the spike-point $x = 4/3$. The results are:

Granularity = 1.0e − 4	
x	Spike(x)
...	...
< 1.326607e + 00, 1.0e − 4 >	< 2.711097e + 00, 1.972222e − 02 >
< 1.328102e + 00, 1.0e − 4 >	< 2.711924e + 00, 2.538571e − 02 >
< 1.329596e + 00, 1.0e − 4 >	< 2.711691e + 00, 3.558000e − 02 >
< 1.331091e + 00, 1.0e − 4 >	< 2.709282e + 00, 5.936667e − 02 >
< 1.332586e + 00, 1.0e − 4 >	< 2.699531e + 00, 1.783000e − 01 > invalid mark
< 1.334081e + 00, 1.0e − 4 >	< 2.703517e + 00, 1.785000e − 01 > invalid mark
< 1.335575e + 00, 1.0e − 4 >	< 2.721241e + 00, 5.956667e − 02 >
< 1.337070e + 00, 1.0e − 4 >	< 2.731621e + 00, 3.578000e − 02 >
< 1.338565e + 00, 1.0e − 4 >	< 2.739826e + 00, 2.558571e − 02 >
< 1.340060e + 00, 1.0e − 4 >	< 2.746972e + 00, 1.992222e − 02 >
...	...

Although the spike-point is not inside any mark, the group of invalid marks, indicating that the results for $Spike(x)$ are possibly wrong, has been reduced to two marks $< 1.332586e + 00, 1.0e − 4 >$ and $< 1.334081e + 00, 1.0e − 3 >$. For any greater tolerance, for example $t = 0.2$, both marks would become valid and the spike would have been hidden.

For a granularity of $1.0e − 5$, the results are:

Granularity = 1.0e − 5	
x	Spike(x)
...	...
< 1.326607e + 00, 1.0e − 5 >	< 2.711097e + 00, 1.972222e − 03 >
< 1.328102e + 00, 1.0e − 5 >	< 2.711924e + 00, 2.538571e − 03 >
< 1.329596e + 00, 1.0e − 5 >	< 2.711691e + 00, 3.558000e − 03 >
< 1.331091e + 00, 1.0e − 5 >	< 2.709282e + 00, 5.936667e − 03 >
< 1.332586e + 00, 1.0e − 5 >	< 2.699531e + 00, 1.783000e − 02 >
< 1.334081e + 00, 1.0e − 5 >	< 2.703517e + 00, 1.785000e − 02 >
< 1.335575e + 00, 1.0e − 5 >	< 2.721241e + 00, 5.956667e − 03 >
< 1.337070e + 00, 1.0e − 5 >	< 2.731621e + 00, 3.578000e − 03 >
< 1.338565e + 00, 1.0e − 5 >	< 2.739826e + 00, 2.558571e − 03 >
...	...

Now the group of invalid marks indicating wrong results for $Spike(x)$ does not exist and the spike has been hidden. This is the same problem which existed for computations with the real numbers: the spike-point $x = 4/3$ has not been taken into account.

The next examples illustrate the importance of the digital scale for the validity of the results.

8.7 Arithmetic Operators

Example 8.7.7 Given the following systems of equations

$$\begin{cases} 0.003x_1 + 59.14x_2 = 59.17 \\ 5.291x_1 - 6.13x_2 \ \ = 46.78 \end{cases}$$

let us study its solution from the standpoint of marks, using the pivot method. Let us suppose that the coefficients and the right hand sides are marks of a digital scale DI_3 with a granularity $g = 1.0e - 3$, and let us also consider a tolerance $t = 0.01$. The system would be expressed in terms of marks as

$$\begin{cases} \langle 3.000e - 3, 1e - 3 \rangle x_1 + \langle 5.914e + 1, 1e - 3 \rangle x_2 = \langle 5.917e + 1, 1e - 3 \rangle \\ \langle 5.291, 1e - 3 \rangle x_1 + \langle -6.130e0, 1e - 3 \rangle x_2 \ \ \ \ \ \ \ \ \ \ \ = \langle 4.678e + 1, 1e - 3 \rangle. \end{cases}$$

Solution 1 Using a scale DI_3 with three decimal digits and taking a bad pivot, e.g.,

$$p = \frac{\langle 5.291e0, 1e - 3 \rangle}{\langle 3.000e - 3, 1e - 3 \rangle} = \langle 1.764e3, 2.e - 3 \rangle,$$

the second equation becomes

$$(\langle -6.130e0, 1.e - 3 \rangle + \langle -5.914e + 1, 1.e - 3 \rangle \cdot p)x_2$$
$$= \langle 4.678e + 1, 1.e - 3 \rangle + \langle -5.917e + 1, 1.e - 3 \rangle \cdot p$$

then

$$x_2 = \langle 1.001, 5.e - 3 \rangle$$

and

$$x_1 = \frac{\langle 5.917e + 1, 1.e - 3 \rangle + \langle -5.914e + 1, 1.e - 3 \rangle \cdot x_2}{\langle 3.000e - 3, 1.e - 3 \rangle}$$
$$= \langle 1.000e + 1, 1.184e1 \rangle.$$

In this case the granularity of the mark x_1 is bigger than 1, therefore this mark is not valid no matter what the tolerance is.

Solution 2 Using a scale DI_6 with six decimal digits, with the same bad pivot.

$$p = \frac{\langle 5.291000e0, 1.e - 3 \rangle}{\langle 3.000000e - 3, 1.e - 3 \rangle} = \langle 1.763667e + 3, 1.002001e - 3 \rangle,$$

the second equation becomes

$$(\langle -6.130000e0, 1.e - 3 \rangle + \langle -5.914000e + 1, 1.e - 3 \rangle \cdot p)x_2$$
$$= \langle 4.678000e + 1, 1.e - 3 \rangle + \langle -5.917000e + 1, 1.e - 3 \rangle \cdot p$$

The solution in this case is

$$x_2 = \langle 1.000000e0, 1.04\ldots e-3\rangle$$

and

$$x_1 = \frac{\langle 5.917000e+1, 1.e-3\rangle + \langle -5.914000e+1, 1.e-3\rangle \cdot x_2}{\langle 3.000000e-3, 1.e-3\rangle}$$
$$= \langle 1.00000e+1, 2.06\ldots e0\rangle.$$

Again the mark x_1 is not valid, since the granularity is bigger than 1. Paradoxically, as the "exact" resolution of the initial real system is $x_1 = 10$ and $x_2 = 1$, apparently a "good" solution is being rejected, i.e., a computation with marks can apparently provokes sometimes a "false alarm" about the correctness of a result. This is not true and the computation with marks warns about what is happening. On the one hand, solving the real system in an exact way, is made under the hypothesis that the coefficients have more than three significant digits, i.e., they have as many digits as the computer provides. This is equivalent to taking the granularity of the data to be much less than 10^{-3}, which is wrong because the data has only three digits, and to suppose that the digits to the right of the third digit are zeros is as arbitrary as to suppose that they are any other numbers.

Solution 3 Using a scale DI_3 with three decimal digits, with the good pivot

$$p = \frac{\langle 3.000e-3, 1.e-3\rangle}{\langle 5.291e0, 1.e-3\rangle} = \langle 5.670e-4, 2.e-3\rangle.$$

the first equation becomes

$$(\langle -5.914e+1, 1.e-3\rangle + \langle 6.130e0/, 1.e-3\rangle \cdot p)x_2$$
$$= \langle -5.917e+1, 1.e-3\rangle + \langle -4.678e+1, 1.e-3\rangle \cdot p$$

then

$$x_2 = \langle 1.001e0, 5.e-3\rangle$$

and starting from the second equation

$$\langle 5.291e0, 1.e-3\rangle x_1 = \langle 6.130e0, 1.e-3\rangle \cdot \langle 1.001e0, 5.e-3\rangle + \langle 4.678e+1, 1.e-3\rangle$$
$$x_1 = \frac{\langle 5.292e+1, 7.e-3\rangle}{\langle 5.291e0, 1.e-3\rangle} = \langle 1.000e+1, 8.e-3\rangle.$$

Marks x_1 and x_2 are valid under the minimal approach since $g < t$. One could think that these valid results reinforce the reasoning that, in solution 2, good solutions have been rejected. It is necessary to insist on the falsity of this reasoning. The pivot for this third process does guarantee the validity. The pivot chosen in the process 2 does not.

With similar proofs, the main term of the granularity when the maximal or maximum granularity approaches are used can be obtain. The following table summarizes the resulting formulas.

Operator	Minimal	Maximal
Product	$g^M_{1,2}$	$g_1 + g_2 + g_1g_2$
Quotient	$\max\{g_1, \dfrac{g_2}{1-g_2}\}$	$\dfrac{g_1+g_2}{1-g_2}$
Sum	$g^M_{1,2}$	$\max\left\{g^M_{1,2}, \left\|\dfrac{c_1g_1+c_2g_2}{c_1+c_2}\right\|\right\}$
Substraction	$\max\left\{g^M_{1,2}, \left\|\dfrac{\|c_1\|g_1}{\|c_1\|-\|c_2\|}\right\|, \left\|\dfrac{\|c_2\|g_2}{\|c_1\|-\|c_2\|}\right\|\right\}$	$\max\left\{g^M_{1,2}, \left\|\dfrac{\|c_1\|g_1+\|c_2\|g_2}{\|c_1\|-\|c_2\|}\right\|\right\}$

The granularities obtained with the maximal approach are bigger than the granularities obtained with the minimal approach, i.e., the maximal approach describes as invalid some marks which, with other approaches, could be valid marks. This is not inconsistent, but simply means that it is necessary to know the reason for choosing one or another approach.

8.8 Semantic Interpretations

In this section an important feature is presented: the logical meaning of the result of any mark arithmetic operator related with the indistinguishable points of the operands. A previous basic property is the following theorem.

Theorem 8.8.1 *If $f_{M(t,n)}$ is a mark operator and $\mathfrak{z} = f_{M(t,n)}(\mathfrak{x}_1, \mathfrak{x}_2)$, then*

$$f^*(Iv(\mathfrak{x}_1), Iv(\mathfrak{x}_2)) \subseteq Exsh(\mathfrak{z}).$$

Proof As $f^*(Iv(\mathfrak{x}_1), Iv(\mathfrak{x}_2)) \subseteq f(c_1, c_2) * [1+t, 1-t]$ and

$$f(c_1, c_2) \in di(f(c_1, c_2)) * \left[1 - \frac{b^{-n}}{2}, 1 + \frac{b^{-n}}{2}\right]$$

the inclusion

$$f(c_1,c_2) * [1+t, 1-t] \subseteq di(f(c_1,c_2)) * [1+t, 1-t] * \left[1 - \frac{b^{-n}}{2}, 1 + \frac{b^{-n}}{2}\right].$$

is true. Moreover and because $g_z \geq \frac{b^{-n}}{2}$, then

$$[1 - \frac{b^{-n}}{2}, 1 + \frac{b^{-n}}{2}] \subseteq [1 - g_z, 1 + g_z]$$

therefore

$$f(c_1,c_2) * [1+t, 1-t] \subseteq di(f(c_1,c_2)) * [1+t, 1-t] * [1 - g_z, 1 + g_z]$$

and the inclusion $f^*(Iv(\mathfrak{x}_1), Iv(\mathfrak{x}_2)) \subseteq Exsh(\mathfrak{z})$ is true. ∎

Corollary 8.8.1 (Interval-semantics of the mark operators) *Under the hypotheses of the previous theorem, Theorem 8.8.1, we have*

$$\forall (z, Exsh'(\mathfrak{z}))\ \exists (x_1, Iv'(\mathfrak{x}_1))\ \exists (x_2, Iv'(\mathfrak{x}_2))\ z = f(x_1, x_2).$$

Proof Apply the interval *-semantic theorem to the previous inclusion, taking into account the improper modality of the intervals $Iv(\mathfrak{x}_1)$, $Iv(\mathfrak{x}_2)$ and $Exsh(\mathfrak{z})$. ∎

Theorem 8.8.2 (Mark-semantics of the mark operators) *If $f_{\mathbb{M}(t,n)}$ is a mark operator on $\mathbb{M}(t,n)$, and $\mathfrak{z} = f_{\mathbb{M}(t,n)}(\mathfrak{x}_1, \mathfrak{x}_2)$, then*

$$\forall (z, Exsh'(\mathfrak{z}))\ z \in (f(c_1,c_2) * [1-t, 1+t])'.$$

Proof From the inclusion

$$[f(c_1,c_2), f(c_1,c_2)] \subseteq di(f(c_1,c_2)) * \left[1 - \frac{b^{-n}}{2}, 1 + \frac{b^{-n}}{2}\right],$$

multiplying both members by $[1+t, 1-t]$, the inclusivity of the product leads to

$$f(c_1,c_2) * [1+t, 1-t] \subseteq di(f(c_1,c_2)) * [1+t, 1-t] * [1 - \frac{b^{-n}}{2}, 1 + \frac{b^{-n}}{2}]$$
$$\subseteq di(f(c_1,c_2)) * [1+t, 1-t] * [1 - g_z, 1 - g_z]$$
$$= Exsh(\mathfrak{z}).$$

As the modality of the involved intervals is improper,

$$Exsh(\mathfrak{z})' \subseteq (f(c_1,c_2) * [1+t, 1-t])'.$$

This means the logical formula is true. ∎

Example 8.8.1 For $\mathfrak{x} = \langle 3.2e0, 5.0e-2, 1.0e-3, 4, 10 \rangle$ and $\mathfrak{y} = \langle 2.9e0, 5.0e-2, 1.0e-3, 4, 10 \rangle$, marks on DI_4,

$$\mathfrak{z} = \mathfrak{x} - \mathfrak{y} = \langle 0.3000e0, 5.0e-2, 1.0667e-2, 4, 10 \rangle$$

which is a valid mark. As

$$Exsh'(\mathfrak{z}) = \text{Prop}(0.3000 * [1+0.05, 1-0.05] * [1-0.010667, 1+0.010667])$$
$$= \text{Prop}([0.3116, 0.2880])$$
$$= [0.2880, 0.3116]'$$

and

$$f(c_1, c_2) * [1-t, 1+t] = 0.3000 * [1-0.05, 1+0.05]) = [0.2850, 0.3150],$$

it is true that

$$\forall (z, [0.2880, 0.3116]') \; z \in [0.2850, 0.3150]'.$$

Remark 8.8.1 The minimum condition of significance $h^{-n} \le g < t < 1$, which decides the validity of the mark, makes $Exsh(\mathfrak{z})$ an improper interval. If the mark is not valid, $Exsh(\mathfrak{z})$ is a proper interval and the semantic is

$$\exists (z, Exsh'(\mathfrak{z})) \; \exists (x_1, Iv'(\mathfrak{x}_1)) \; \exists (x_2, Iv'(\mathfrak{x}_2)) \; z = f(x_1, x_2)$$

and no particular point of $Exsh(\mathfrak{z})'$ can be considered as a "good" result of the computation.

8.9 Functions of Marks

Definition 8.9.1 (Function of marks) Let $f : \mathbb{R}^k \to \mathbb{R}$ be a real continuous function such that the operators of its syntactic tree admit associated mark operators (see Definition 8.5.1), and $\mathfrak{x}_1, \ldots, \mathfrak{x}_k \in \mathbb{M}(t, n)$. The *function of marks* associated to a real continuous function f with arguments $x_1, \ldots, x_k$ will be represented by $f_{\mathbb{M}(t,n)}(\mathfrak{x}_1, \ldots, \mathfrak{x}_k)$ and is a function $f_{\mathbb{M}(t,n)} : \mathbb{M}(t, n)^k \to \mathbb{M}(t, n)$ such that

1. each variable x_i of f is replaced by the corresponding mark $\mathfrak{x}_i$, considering as independent every incidence of $\mathfrak{x}_i$,
2. each operator of the syntactic tree of f is replaced by the corresponding mark operator over $\mathbb{M}(t, n)$.

Important functions of marks are power series, which allows of defining the extension to marks of the main one-variable operators such as log, exp, pow, sin, cos, tan, arcsin, arccos, arctan, etc.

Lemma 8.9.1 *Let* $\mathfrak{x} = \langle c, t, g, n, b \rangle \in \mathbb{M}(t, n)$ *be a mark and let*

$$\tilde{\mathfrak{x}} = \langle c, t - g(1+t), b^{-n}, n, b \rangle \in \mathbb{M}(t - g(1+t), n)$$

be the associated mark to the external shadow of $\mathfrak{x}$. *The inclusion*

$$Exsh(\mathfrak{x}) \subseteq Iv(\tilde{\mathfrak{x}})$$

is true and, if $\mathfrak{y}$ *is a mark with center c and tolerance* ξ *such that* $\xi > \tilde{t}$, *then*

$$Exsh(\mathfrak{x}) \not\subseteq Iv(\mathfrak{y}).$$

Proof By definition,

$$Exsh(\mathfrak{x}) = c * [1 + t - g - gt, 1 - t + g - gt]$$
$$Iv(\tilde{\mathfrak{x}}) = c * [1 + t - g(1+t), 1 - t + g(1+t)]$$
$$= c * [1 + t - g - gt, 1 - t + g + gt]$$

and since

$$1 + t - g - gt \geq 1 + t - g - gt$$
$$1 - t + g - gt \leq 1 - t + g + gt$$

then $Exsh(\mathfrak{x}) \subseteq Iv(\tilde{\mathfrak{x}})$. On the other hand, if $\mathfrak{y} = \langle c, \xi, g', n', b' \rangle$ is a mark satisfying $\xi > \tilde{t}$, then

$$Exsh(\mathfrak{x}) = c * [1 + t - g - gt, 1 - t + g - gt]$$
$$= c * [1 + \tilde{t}, 1 - t + g - gt]$$
$$\not\subseteq c * [1 + \xi, 1 - \xi]. \qquad \blacksquare$$

The next theorem provides a semantic interpretation to the result of a function of marks. It is interesting to emphasize that the semantics of the mark operators, see Theorem 8.8.2 and Corollary 8.8.1, can not be applied recursively, since the universal quantifier of these semantics corresponds to the external shadow of the result, which is not a mark. So it is not possible to continue with the semantics starting from this external shadow, but with the mark associated to this external shadow. For that reason, the following semantic theorem, in the particular case when the syntax tree of the function f has a unique operator, does not provide the same semantic obtained for a mark operator.

8.9 Functions of Marks

Theorem 8.9.1 (Semantic for a function of marks) *Let us consider the function of marks* $f_{\mathbb{M}(t,n)} : \mathbb{M}(t,n)^k \longrightarrow \mathbb{M}(t,n)$. *If* $\mathfrak{z} = f_{\mathbb{M}(t,n)}(\mathfrak{x}_1, \ldots, \mathfrak{x}_k)$ *and* $\tilde{\mathfrak{z}}$ *is the associated mark to the shadow of* $\mathfrak{z}$, *supposing that all the involved marks are valid, then*

$$(\forall z \in Iv'(\tilde{\mathfrak{z}})) \ (\exists x_1 \in Iv'(\mathfrak{x}_1)) \ldots (\exists x_k \in Iv'(\mathfrak{x}_k)) \ z = f(x_1, \ldots, x_k).$$

Proof Only the proof for one step of the syntactic tree of f will be developed, since the general proof is a simple induction process based on this step. If $\mathfrak{z} = f_{\mathbb{M}(t,n)}(\mathfrak{x}_1, \mathfrak{x}_2)$ and $\tilde{\mathfrak{x}}_1, \tilde{\mathfrak{x}}_2, \tilde{\mathfrak{z}}$ are the associated marks to the shadows of $\mathfrak{x}_1, \mathfrak{x}_2$ and $\mathfrak{z}$, then denoting $(t - \max(g_1, g_2))(1+t))$ by $\tilde{t}$, and considering the following inclusion

$$f(c_1, c_2) * [1 + \tilde{t}, 1 - \tilde{t}] \subseteq di(f(c_1, c_2)) * [1 + \tilde{t}, 1 - \tilde{t}] * [1 - \tfrac{b^{-n}}{2}, 1 + \tfrac{b^{-n}}{2})$$
$$= di(f(c_1, c_2)) * [(1 + \tilde{t})(1 - \tfrac{b^{-n}}{2}), (1 - \tilde{t})(1 + \tfrac{b^{-n}}{2})],$$

since $g_z \geq \max\{g_1, g_2\} + b^{-n}$, then

$$(1 + \tilde{t})(1 - \frac{b^{-n}}{2}) \geq 1 + t - g_z(1 + t)$$

and

$$(1 - \tilde{t})(1 + \frac{b^{-n}}{2}) \leq 1 - t + g_z(1 + t)$$

and therefore

$$f(c_1, c_2) * [1 + \tilde{t}, 1 - \tilde{t}] \subseteq Iv(\tilde{\mathfrak{z}}).$$

Thus

$$f^*(Iv(\tilde{\mathfrak{x}}_1), Iv(\tilde{\mathfrak{x}}_2)) \subseteq f^*(c_1 * [1 + \tilde{t}, 1 - \tilde{t}], c_2 * [1 + \tilde{t}, 1 - \tilde{t}])$$
$$\subseteq f(c_1, c_2) * [1 + \tilde{t}, 1 - \tilde{t}]$$
$$\subseteq Iv(\tilde{\mathfrak{z}}).$$

Since $Iv(\tilde{\mathfrak{x}}_1)' \subseteq Iv(\mathfrak{x}_1)'$ and $Iv(\tilde{\mathfrak{x}}_2)' \subseteq Iv(\mathfrak{x}_2)'$, an induction process provides the desired semantic. ∎

Remark 8.9.1 For the mark $\tilde{\mathfrak{z}} = \langle c_z, \tilde{t}, b^{-n}, n, b \rangle$, where $\tilde{t} = t - g(1+t)$ is the effective tolerance, the condition of significance $b^{-n} < \tilde{t}$, which decides the validity of the mark $\tilde{\mathfrak{z}}$, makes $Iv(\tilde{\mathfrak{z}})$ an improper interval because $Iv(\tilde{\mathfrak{z}}) = c_z * [1 + t - g(1 + t), 1 - t + g(1+t)]$ improper is equivalent to

$$1 + t - g(1+t) > 1 - t + g(1+t) \Leftrightarrow g < \frac{t}{1+t}.$$

If the mark is not valid, $Iv(\tilde{\mathfrak{z}})$ is a proper interval and the semantics is

$$(\exists z \in Iv'(\tilde{\mathfrak{z}})) \, (\exists x_1 \in Iv'(\mathfrak{x}_1)) \ldots (\exists x_k \in Iv'(\mathfrak{x}_k)) \quad z = f(x_1, \ldots, x_k).$$

and no particular point of $Iv(\tilde{\mathfrak{z}})'$ can be consider as a "good" result of the computation.

Example 8.9.1 The mark extension of the continuous function

$$f(x, y) = \frac{x+y}{x-y}$$

to the marks

$$\mathfrak{x} = \langle 3.2e0, 1.0e-2, 1.0e-3, 5, 10 \rangle$$

and

$$\mathfrak{y} = \langle 3.9e0, 1.0e-2, 1.0e-3, 5, 10 \rangle,$$

using a scale DI_5 with five decimal digits, is

$$\tilde{\mathfrak{z}} = \langle -1.01429e+1, 1.0e-2, 5.60466e-3, 5, 10 \rangle$$

which is a valid mark because its granularity is less than its effective tolerance. As

$$Iv(\mathfrak{x}) = [3.232, 3.168]$$
$$Iv(\mathfrak{y}) = [3.939, 3.861]$$
$$Iv(ExshM(\tilde{\mathfrak{z}})) = [-10.0988, -10.1869],$$

the semantics of this result is

$$(\forall z \in [-10.1869, -10.0988]') \, (\exists x \in [3.168, 3.232]') \, (\exists y \in [3.861, 3.939]')$$
$$z = \frac{x+y}{x-y}$$

and, therefore, every point of the interval $[-10.1869, -10.0988]'$ is a valid result. The mark extension of the same function to the marks

$$\mathfrak{x} = \langle 3.2e0, 1.0e-2, 1.0e-3, 5, 10 \rangle$$

and

$$\mathfrak{y} = \langle 3.5e0, 1.0e-2, 1.0e-3, 5, 10 \rangle,$$

8.9 Functions of Marks

using a scale DI_5 with five decimal digits, is

$$\mathfrak{z} = \langle -2.23333e + 1, 1.0e - 2, 1.18064e - 2, 5, 10 \rangle$$

which is an invalid mark because its granularity is greater than its effective tolerance. The consequence of this fact is that

$$Iv(\mathfrak{x}) = [3.232, 3.168]$$
$$Iv(\mathfrak{y})) = [3.535, 3.465]$$
$$Iv(ExshM(\mathfrak{z})) = [-22.2904, -22.3763]$$

and, due to the proper modality of the interval $Iv(ExshM(\mathfrak{z}))$, the semantics of this result is

$$(\exists z \in [-22.2904, -22.3763]')\ (\exists x \in [3.168, 3.232]')\ (\exists y \in [3.465, 3.535]')$$
$$z = \frac{x+y}{x-y}$$

and not every value of $[-22.2904, -22.3763]'$ is valid, but only some unknown value, i.e., it is not possible to know the result of the computation.

It is important to observe that the semantic interpretation is valid whatever the used approach to compute the granularity, although the interval to which this semantic interpretation is applied is modified.

Lemma 8.9.2 *Under the maximal approach of the computation of the granularity, given* $f : \mathbb{R}^k \to \mathbb{R}$ *a real continuous centered function, for the marks* $\mathfrak{x}_1, \ldots, \mathfrak{x}_k \in \mathbb{M}(t, n)$, *if* $\langle c_z, g_z \rangle = f_{\mathbb{M}(t,n)}(\mathfrak{x}_1, \ldots, \mathfrak{x}_k)$, *and* g_z *is compatible with* αt, *we have that*

$$\langle c_z, g_z \rangle \approx_\alpha f_{\mathbb{M}(t,\infty)}(\mathfrak{x}_1, \ldots, \mathfrak{x}_k).$$

Proof Let us consider the function f that has as syntactic tree

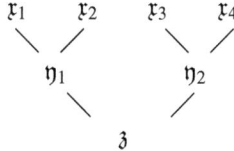

since in a general case we will proceed by induction. If

$$\left.\begin{array}{l}\mathfrak{x}_1 = \langle c_1, g_1\rangle \\ \mathfrak{x}_2 = \langle c_2, g_2\rangle\end{array}\right\} \Rightarrow \mathfrak{y}_1 = f_{1\mathbb{M}(t,n)}(\mathfrak{x}_1, \mathfrak{x}_2) = \langle dif_1(c_1, c_2), g_{y1}\rangle$$

$$\left.\begin{array}{l}\mathfrak{x}_3 = \langle c_3, g_3\rangle \\ \mathfrak{x}_4 = \langle c_4, g_4\rangle\end{array}\right\} \Rightarrow \mathfrak{y}_2 = f_{2\mathbb{M}(t,n)}(\mathfrak{x}_3, \mathfrak{x}_4) = \langle dif_2(c_3, c_4), g_{y2}\rangle$$

$$\mathfrak{z} = \langle c_z, g_z\rangle \Rightarrow \mathfrak{z} = f_{\mathbb{M}(t,n)}(\mathfrak{y}_1, \mathfrak{y}_2) = \langle dif(dif_1(c_1, c_2), dif_2(c_3, c_4)), g_z\rangle,$$

then

$$dif(dif_1(c_1, c_2), dif_2(c_3, c_4))$$
$$\supseteq f(dif_1(c_1, c_2), dif_2(c_3, c_4)) * \text{Impr}(1 \pm \frac{b^{-n}}{2})$$
$$\supseteq fR(f_1(c_1, c_2) * \text{Impr}(1 \pm \frac{b^{-n}}{2}), f_2(c_3, c_4) * \text{Impr}(1 \pm \frac{b^{-n}}{2})) * \text{Impr}(1 \pm \frac{b^{-n}}{2})$$
$$\supseteq fR(f_1(c_1, c_2) * \text{Impr}(1 \pm g_{y1}), f_2(c_3, c_4) * \text{Impr}(1 \pm g_{y2})) * \text{Impr}(1 \pm \frac{b^{-n}}{2})$$

and, from the maximal approach in the computation of the granularity,

$$fR(f_1(c_1, c_2) * \text{Impr}(1 \pm g_{y1}), f_2(c_3, c_4) * \text{Impr}(1 \pm g_{y2}))$$
$$\supseteq f(f_1(c_1, c_2), f_2(c_3, c_4)) * \text{Impr}(1 \pm \gamma_{y1,y2}). \tag{8.11}$$

Applying this last inclusion (8.11) to the previous result, one has

$$dif(dif_1(c_1, c_2), dif_2(c_3, c_4))$$
$$\supseteq f(f_1(c_1, c_2), f_2(c_3, c_4)) * \text{Impr}(1 \pm \gamma_{y1,y2}) * \text{Impr}(1 \pm \frac{b^{-n}}{2})$$
$$\supseteq f(f_1(c_1, c_2), f_2(c_3, c_4)) * \text{Impr}(1 \pm g_z),$$

and therefore, since $\alpha t > g_z$,

$$dif(dif_1(c_1, c_2), dif_2(c_3, c_4)) \in f(f_1(c_1, c_2), f_2(c_3, c_4)) * \text{Prop}(1 \pm \alpha t). \qquad \blacksquare$$

Remark 8.9.2 The result of this theorem can be expressed as

$$\exists (y, Ind_\alpha(f_{\mathbb{M}(t,\infty)}(\mathfrak{x}_1, \ldots, \mathfrak{x}_k))) \, c_z = y.$$

8.10 Remarks About Granularity

Among the elements that constitute a mark, granularity deserves a separate comment and especially its evolution through a computation.

Initially a mark comes from a reading made in a measuring device and its granularity is the numeric value that reflects the inaccuracy of the measurement. So it is natural to identify the granularity with the relative error of the reading, based fundamentally on the error of the measuring device but being able to include relative aspects of the same phenomenon.

A second step is the computation process with marks. The resulting granularity determines the validity of the resulting mark, and the quotient between the granularity and the tolerance measures the imprecision of the mark. During the computation process, the granularity will increase in a fashion depending on the approach with which the granularity is computed. In this computation, the phenomenon being studied will dictate the approach to calculating the granularity, with the only imposition that the resulting granularity will never be smaller than any granularity of the data. Accepting this norm, different approaches for the calculation of the granularity can be defined, for example:

- Minimal approach: The resulting granularity reflects the biggest projection of the granularity of each data in the result. This approach reflects situations of the "chain" type in that the imprecision of the process comes from the biggest imprecisions of the data. A situation of this type appears when, for example the resistance of a beam structure is modelled, since the final resistance will become determined by the incidence of the beam less resistant in the structure or, for example the study of string resistances, since the resistance of one rope is the smallest resistance of each one of its points, etc.
- Maximal approach: This is based on a interval treatment of the error propagation. The granularity is a relative error and it follows the rules of error propagation (however, from a interval point of view). This approach will be used when it is important that the granularity of the result takes into account all the granularities of the data and their propagation.
- Maximum granularity approach: The resultant granularity is the maximum of the granularities of the data. This approach is applicable when the marks are simple objects of a calculation scale.
- Statistical approach: This consists in considering the granularities as statistical distribution functions. The operations on marks would give a granularity obtained from operating on those distribution functions.

The system of marks has its own entity to reflect the evolution of the computations starting from readings of numeric scales. In fact, any computation yields a center and a granularity. Computations performed using marks reflect the gradual loss of information due to numerical errors and truncations and they give warnings relevant to taking decisions either on the acceptance of the results or about the utility of searching for more precision to attain the required validity.

A third step is the semantic interpretation of the result of any mark arithmetic operator related with the indistinguishable points of the operands. For any continuous centered function, the corresponding associated function of marks has been defined by substitution of each variable by the corresponding mark and each operator of the syntactic tree of f by the corresponding mark operator. Results about the semantic of a centered function complete the general development of the theory of marks. The semantic interpretation is valid whichever approach to compute the granularity is used, although the interval that this semantic interpretation is applied to may be modified. It is interesting to note that the interpretation is not referred to a centered interval with relative width similar to the value of the granularity. Moreover it is possible to make different studies of possible semantics that can be obtained by applying different approaches to computing the propagation of the granularity.

The theory of mark also opens a door to extend the traditional interval treatment considering marks as bounds of intervals, a fact which will allow ignoring the process of controlling the unavoidable truncations.

Chapter 9
Intervals of Marks

9.1 Introduction

Intervals, whether classical or modal, pretend to represent numerical information in a coherent way and, for that, one of the main problems is rounding. Indeed, using a digital scale with a finite number of digits, computations will have to be rounded in a convenient way. Working with non-interval numeric values, the best rounding is that which guarantees that the obtained value is "the closest" to the theoretical solution. Working with modal intervals the rule of rounding cannot be the same. Traditionally, the rounding process has been always a nuisance inherent in interval computation, but necessary to keep the semantic interpretations that these computations provide. Until now, in a modal interval computation, the best rounding rule has been the outer one, that is to say, if $Y \in I^*(\mathbb{R})$ represents the exact solution, and $Z \in I^*(\mathbb{R})$ the computed solution, it is necessary that $Y \subseteq Z$, which is compatible with the semantics of the extension f^*.

Although the apparent correct rounding is this outer one, there are problems in which outer rounding is not the appropriate solution. For example problems where one interval variable appears with two modalities: proper and improper. In this case no rounding, whether outer or inner, can solve this problem, as Example 8.2.1 proves. Obviously, in classical interval analysis these problems do not appear because the interval variables can not act from two different modal points of view.

In this chapter a system of intervals with mark bounds, closely related to the modal intervals, will be built to allow computations with a finite system of digits without any truncation. These intervals will be called *intervals of marks*, and in their construction marks, as indiscernibility intervals around a point, will be used. The main concepts and construction will be parallel to the construction of the modal intervals.

9.2 Intervals of Marks

Similarly to the construction of modal intervals, where the logical context is formed by the set of real numbers $\mathbb{R}$, the set of set-theoretical intervals $I(\mathbb{R})$ and the set of predicates on the real line $P : \mathbb{R} \to \{0, 1\}$, the logical basis for the construction of intervals of marks will be the set of marks, the set of set-theoretical intervals of marks, and the set of predicates on the set of marks.

Many concepts, properties, and proofs in the construction of intervals of marks will be the same, *mutatis mutandis*, as those for the modal intervals. So, in the following development such proofs will not be given in detail and will be replaced by a reference to the corresponding property of modal intervals.

Definition 9.2.1 (Set-theoretical intervals of marks) Let $\underline{\mathfrak{a}} = \langle \underline{a}, g \rangle$ and $\overline{\mathfrak{a}} = \langle \overline{a}, g \rangle$ be two comparable marks of $\mathbb{M}(t, n, b)$ with the same granularity. A *set-theoretical interval of marks*, with bounds $\underline{\mathfrak{a}}, \overline{\mathfrak{a}} \in \mathbb{M}(t, n, b)$ is

$$\mathfrak{A}' = [\underline{\mathfrak{a}}, \overline{\mathfrak{a}}]' = \{\mathfrak{a} \in \mathbb{M}(t, \infty, b) \mid \mathfrak{a} = \langle a, g' \rangle,\ a \in \mathbb{R},\ g' \in]0, 1],\ \underline{\mathfrak{a}} \leq \mathfrak{a} \leq \overline{\mathfrak{a}}\}.$$

As to specify the value of b is often irrelevant, the set of marks will continue being denoted by $\mathbb{M}(t, n)$, omitting the value of b.

The set of proper intervals of marks will be denoted by $I(\mathbb{M}(t, n))$, abridged to $I(\mathbb{M})$ when the type of the marks is arranged in advance.

In the same way as $\text{Pred}(\mathbb{R})$ denotes the set of the predicates over the real numbers,

$$\text{Pred}(\mathbb{R}) = \{P \mid P : \mathbb{R} \to \{0, 1\}\},$$

$\text{Pred}(\mathbb{M})$ will denotes the set of predicates over the set of marks of the same type $\mathbb{M}(t, n)$.

Definition 9.2.2 (Mark predicates)

$$\text{Pred}(\mathbb{M}) = \{P \mid P : \mathbb{M}(t, n) \to \{0, 1\}\}.$$

A mark $\mathfrak{a} \in \mathbb{M}(t, n)$ can be identified with the set of predicates that it satisfies, that is to say,

$$\mathfrak{a} \leftrightarrow \{P \in \text{Pred}(\mathbb{M}) \mid P(\mathfrak{a}) = 1\}.$$

Since these systems of predicates are defined over a given type of marks, the structure $\text{Pred}(\mathbb{M})$ is isomorphic to the structure of $\text{Pred}(\mathbb{R})$. For this reason the predicates defined over a set of marks of the same type can be simply considered as functions of the centers of the marks.

9.2 Intervals of Marks

A classical interval $X' \in I(\mathbb{R})$ can be identified with the set of the properties that its points satisfy, in any of the forms

$$X' \leftrightarrow \bigcup_{x \in X'} \{P \in \text{Pred}(\mathbb{R}) \mid P(x) = 1\}$$

or

$$X' \leftrightarrow \bigcap_{x \in X'} \{P \in \text{Pred}(\mathbb{R}) \mid P(x) = 1\}.$$

In parallel with these results, a proper interval of marks $\mathfrak{A}' \in I(\mathbb{M})$ can also be identified with the set of predicates that its marks satisfy, according to one of the constructions

$$\mathfrak{A}' \leftrightarrow \bigcup_{\mathfrak{a} \in \mathfrak{A}'} \{P \in \text{Pred}(\mathbb{M}) \mid P(\mathfrak{a}) = 1\}$$

or

$$\mathfrak{A}' \leftrightarrow \bigcap_{\mathfrak{a} \in \mathfrak{A}'} \{P \in \text{Pred}(\mathbb{M}) \mid P(\mathfrak{a}) = 1\}.$$

To relate intervals of marks with the sets $\text{Pred}(\mathfrak{a})$ corresponding to their marks, we define a modal interval of marks.

Definition 9.2.3 (Modal interval of marks) A *modal interval of marks* of the type $\mathbb{M}(t, n)$ is a couple formed by a proper interval of marks $\mathfrak{A}'$, called the *domain* of the modal interval $\text{Set}(\mathfrak{A}) = \mathfrak{A}'$, together with a quantifier $Q_\mathfrak{A}$, called the *modality* $\text{Mod}(\mathfrak{A}) = Q_\mathfrak{A}$,

$$\mathfrak{A} = (\mathfrak{A}', Q_\mathfrak{A}),$$

where $\mathfrak{A}' \in I(\mathbb{M})$, $Q_\mathfrak{A} \in \{\exists, \forall\}$.

The set of modal intervals of marks of the type $\mathbb{M}(t, n)$ is denoted by $I^*(\mathbb{M}(t, n))$, abridged to $I^*(\mathbb{M})$ when the type of the marks is arranged in advance.

This fundamental definition leads naturally to considering the following subsets:

Definition 9.2.4 (Subsets of intervals of marks)

$I_e(\mathbb{M}) = \{(\mathfrak{A}', \exists) \mid \mathfrak{A}' \in I(\mathbb{M})\}$, set of proper intervals of marks

$I_u(\mathbb{M}) = \{(\mathfrak{A}', \forall) \mid \mathfrak{A}' \in I(\mathbb{M})\}$, set of improper intervals of marks

$I_p(\mathbb{M}) = \{[\mathfrak{A}, \mathfrak{A}] \mid \mathfrak{A} \in \mathbb{M}(t, n)\}$, set of degenerate intervals of marks.

An important concept in the study of modal intervals was the modal quantifier Q which associated to every real predicate $P \in \text{Pred}(\mathbb{R})$ a unique hereditary interval predicate $P^*(X) \in \text{Pred}(I^*(\mathbb{R}))$ over $I^*(\mathbb{R})$. Similarly

Definition 9.2.5 (Modal quantifier) For a modal interval of marks $\mathfrak{A} = (\mathfrak{A}', Q_\mathfrak{A}) \in I^*(\mathbb{M})$, the *modal quantifier* associates to each predicate $P \in \text{Pred}(\mathbb{M})$ a unique predicate over $I^*(\mathbb{M})$ through the construction

$$Q(\mathfrak{a}, \mathfrak{A}) P(\mathfrak{a}) \Leftrightarrow \begin{cases} (\exists \mathfrak{a} \in \mathfrak{A}') \ P(\mathfrak{a}) \text{ if } \text{mod}(\mathfrak{A}) = \exists \\ (\forall \mathfrak{a} \in \mathfrak{A}') \ P(\mathfrak{a}) \text{ if } \text{mod}(\mathfrak{A}) = \forall. \end{cases}$$

For instance,

$$Q(\langle x, g \rangle, ([\langle -3, 0.001 \rangle, \langle 1, 0.001 \rangle]', \exists)) \langle x, g \rangle \geq 0$$

means

$$(\exists \langle x, g \rangle \in [\langle -3, 0.001 \rangle, \langle 1, 0.001 \rangle]') \langle x, g \rangle \geq 0$$

and

$$Q(\langle x, g \rangle, ([\langle 1, 0.001 \rangle, \langle 2, 0.001 \rangle]', \forall)) \langle x, g \rangle \geq 0$$

means

$$(\forall \langle x, g \rangle \in [\langle 1, 0.001 \rangle, \langle 2, 0.001 \rangle]') \langle x, g \rangle \geq 0.$$

Definition 9.2.6 (Set of predicates validated by a modal interval of marks) For an interval of marks $\mathfrak{A}$ the set of *predicates accepted* by $\mathfrak{A}$ is

$$\text{Pred}(\mathfrak{A}) = \{P \in \text{Pred}(\mathbb{M}) \mid Q(\mathfrak{a}, \mathfrak{A}) P(\mathfrak{a})\}.$$

Lemma 9.2.1 (Predicate of modal intervals of marks) *If $\bigcup_{\mathfrak{a} \in \mathfrak{A}'}$ is the "union operator" of a family of set of marks indexed by $\mathfrak{a}$ ranging over $\mathfrak{A}'$, and $\bigcap_{\mathfrak{a} \in \mathfrak{A}'}$ is the corresponding "intersection operator", then*

$$\text{Pred}(\mathfrak{A}', \exists) = \{P \in \text{Pred}(\mathbb{M}) \mid (\exists \mathfrak{a} \in \mathfrak{A}') P(\mathfrak{a})\} = \bigcup_{\mathfrak{a} \in \mathfrak{A}'} \text{Pred}(\mathfrak{a})$$

$$\text{Pred}(\mathfrak{A}', \forall) = \{P \in \text{Pred}(\mathbb{M}) \mid (\forall \mathfrak{a} \in \mathfrak{A}') P(\mathfrak{a})\} = \bigcap_{\mathfrak{a} \in \mathfrak{A}'} \text{Pred}(\mathfrak{a})$$

Proof See Lemma 2.2.1. ∎

Definition 9.2.7 (Set of co-predicates rejected by a modal interval of marks) The set of *co-predicates* is the set of all the predicates $P \in \text{Pred}(\mathbb{M})$ rejected by $\mathfrak{A}$,

$$\text{Copred}(\mathfrak{A}) = \{P \in \text{Pred}(\mathbb{M}) \mid \neg Q(\mathfrak{a}, \mathfrak{A}) P(\mathfrak{a})\}.$$

9.2 Intervals of Marks

To represent an interval of marks it is very convenient to use its *canonical notation*, defined by its bounds. If $\underline{a} = \langle \underline{a}, g \rangle$ and $\overline{a} = \langle \overline{a}, g \rangle$ are the bounds of the interval of marks $\mathfrak{A} = [\underline{a}, \overline{a}]$, then

$$\mathfrak{A} = [\underline{a}, \overline{a}] = [\langle \underline{a}, g \rangle, \langle \overline{a}, g \rangle] = \begin{cases} ([\underline{a}, \overline{a}]', \exists) & \text{if } \underline{a} \leq \overline{a} \\ ([\overline{a}, \underline{a}]', \forall) & \text{if } \underline{a} \geq \overline{a}. \end{cases}$$

The treatment of the duality of modal intervals of marks is a translation of the treatment of the duality of the real modal intervals.

Definition 9.2.8 (Dual operator) For $\mathfrak{A} = (\mathfrak{A}', Q_\mathfrak{A}) \in I^*(\mathbb{M})$, the *dual operator* over $\mathfrak{A}$, denoted by $\text{Dual}(\mathfrak{A})$ is defined as

$$\text{Dual}(\mathfrak{A}) = (\mathfrak{A}', \text{Dual}(Q_\mathfrak{A})),$$

where

$$\text{Dual}(Q_\mathfrak{A}) = \begin{cases} \forall & \text{if } Q_\mathfrak{A} = \exists \\ \exists & \text{if } Q_\mathfrak{A} = \forall. \end{cases}$$

In terms of the canonical notation

$$\mathfrak{A} = [\underline{a}, \overline{a}] \Leftrightarrow \text{Dual}(\mathfrak{A}) = [\overline{a}, \underline{a}].$$

Definition 9.2.9 (Interval projection) The *interval projection*, $P_\mathbb{R}(\mathfrak{A})$, of the interval of marks $\mathfrak{A} = [\langle \underline{a}, g \rangle, \langle \overline{a}, g \rangle]$ is the interval $[\underline{a}, \overline{a}] \in I^*(\mathbb{R})$, i.e.,

$$P_\mathbb{R}([\langle \underline{a}, g \rangle, \langle \overline{a}, g \rangle]) = [\underline{a}, \overline{a}].$$

Sometimes it will be necessary to make an *immersion* of a modal interval $[\underline{a}, \overline{a}] \in I^*(\mathbb{R})$ inside a mark scale, for a given granularity g, distinguishing between

- *Immersion* of the interval $[\underline{a}, \overline{a}]$ into the mark scale $\mathbb{M}(t, \infty)$ is the interval of marks $[\langle \underline{a}, g \rangle, \langle \overline{a}, g \rangle]$ of the type $\mathbb{M}(t, \infty)$.
- *Digital immersion* of the interval $[\underline{a}, \overline{a}]$ into the mark scale $\mathbb{M}(t, n)$ is the interval of marks

$$[\langle DI_n(\underline{a}), g + b^{-n} \rangle, \langle DI_n(\overline{a}), g + b^{-n} \rangle],$$

where b^{-n} is the maximum relative separation between two points of the scale.

As the digital immersion involves an increase in the granularity, it only makes sense if $\langle DI_n(\underline{a}), g + b^{-n} \rangle$ and $\langle DI_n(\overline{a}), g + b^{-n} \rangle$ are valid marks.

9.3 Relations in the Set of Intervals of Marks

Any relation between intervals of marks must be considered under two different points of view, material or weak, defined only between *comparable intervals of marks*, that is, when their bounds are marks with the same tolerance and expressed in the same basis.

9.3.1 Material Relations

Formally any material relation between intervals of marks is defined in a similar way to the parallel modal interval relation.

Definition 9.3.1 (Material inclusion and material equality) If $\mathfrak{A}$ and $\mathfrak{B}$ are comparable intervals of marks, $\mathfrak{A}$ is *materially included* in $\mathfrak{B}$, when

$$\mathfrak{A} \subseteq \mathfrak{B} : (\text{Pred}(\mathfrak{A}) \subseteq \text{Pred}(\mathfrak{B})).$$

and $\mathfrak{A}$ is *materially equal* to $\mathfrak{B}$, when

$$\mathfrak{A} = \mathfrak{B} : (\mathfrak{A} \subseteq \mathfrak{B}, \mathfrak{A} \supseteq \mathfrak{B}) \Leftrightarrow \text{Pred}(\mathfrak{A}) = \text{Pred}(\mathfrak{B}).$$

The relation of material inclusion between intervals of marks is, obviously, an order relation, and material equality is an equivalence relation.

Material inclusion of modal intervals of marks can be related with the inclusions of their corresponding ranges.

Theorem 9.3.1 *If* $\mathfrak{A} = (\mathfrak{A}', Q_\mathfrak{A})$ *and* $\mathfrak{B} = (\mathfrak{B}', Q_\mathfrak{B})$ *then*

$$\mathfrak{A} \subseteq \mathfrak{B} \Leftrightarrow \begin{cases} \mathfrak{A}' \subseteq \mathfrak{B}' & \text{if } Q_\mathfrak{A} = Q_\mathfrak{B} = \exists \\ \mathfrak{A}' \supseteq \mathfrak{B}' & \text{if } Q_\mathfrak{A} = Q_\mathfrak{B} = \forall \\ \mathfrak{A}' \cap \mathfrak{B}' \neq \emptyset & \text{if } Q_\mathfrak{A} = \forall, Q_\mathfrak{B} = \exists \\ \mathfrak{A}' = \mathfrak{B}' = [\mathfrak{a}, \mathfrak{a}] & \text{if } Q_\mathfrak{A} = \exists, Q_\mathfrak{B} = \forall. \end{cases}$$

Proof See Lemma 2.2.3. ∎

From these equivalences, material equality and inclusion between intervals of marks can be related with equalities and inclusions between their respective bounds. So

$$[\underline{\mathfrak{a}}, \overline{\mathfrak{a}}] \subseteq [\underline{\mathfrak{b}}, \overline{\mathfrak{b}}] \Leftrightarrow (\underline{\mathfrak{a}} \geq \underline{\mathfrak{b}}, \overline{\mathfrak{a}} \leq \overline{\mathfrak{b}}),$$

$$[\underline{\mathfrak{a}}, \overline{\mathfrak{a}}] = [\underline{\mathfrak{b}}, \overline{\mathfrak{b}}] \Leftrightarrow (\underline{\mathfrak{a}} = \underline{\mathfrak{b}}, \overline{\mathfrak{a}} = \overline{\mathfrak{b}}).$$

9.3 Relations in the Set of Intervals of Marks

Also, in a parallel way to modal intervals with real bounds, the material inequalities between two comparable intervals of marks are defined:

Definition 9.3.2 (Material inequality) If $\mathfrak{A} = [\underline{a}, \overline{a}], \mathfrak{B} = [\underline{b}, \overline{b}]$ are two comparable intervals of marks, $\mathfrak{A}$ is *materially less than or equal to* $\mathfrak{B}$, denoted by $\mathfrak{A} \leq \mathfrak{B}$, when

$$\mathfrak{A} \leq \mathfrak{B} \Leftrightarrow (\underline{a} \leq \underline{b}, \overline{a} \leq \overline{b}).$$

and $\mathfrak{A}$ is *materially greater than or equal to* $\mathfrak{B}$ when

$$\mathfrak{A} \geq \mathfrak{B} \Leftrightarrow \mathfrak{B} \leq \mathfrak{A}.$$

From the properties of these relations between marks, the parallel relations between intervals of marks are reflexive, antisymmetric and transitive.

9.3.2 Weak Relations

Suppose $\alpha \in [0, 1]$ and $\mathfrak{A}$ and $\mathfrak{B}$ are comparable mark intervals with granularities compatible with αt.

Definition 9.3.3 (Weak equality) If $\mathfrak{A}$ and $\mathfrak{B}$ are comparable intervals of marks, $\mathfrak{A}$ is *weakly equal* to $\mathfrak{B}$ when

$$\mathfrak{A} \approx_\alpha \mathfrak{B} \Leftrightarrow (\underline{a} \approx_\alpha \underline{b}, \overline{a} \approx_\alpha \overline{b}).$$

Definition 9.3.4 (Weak inclusion) If $\mathfrak{A}$ and $\mathfrak{B}$ are comparable intervals of marks, $\mathfrak{A}$ is *weakly included* in $\mathfrak{B}$ when

$$\mathfrak{A} \subseteq_\alpha \mathfrak{B} \Leftrightarrow (\underline{a} \succeq_\alpha \underline{b}, \overline{a} \preceq_\alpha \overline{b}).$$

For $\alpha, \beta \in [0, 1]$ with $\alpha + \beta \leq 1$, if $\mathfrak{A}, \mathfrak{B}$ and $\mathfrak{C}$ are intervals of marks with granularities compatible with αt, βt and $(\alpha + \beta)t$, the following properties hold

- $\mathfrak{A} \approx_\alpha \mathfrak{A}$, $\mathfrak{A} \subseteq_\alpha \mathfrak{A}$.
- $\mathfrak{A} \approx_\alpha \mathfrak{B} \Leftrightarrow \mathfrak{B} \approx_\alpha \mathfrak{A}$, $(\mathfrak{A} \subseteq_\alpha \mathfrak{B}, \mathfrak{B} \subseteq_\alpha \mathfrak{A}) \Leftrightarrow \mathfrak{A} \approx_\alpha \mathfrak{B}$.
- $(\mathfrak{A} \approx_\alpha \mathfrak{B}, \mathfrak{B} \approx_\beta \mathfrak{C}) \Rightarrow \mathfrak{A} \approx_{\alpha+\beta} \mathfrak{C}$, $(\mathfrak{A} \subseteq_\alpha \mathfrak{B}, \mathfrak{B} \subseteq_\beta \mathfrak{C}) \Rightarrow \mathfrak{A} \subseteq_{\alpha+\beta} \mathfrak{C}$.

Definition 9.3.5 (Weak inequality) If $\mathfrak{A}$ and $\mathfrak{B}$ are comparable intervals of marks, $\mathfrak{A}$ is *weakly less than or equal to* $\mathfrak{B}$ when

$$\mathfrak{A} \preceq_\alpha \mathfrak{B} \Leftrightarrow (\underline{a} \preceq_\alpha \underline{b}, \overline{a} \preceq_\alpha \overline{b}).$$

and $\mathfrak{A}$ is *weakly greater than or equal to* $\mathfrak{B}$ when

$$\mathfrak{A} \succeq_\alpha \mathfrak{B} \Leftrightarrow \mathfrak{B} \preceq_\alpha \mathfrak{A}$$

From the properties of these relations between marks, these parallel relations between intervals of marks are reflexive, antisymmetric, and $(\alpha + \beta)$-transitive.

Lemma 9.3.1 *If $\mathfrak{A}, \mathfrak{B}, \mathfrak{C} \in I^*(\mathbb{M})$ are comparable intervals of marks with granularities g_a, g_b and g_c compatible with αt, then*

1. $(\mathfrak{A} \subseteq \mathfrak{B} \Rightarrow \mathfrak{A} \subseteq_\alpha \mathfrak{B})$, $(\mathfrak{A} = \mathfrak{B} \Rightarrow \mathfrak{A} \approx_\alpha \mathfrak{B})$.
2. $(\mathfrak{A} \subseteq \mathfrak{B}, \mathfrak{B} \subseteq_\alpha \mathfrak{C} \Rightarrow \mathfrak{A} \subseteq_\alpha \mathfrak{C})$, $(\mathfrak{A} \subseteq \mathfrak{B}, \mathfrak{B} \approx_\alpha \mathfrak{C} \Rightarrow \mathfrak{A} \subseteq_\alpha \mathfrak{C})$.

Proof From the properties of the weak relationship between marks. ∎

Lemma 9.3.2 *Let $\mathfrak{A}'$ and $\mathfrak{B}'$ be proper intervals of marks. If they are comparable and with granularities compatible with αt, then*

$$\mathfrak{A}' \subseteq_\alpha \mathfrak{B}' \Leftrightarrow (\forall \mathfrak{a} \in \mathfrak{A}')\, (\exists \mathfrak{b} \in \mathfrak{B}')\, \mathfrak{a} \approx_\alpha \mathfrak{b}.$$

Proof For $\mathfrak{A}' = [\langle \underline{a}, g_a \rangle, \langle \overline{a}, g_a \rangle]$ and $\mathfrak{B}' = [\langle \underline{b}, g_b \rangle, \langle \overline{b}, g_b \rangle]$,

⇐: From the hypothesis, if $\underline{\mathfrak{a}} \in \mathfrak{A}'$, then $(\exists \mathfrak{b}_1 \in \mathfrak{B}')\, \underline{\mathfrak{a}} \approx_\alpha \mathfrak{b}_1$. By definition of an interval of marks, $\underline{b} \leq \mathfrak{b}_1 \leq \overline{b}$. Using the properties of the weak relationships between marks

$$\underline{\mathfrak{a}} \approx_\alpha \mathfrak{b}_1 \geq \underline{\mathfrak{b}} \Rightarrow \underline{\mathfrak{a}} \succeq_\alpha \underline{\mathfrak{b}}.$$

Repeating this process for $\overline{\mathfrak{a}} \in \mathfrak{A}'$ the analogous result $\overline{\mathfrak{a}} \preceq_\alpha \overline{\mathfrak{b}}$ is obtained.

⇒: Let $\underline{\mathfrak{a}} = \langle \underline{a}, g_a \rangle$, $\overline{\mathfrak{a}} = \langle \overline{a}, g_a \rangle$, $\underline{\mathfrak{b}} = \langle \underline{b}, g_b \rangle$, $\overline{\mathfrak{b}} = \langle \overline{b}, g_b \rangle$. If $\mathfrak{a} \in \mathfrak{A}'$, then $\mathfrak{a} = \langle a, \tilde{g} \rangle$ with $\underline{a} \leq a \leq \overline{a}$. As $\mathfrak{A}' \subseteq_\alpha \mathfrak{B}'$ is supposed true, the following situations can occur:

1. $(\underline{a} > \underline{b})$ and $(\overline{a} < \overline{b})$. In this case, as $\underline{b} \leq \underline{a} \leq a \leq \overline{a} \leq \overline{b}$, then $a \in [\underline{b}, \overline{b}]'$ and defining $\mathfrak{b} = \langle a, g \rangle$, it is true that $\mathfrak{b} \in \mathfrak{B}'$ and $\mathfrak{b} \approx_\alpha \mathfrak{a}$.
2. $(\underline{a} > \underline{b})$ and $(\overline{a} \in Ind\langle \overline{b}, g_b, \alpha t, n, b \rangle$ or $\overline{b} \in Ind\langle \overline{a}, g_a, \alpha t, n, b \rangle)$. Let us suppose $\overline{a} \geq \overline{b}$ (otherwise it would be the previous situation),

   - $a \in [\underline{a}, \overline{b}]'$ implies $\underline{b} \leq \underline{a} \leq a \leq \overline{a} \leq \overline{b}$ and the same reasoning as for case (1) leads to the conclusion.
   - $a \in [\overline{b}, \overline{a}]'$. This situation allows taking $\mathfrak{b} = \langle \overline{b}, g \rangle$ and, in this case,

     (i) if $\overline{a} \in Ind\langle \overline{b}, g_b, \alpha t, n, b \rangle$,

     $$\overline{b} \leq a \leq \overline{a} \leq \max\left\{\overline{b}(1 + \alpha t), \overline{b}(1 - \alpha t)\right\},$$

     therefore $a \in Ind\langle \overline{b}, g_b, \alpha t, n, b \rangle$ and $\overline{\mathfrak{b}} \approx_\alpha \mathfrak{a}$.

     (ii) if $\overline{b} \in Ind\langle \overline{a}, g_a, \alpha t, n, b \rangle$,

     $$\min\{\overline{a}(1 - \alpha t), \overline{a}(1 + \alpha t)\} \leq \overline{b} \leq \overline{a},$$

9.3 Relations in the Set of Intervals of Marks

since $\overline{b} \leq a \leq \overline{a}$, then

$$a(1 - \alpha t) \leq \overline{a}(1 - \alpha t), a(1 + \alpha t) \leq \overline{a}(1 + \alpha t)$$

and

$$\min\{a(1 - \alpha t), a(1 + \alpha t)\} \leq \min\{\overline{a}(1 - \alpha t), \overline{a}(1 + \alpha t)\},$$

therefore

$$\min\{a(1 - \alpha t), a(1 + \alpha t)\} \leq \overline{b} \leq a,$$

which implies $\overline{b} \in Ind_\alpha(\mathfrak{a})$ and consequently $\overline{b} \approx_\alpha \mathfrak{a}$.

3. ($\underline{a} \in Ind\langle \underline{b}, g_b, \alpha t, n, b\rangle$ or $\underline{b} \in Ind\langle \underline{a}, g_a, \alpha t, n, b\rangle$) and ($\overline{a} < \overline{b}$). Analogous to 2.
4. ($\underline{a} \in Ind\langle \underline{b}, g_b, \alpha t, n, b\rangle$ or $\underline{b} \in Ind\langle \underline{a}, g_a, \alpha t, n, b\rangle$) and
   ($\overline{a} \in Ind\langle \overline{b}, g_b, \alpha t, n, b\rangle$ or $\overline{b} \in Ind\langle \overline{a}, g_a, \alpha t, n, b\rangle$).
   Combine the possibilities examined in 2 and in 3. ∎

Lemma 9.3.3 *Suppose that* $\mathfrak{A}, \mathfrak{B}, \mathfrak{C} \in I^*(\mathbb{M})$ *are comparable intervals of marks with granularities compatible with* αt. *Then*

1. *If* $\mathfrak{A}$ *and* $\mathfrak{B}$ *are proper,*

$$\mathfrak{A} \subseteq_\alpha \mathfrak{B} \Leftrightarrow (\forall \mathfrak{a} \in \mathfrak{A}')\,(\exists \mathfrak{b} \in \mathfrak{B}')\,\mathfrak{a} \approx_\alpha \mathfrak{b}.$$

2. *If* $\mathfrak{A}$ *and* $\mathfrak{B}$ *are improper,*

$$\mathfrak{A} \subseteq_\alpha \mathfrak{B} \Leftrightarrow (\forall \mathfrak{b} \in \mathfrak{B}')\,(\exists \mathfrak{a} \in \mathfrak{A}')\,\mathfrak{b} \approx_\alpha \mathfrak{a}.$$

3. *If* $\mathfrak{A}$ *is improper and* $\mathfrak{B}$ *is proper,*

$$\mathfrak{A} \subseteq_\alpha \mathfrak{B} \Leftrightarrow (\exists \mathfrak{a} \in \mathfrak{A}')\,(\exists \mathfrak{b} \in \mathfrak{B}')\,\mathfrak{a} \approx_\alpha \mathfrak{b}.$$

4. *If* $\mathfrak{A}$ *is proper and* $\mathfrak{B}$ *is improper,*

$$\mathfrak{A} \subseteq_\alpha \mathfrak{B} \Leftrightarrow (\forall \mathfrak{a} \in \mathfrak{A}')\,(\forall \mathfrak{b} \in \mathfrak{B}')\,\mathfrak{a} \approx_\alpha \mathfrak{b}.$$

Proof For $\mathfrak{A} = [\underline{a}, \overline{a}]$ and $\mathfrak{B} = [\underline{b}, \overline{b}]$:

1. If $\mathfrak{A}$ and $\mathfrak{B}$ are proper, Lemma 9.3.2 leads to the result.
2. If $\mathfrak{A}$ and $\mathfrak{B}$ are improper and $\mathfrak{A} \subseteq_\alpha \mathfrak{B}$, then $\mathfrak{A}' = [\overline{a}, \underline{a}]$ and $\mathfrak{B}' = [\overline{b}, \underline{b}]$, $\overline{b} \succeq_\alpha \overline{a}$ and $\underline{b} \preceq_\alpha \underline{a}$. It only remains to apply Lemma 9.3.2 to the inclusion $\mathfrak{B}' \subseteq_\alpha \mathfrak{A}'$.
3. If $\mathfrak{A}$ is improper and $\mathfrak{B}$ proper, then on splitting the equivalence into two implications, we reason as follows.

$\Rightarrow$: Let $\underline{a}, \overline{a}, \underline{b}, \overline{b}$ be the centers of the marks $\mathfrak{a}, \overline{\mathfrak{a}}, \mathfrak{b}, \overline{\mathfrak{b}}$, respectively. The inequalities $\underline{a} \succeq_\alpha \underline{b}$ and $\overline{a} \preceq_\alpha \overline{b}$ allow distinguishing between the following cases:

- $\underline{a} \geq \underline{b}$ and $\overline{a} \leq \overline{b}$. In this case $[\underline{a}, \overline{a}] \subseteq [\underline{b}, \overline{b}]$ and therefore

$$(\exists a \in [\underline{a}, \overline{a}]')(\exists b \in [\underline{b}, \overline{b}]')\, a = b$$

  and the marks $\langle a, g \rangle$ y $\langle b, g \rangle$ accomplish the wanted weak equality.
- $\underline{a} \geq \underline{b}$ and $\overline{a} \approx_\alpha \overline{b}$. Defining $\mathfrak{a} = \overline{a}$ and $\mathfrak{b} = \overline{b}$, then $\mathfrak{a} \approx_\alpha \mathfrak{b}$.
- The two remaining cases are similar to the previous one.

$\Leftarrow$: Let $\mathfrak{a} \in \mathfrak{A}'$ and $\mathfrak{b} \in \mathfrak{B}'$ be marks such that $\mathfrak{a} \approx_\alpha \mathfrak{b}$. As $\overline{a} \leq a \leq \underline{a}$ and $\underline{b} \leq b \leq \overline{b}$,

$$\underline{b} \leq b \approx_\alpha a \leq \underline{a} \Rightarrow \underline{b} \preceq_\alpha \underline{a}$$
$$\overline{a} \leq a \approx_\alpha b \leq \overline{b} \Rightarrow \overline{a} \preceq_\alpha \overline{b}.$$

4. If $\mathfrak{A}$ is proper and $\mathfrak{B}$ is improper then:

$\Rightarrow$: Let be $\mathfrak{a} \in \mathfrak{A}'$ and $\mathfrak{b} \in \mathfrak{B}'$. Then

$$\underline{a} \leq a \leq \overline{a} \preceq_\alpha \overline{b} \leq b \Rightarrow a \preceq_\alpha b$$
$$\overline{b} \leq b \leq \underline{b} \preceq_\alpha \underline{a} \leq a \Rightarrow b \preceq_\alpha a,$$

and therefore $\mathfrak{a} \approx_\alpha \mathfrak{b}$.

$\Leftarrow$: For $\mathfrak{a} \in \mathfrak{A}'$ and $\mathfrak{b} \in \mathfrak{B}'$,

$$\underline{b} \leq \underline{b} \approx_\alpha \underline{a} \leq \underline{a} \Rightarrow \underline{b} \preceq_\alpha \underline{a}$$
$$\overline{a} \leq \overline{a} \approx_\alpha \overline{b} \leq \overline{b} \Rightarrow \overline{a} \preceq_\alpha \overline{b}.$$

Lemma 9.3.4 *Suppose $\mathfrak{A}, \mathfrak{B}, \mathfrak{C}$ are comparable intervals of marks with granularities compatible with αt. Then:*

1. $\mathfrak{A} \leq \mathfrak{B} \Rightarrow (\forall \alpha,]0, 1]) \, \mathfrak{A} \preceq_\alpha \mathfrak{B}$.
2. $\mathfrak{A} \approx_\alpha \mathfrak{B}, \mathfrak{B} \leq \mathfrak{C} \Rightarrow \mathfrak{A} \preceq_\alpha \mathfrak{C}$.

Proof From the properties of the weak inequalities between marks. ∎

9.3.3 Interval Lattices

The lattice operations "meet" and "join" on $I^*(\mathbb{M})$ are defined for bounded families of modal intervals as the infimum and supremum of the modal inclusion order relation on $I^*(\mathbb{M})$.

9.3 Relations in the Set of Intervals of Marks 239

Definition 9.3.6 ("Meet" and "join" on $(I^*(\mathbb{M}), \subseteq)$) For a indexed family $\mathfrak{A}(I) = \{\mathfrak{A}(i) \in I^*(\mathbb{M}) \mid i \in I\}$ (I is the domain of the index):

$$\bigwedge_{i \in I} \mathfrak{A}(i) = \mathfrak{A} \in I^*(\mathbb{M}) \text{ is such that } (\forall i \in I) \, (\mathfrak{X} \subseteq \mathfrak{A}(i) \Leftrightarrow \mathfrak{X} \subseteq \mathfrak{A}),$$

$$\bigvee_{i \in I} \mathfrak{A}(i) = \mathfrak{B} \in I^*(\mathbb{M}) \text{ is such that } (\forall i \in I) \, (\mathfrak{X} \supseteq \mathfrak{A}(i) \Leftrightarrow \mathfrak{X} \supseteq \mathfrak{B}),$$

writing $(\mathfrak{A} \wedge \mathfrak{B})$ and $(\mathfrak{A} \vee \mathfrak{B})$ for the corresponding case of two operands.

The lattice operations "min" and "max" on $I^*(\mathbb{M})$ are defined on bounded families of modal intervals as the infimum and supremum of the modal inequality order relation on $I^*(\mathbb{M})$.

Definition 9.3.7 ("Min" and "max" on $(I^*(\mathbb{M}), \subseteq)$) For a bounded family $\mathfrak{A}(I) = \{\mathfrak{A}(i) \in I^*(\mathbb{M}) \mid i \in I\}$:

$$\operatorname*{Min}_{i \in I} \mathfrak{A}(i) = \mathfrak{A} \in I^*(\mathbb{M}) \text{ is such that } (\forall i \in I) \, (\mathfrak{X} \leq \mathfrak{A}(i) \Leftrightarrow \mathfrak{X} \leq \mathfrak{A}),$$

$$\operatorname*{Max}_{i \in I} \mathfrak{A}(i) = \mathfrak{B} \in I^*(\mathbb{M}) \text{ is such that } (\forall i \in I) \, (\mathfrak{X} \geq \mathfrak{A}(i) \Leftrightarrow \mathfrak{X} \geq \mathfrak{B}).$$

In terms of the bounds, for a indexed family of $\mathfrak{A}(I) \subseteq I^*(\mathbb{M})$, if $\mathfrak{A}(i) = [\underline{a}(i), \bar{a}(i)]$:

(1) $\bigwedge_{i \in I} \mathfrak{A}(i) = [\max_{i \in I} \underline{a}(i), \min_{i \in I} \bar{a}(i)]$

(2) $\bigvee_{i \in I} \mathfrak{A}(i) = [\min_{i \in I} \underline{a}(i), \max_{i \in I} \bar{a}(i)]$

(3) $\operatorname*{Min}_{i \in I} \mathfrak{A}(i) = [\min_{i \in I} \underline{a}(i), \min_{i \in I} \bar{a}(i)]$

(4) $\operatorname*{Max}_{i \in I} \mathfrak{A}(i) = [\max_{i \in I} \underline{a}(i), \max_{i \in I} \bar{a}(i)]$.

The set of comparable intervals of marks is a $\subseteq$-reticle and a $\leq$-reticle, in a similar way to $I^*(\mathbb{R})$.

Definition 9.3.8 (Meet–join operator on $(I^*(\mathbb{M}), \subseteq)$) For an interval of marks $\mathfrak{A} \in I^*(\mathbb{M})$

$$\Omega_{(\mathfrak{a}, \mathfrak{A})} = \bigwedge_{\mathfrak{a} \in \mathfrak{A}'} \quad \text{if } \mathfrak{A} \text{ is improper}$$

$$\Omega_{(\mathfrak{a}, \mathfrak{A})} = \bigvee_{\mathfrak{a} \in \mathfrak{A}'} \quad \text{if } \mathfrak{A} \text{ is proper.}$$

Lemma 9.3.5 *For any interval of marks* $\mathfrak{A} \in I^*(\mathbb{M})$,

$$\mathfrak{A} = \Omega_{(\mathfrak{a}, \mathfrak{A})} [\mathfrak{a}, \mathfrak{a}].$$

Proof

1) If $\mathfrak{A}$ is improper, then $\bigwedge_{\mathfrak{a} \in \mathfrak{A}'} [\mathfrak{a}, \mathfrak{a}] = [\max_{\mathfrak{a} \in \mathfrak{A}'} \mathfrak{a}, \min_{\mathfrak{a} \in \mathfrak{A}'} \mathfrak{a}]$.

2) If $\mathfrak{A}$ is proper, then $\bigvee_{\mathfrak{a} \in \mathfrak{A}'} [\mathfrak{a}, \mathfrak{a}] = [\min_{\mathfrak{a} \in \mathfrak{A}'} \mathfrak{a}, \max_{\mathfrak{a} \in \mathfrak{A}'} \mathfrak{a}]$.

∎

Lemma 9.3.6 *If $\mathfrak{A}, \mathfrak{B} \in I^*(\mathbb{M})$, then*

1. $\mathrm{Pred}(\mathfrak{A} \wedge \mathfrak{B}) \subseteq \mathrm{Pred}(\mathfrak{A}) \cap \mathrm{Pred}(\mathfrak{B})$.
2. $\mathrm{Pred}(\mathfrak{A} \vee \mathfrak{B}) \supseteq \mathrm{Pred}(\mathfrak{A}) \cup \mathrm{Pred}(\mathfrak{B})$.
3. $\mathrm{Copred}(\mathfrak{A} \wedge \mathfrak{B}) \supseteq \mathrm{Copred}(\mathfrak{A}) \cup \mathrm{Copred}(\mathfrak{B})$.
4. $\mathrm{Copred}(\mathfrak{A} \vee \mathfrak{B}) \subseteq \mathrm{Copred}(\mathfrak{A}) \cap \mathrm{Copred}(\mathfrak{B})$.

Proof See Lemma 2.2.12. ∎

These relations are not equalities because the meet of two intervals of marks is not identifiable with the intersection of sets of predicates and the join can not be identified with the union.

9.3.4 Interval Predicates and Co-predicates

Definition 9.3.9 (Set of interval predicates)

$$\mathrm{Pred}^*(\mathbb{M}) = \{\mathfrak{x} \in \mathcal{X}' \mid \mathfrak{X} \in I^*(\mathbb{M})\}.$$

Definition 9.3.10 (Set of interval co-predicates)

$$\mathrm{Copred}^*(\mathbb{M}) = \{\mathfrak{x} \notin \mathcal{X}' \mid \mathfrak{X} \in I^*(\mathbb{M})\}.$$

Definition 9.3.11 (Set of accepted interval predicates) For a interval of marks $\mathfrak{A}$ the set of *predicates accepted* by $\mathfrak{A}$ is

$$\mathrm{Pred}^*(\mathfrak{A}) = \{(\mathfrak{x} \in \mathcal{X}') \in \mathrm{Pred}^*(\mathbb{M}) \mid Q(\mathfrak{x}, \mathfrak{A}) \: \mathfrak{x} \in \mathcal{X}'\}.$$

Definition 9.3.12 (Set of rejected interval co-predicates) For a interval of marks $\mathfrak{A}$ the set of *co-predicates rejected* by $\mathfrak{A}$ is

$$\mathrm{Copred}^*(\mathfrak{A}) = \{(\mathfrak{x} \notin \mathcal{X}') \in \mathrm{Copred}^*(\mathbb{M}) \mid \neg Q(\mathfrak{x}, \mathfrak{A}) \: \mathfrak{x} \notin \mathcal{X}'\}.$$

Definition 9.3.13 (Proper and improper operators) For $\mathfrak{A} = (\mathfrak{A}', Q_{\mathfrak{A}}) \in I^*(\mathbb{M})$, the *proper operator* over $\mathfrak{A}$, denoted by $\mathrm{Prop}(\mathfrak{A})$, is defined as

$$\mathrm{Prop}(\mathfrak{A}) = \mathrm{Prop}(\mathfrak{A}') = (\mathfrak{A}', \exists)$$

and the *improper operator*, denoted by $\text{Impr}(\mathfrak{X})$ as

$$\text{Impr}(\mathfrak{A}) = \text{Impr}(\mathfrak{A}) = (\mathfrak{A}', \forall).$$

The operators meet and join produce the equivalences contained in the following lemma.

Lemma 9.3.7 *If* $\mathfrak{A}, \mathfrak{X} \in I^*(\mathbb{M})$, *then*

1. $(\mathfrak{x} \in \mathfrak{X}') \in \text{Pred}^*(\mathfrak{A}) \Leftrightarrow \text{Impr}(\mathfrak{X}) \subseteq \mathfrak{A}$.
2. $(\neg(\mathfrak{x} \in \mathfrak{X}')) \in \text{Copred}^*(\mathfrak{A}) \Leftrightarrow \text{Prop}(\mathfrak{X}) \supseteq \mathfrak{A}$.

Proof See Lemma 2.2.13. ■

Lemma 9.3.8 *If* $\mathfrak{A}, \mathfrak{B} \in I^*(\mathbb{M})$, *then*

1. $\text{Pred}^*(\mathfrak{A} \wedge \mathfrak{B}) = \text{Pred}^*(\mathfrak{A}) \cap \text{Pred}^*(\mathfrak{B})$.
2. $\text{Pred}^*(\mathfrak{A} \vee \mathfrak{B}) \supseteq \text{Pred}^*(\mathfrak{A}) \cup \text{Pred}^*(\mathfrak{B})$.
3. $\text{Copred}^*(\mathfrak{A} \wedge \mathfrak{B}) \supseteq \text{Copred}^*(\mathfrak{A}) \cup \text{Copred}^*(\mathfrak{B})$.
4. $\text{Copred}^*(\mathfrak{A} \vee \mathfrak{B}) = \text{Copred}^*(\mathfrak{A}) \cap \text{Copred}^*(\mathfrak{B})$.

Proof See Lemma 2.2.14. ■

9.3.5 k-Dimensional Intervals of Marks

The system of intervals of marks can be extended in a natural way to k-dimensional intervals of marks.

The symbol $I^*(\mathbb{M}(t, n)^k)$ will indicate the set of k-dimensional modal intervals of marks, where (t, n) is the type of the involved marks, sometimes abridged to $I^*(\mathbb{M}^k)$ when the type of the marks is set in advance or is implicit.

Definition 9.3.14 (Set of k-dimensional modal intervals of marks)

$$I^*(\mathbb{M}^k) = \{([a_1, b_1], \ldots, [a_k, b_k]) \mid (\forall i \in I) [a_i, b_i] \in I^*(\mathbb{M}(t, n))\}.$$

Definitions and relationships in $I^*(\mathbb{M})$ are easily generalized to $I^*(\mathbb{M}^k)$.

Definition 9.3.15 (k-dimensional inclusion and equality) For $\mathfrak{A} = (\mathfrak{A}_1, \ldots, \mathfrak{A}_k) \in I^*(\mathbb{M}^k)$, $\mathfrak{B} = (\mathfrak{B}_1, \ldots, \mathfrak{B}_k) \in I^*(\mathbb{M}^k)$,

$$\mathfrak{A} \subseteq \mathfrak{B} : (\mathfrak{A}_1 \subseteq \mathfrak{B}_1, \ldots, \mathfrak{A}_k \subseteq \mathfrak{B}_k)$$
$$\mathfrak{A} = \mathfrak{B} : (\mathfrak{A}_1 = \mathfrak{B}_1, \ldots, \mathfrak{A}_k = \mathfrak{B}_k).$$

Definition 9.3.16 (Proper and Improper operators) For $\mathfrak{X} \in I^*(\mathbb{M}^k)$, $\mathfrak{X}' \in I(\mathbb{M}^k)$,

$$\text{Prop}(\mathfrak{X}) = \text{Prop}(\mathfrak{X}') = ((\mathfrak{X}'_1, \exists), \ldots, (\mathfrak{X}'_k, \exists));$$
$$\text{Impr}(\mathfrak{X}) = \text{Impr}(\mathfrak{X}') = ((\mathfrak{X}'_1, \forall), \ldots, (\mathfrak{X}'_k, \forall)).$$

Definition 9.3.17 (Join and meet of interval-indexed families) If $\mathcal{X}' \in I(\mathbb{M}^k)$, $\mathfrak{x} \in \mathbb{M}^k$ and $\mathfrak{F}(\mathfrak{x}) = [\mathfrak{F}_1(\mathfrak{x}), \mathfrak{F}_2(\mathfrak{x})] \in I^*(\mathbb{M})$,

$$\bigwedge_{\mathfrak{x} \in \mathcal{X}'} \mathfrak{F}(\mathfrak{x}) = \bigwedge_{\mathfrak{x}_1 \in \mathcal{X}'_1} \ldots \bigwedge_{\mathfrak{x}_k \in \mathcal{X}'_k} \mathfrak{F}(\mathfrak{x}_1, \ldots, \mathfrak{x}_v) = [\max_{(\mathfrak{x}, \mathcal{X}')} \mathfrak{F}_1(\mathfrak{x}), \min_{\mathfrak{x} \in \mathcal{X}'} \mathfrak{F}_2(\mathfrak{x})]$$

$$\bigvee_{\mathfrak{x} \in \mathcal{X}'} \mathfrak{F}(\mathfrak{x}) = \bigvee_{\mathfrak{x}_1 \in \mathcal{X}'_1} \ldots \bigvee_{\mathfrak{x}_k \in \mathcal{X}'_k} \mathfrak{F}(\mathfrak{x}_1, \ldots, \mathfrak{x}_v) = [\min_{\mathfrak{x} \in \mathcal{X}'} \mathfrak{F}_1(\mathfrak{x}), \max_{\mathfrak{x} \in \mathcal{X}'} \mathfrak{F}_2(\mathfrak{x})]$$

(the order of the component operators is irrelevant in both cases).

Definition 9.3.18 (Sets of k-dimensional interval predicates) For $\mathfrak{A} \in I^*(\mathbb{M}^k)$, $\mathcal{X}' \in I(\mathbb{M}^k)$ and $\mathfrak{x} \in \mathbb{M}^k$

$$(\mathfrak{x} \in \mathcal{X}') = (\mathfrak{x}_1 \in \mathcal{X}'_1, \ldots, \mathfrak{x}_k \in \mathcal{X}'_k);$$

$$\text{Pred}^*(\mathfrak{A}) = \{(\mathfrak{x} \in \mathcal{X}') \mid (\mathfrak{x}_1 \in \mathcal{X}'_1) \in \text{Pred}^*(\mathfrak{A}_1), \ldots, (\mathfrak{x}_k \in \mathcal{X}'_k) \in \text{Pred}^*(\mathfrak{A}_k)\}).$$

Remark 9.3.1 The condition defining the set $\text{Pred}^*(\mathfrak{A})$ is equivalent to

$$Q(\mathfrak{x}_1, \mathfrak{A}_1) \ldots Q(\mathfrak{x}_k, \mathfrak{A}_k) \ (\mathfrak{x}_1 \in \mathcal{X}'_1, \ldots, \mathfrak{x}_k \in \mathcal{X}'_k),$$

where the order of the modal quantifiers does not matter, because of the independence of the arguments of the $\mathfrak{x}_i$ among the predicates $\mathfrak{x}_i \in \mathcal{X}'_i$.

Definition 9.3.19 (Sets of k-dimensional interval co-predicates) For $\mathfrak{A} \in I^*(\mathbb{M}^k)$, $\mathcal{X}' \in I(\mathbb{M}^k)$, and $\mathfrak{x} \in \mathbb{M}^k$

$$(\mathfrak{x} \notin \mathcal{X}') = \neg(\mathfrak{x} \in \mathcal{X}') = (\mathfrak{x}_1 \notin \mathcal{X}'_1 \text{ or} \ldots \text{or } \mathfrak{x}_k \notin \mathcal{X}'_k)$$

$$\text{Copred}^*(\mathfrak{A}) = \{(\mathfrak{x} \notin \mathcal{X}') \mid (\mathfrak{x}_1 \notin \mathcal{X}'_1) \in \text{Copred}^*(\mathfrak{A}_1), \ldots, (\mathfrak{x}_k \notin \mathcal{X}'_k)$$
$$\in \text{Copred}^*(\mathfrak{A}_k)\}.$$

Lemma 9.3.9 (Interval representation of $\text{Pred}^*(\mathfrak{A})$ and $\text{Copred}^*(\mathfrak{A})$) *Under the hypotheses of the previous definitions:*

$$(\mathfrak{x} \in \mathcal{X}') \in \text{Pred}^*(\mathfrak{A}) \Leftrightarrow \text{Impr}(\mathcal{X}') \subseteq \mathfrak{A};$$

$$(\mathfrak{x} \notin \mathcal{X}') \in \text{Copred}^*(\mathfrak{A}) \Leftrightarrow \text{Prop}(\mathcal{X}') \supseteq \mathfrak{A}.$$

Proof See Lemma 2.2.13. ∎

9.4 Interval Extensions of Functions of Marks

Recalling the concepts covered in Chap. 3, in the classical set-theoretical interval analysis, one extension of a $\mathbb{R}^k$ to $\mathbb{R}$ continuous function $z = f(x_1, \ldots, x_k)$ is the *united extension* R_f or *range of f*. For the interval argument $X' = (X'_1, \ldots, X'_k) \in I(\mathbb{R}^k)$ it is defined as the range of f-values on X'

9.4 Interval Extensions of Functions of Marks

$$R_f(X'_1, \ldots, X'_k) = \{f(x_1, \ldots, x_k) \mid x_1 \in X'_1, \ldots, x_k \in X'_k\}$$
$$= [\min\{f(x_1, \ldots, x_k) \mid x_1 \in X'_1, \ldots, x_k \in X'_k\},$$
$$\max\{f(x_1, \ldots, x_k) \mid x_1 \in X'_1, \ldots, x_k \in X'_k\}]$$

which can be considered as a semantic extension of f, since it admits the logical interpretations

$$(\forall x_1 \in X'_1) \cdots (\forall x_k \in X'_k) \, (\exists z \in R_f(X'_1, \ldots, X'_k)) \, z = f(x_1, \ldots, x_k)$$

and

$$(\forall z \in R_f(X'_1, \ldots, X'_k)) \, (\exists x_1 \in X'_1) \cdots (\exists x_k \in X'_k) \, z = f(x_1, \ldots, x_k).$$

In the context of set-theoretical intervals of marks the, corresponding concept is given in the next definition.

Definition 9.4.1 (United extension of a function of marks) The *united extension* or *range of values* of a function of marks $f_{\mathbb{M}(t,n)} : \mathbb{M}(t,n)^k \to \mathbb{M}(t,n)$ associated to the continuous function $f : \mathbb{R}^k \to \mathbb{R}$, for the interval argument $\mathcal{X}' = (\mathcal{X}'_1, \ldots, \mathcal{X}'_k) \in I(\mathbb{M}(t,n)^k)$ is the function

$$R_{f_{\mathbb{M}(t,n)}} : I(\mathbb{M}(t,n)^k) \to I(\mathbb{M}(t,\infty))$$

defined by

$$R_{f_{\mathbb{M}(t,n)}}(\mathcal{X}'_1, \ldots, \mathcal{X}'_k) = \langle R_f(P_{\mathbb{R}}(\mathcal{X}'_1, \ldots, \mathcal{X}'_k)), g_f(\mathcal{X}'_1, \ldots, \mathcal{X}'_k) \rangle,$$

where

1) $P_{\mathbb{R}}(\mathcal{X}'_1, \ldots, \mathcal{X}'_k)$ is the real projection of $(\mathcal{X}'_1, \ldots, \mathcal{X}'_k)$,
2) $R_f(P_{\mathbb{R}}(\mathcal{X}'_1, \ldots, \mathcal{X}'_k))$ is the united extension of the function f over the classical interval $P_{\mathbb{R}}(\mathcal{X}'_1, \ldots, \mathcal{X}'_k)$, and
3) $g_f(\mathcal{X}'_1, \ldots, \mathcal{X}'_k)$ is the greatest of the granularities obtained if the computations with the bounds of $R_{f_{\mathbb{M}}}(\mathcal{X}'_1, \ldots, \mathcal{X}'_k)$ are made with the corresponding marks in a theoretical form, i.e., with $n = \infty$.

Remark 9.4.1 The united extension is just analytic and not computed. For this reason its granularity has to be the greatest granularity of the mark bounds.

The united extension verifies the logical statements

$$(\forall \mathfrak{x}_1 \in \mathcal{X}'_1) \cdots (\forall \mathfrak{x}_k \in \mathcal{X}'_k) \, (\exists \mathfrak{z} \in R_{f_{\mathbb{M}}}(\mathcal{X}'_1, \ldots, \mathcal{X}'_k)) \, \mathfrak{z} = f_{\mathbb{M}(t,\infty)}(\mathfrak{x}_1, \ldots, \mathfrak{x}_k).$$

and

$$(\forall \mathfrak{z} \in R_{f_{\mathbb{M}}}(\mathcal{X}'_1, \ldots, \mathcal{X}'_k)) \, (\exists \mathfrak{x}_1 \in \mathcal{X}'_1) \cdots (\exists \mathfrak{x}_k \in \mathcal{X}'_k) \, \mathfrak{z} = f_{\mathbb{M}(t,\infty)}(\mathfrak{x}_1, \ldots, \mathfrak{x}_k),$$

so it can be considered as a semantic extension of the function of marks $f_{\mathbb{M}(t,\infty)}$.

9.4.1 Semantic Functions

In the context of modal intervals of marks, it may be expected, as a starting point, that the relation $\mathfrak{z} = f_{\mathbb{M}(t,\infty)}(\mathfrak{x}_1, \ldots, \mathfrak{x}_k)$ must become some kind of interval relation $\mathfrak{Z} = F(f)(\mathfrak{X}_1, \ldots, \mathfrak{X}_k)$ guaranteeing some sort of $(n+1)$-dimensional interval predicate of the form

$$Q(\mathfrak{x}_1, \mathfrak{X}_1) \cdots Q(\mathfrak{x}_k, \mathfrak{X}_k) \, Q(\mathfrak{z}, \mathfrak{Z}) \, \mathfrak{z} = f_{\mathbb{M}(t,\infty)}(\mathfrak{x}_1, \ldots, \mathfrak{x}_k),$$

where an ordering problem obviously arises since the quantifying prefixes are not generally commutable.

Definition 9.4.2 (Poor computational extension) The function

$$F_{\mathbb{M}(t,\infty)} : \mathbb{M}(t,\infty)^k \to I(\mathbb{M}(t,\infty))$$

is a "poor computational extension" of the function of marks $f_{\mathbb{M}(t,\infty)}$ associated to a continuous real function $f : \mathbb{R}^k \to \mathbb{R}$, when for a given $\langle a, g \rangle \in \mathbb{M}(t,\infty)^k$ such that $f_{\mathbb{M}(t,\infty)}(\langle a, g \rangle)$ is valid, the existence of $F_{\mathbb{M}(t,\infty)}(\langle a, g \rangle)'$ implies that

$$f_{\mathbb{M}(t,\infty)}(\langle a, g \rangle) \in F_{\mathbb{M}(t,\infty)}(\langle a, g \rangle)'.$$

Lemma 9.4.1 (Semantic formulation of a poor computational extension) *Suppose $F_{\mathbb{M}(t,\infty)} : \mathbb{M}(t,\infty)^k \to I(\mathbb{M}(t,\infty))$ is a poor computational extension of the function of marks $f_{\mathbb{M}(t,\infty)}$ associated to a continuous real function $f : \mathbb{R}^k \to \mathbb{R}$. Then if $\mathfrak{a} \in \mathbb{M}(t,\infty)^k$, $f_{\mathbb{M}(t,\infty)}(\mathfrak{a})$ is a valid mark and $F_{\mathbb{M}(t,\infty)}(\mathfrak{a})'$ exists, then $f_{\mathbb{M}(t,\infty)}(\mathfrak{a}) \in F_{\mathbb{M}(t,\infty)}(\mathfrak{a})'$ is equivalent to*

$$(\forall \mathfrak{X}' \in I(\mathbb{M}(t,\infty)^k))$$

$$((\mathfrak{x} \in \mathfrak{X}') \in \mathrm{Pred}^*([\mathfrak{a}, \mathfrak{a}]) \Rightarrow (\mathfrak{z} \in R_f(\mathfrak{X}')) \in \mathrm{Pred}^*(\mathrm{Prop}(F_{\mathbb{M}(t,\infty)}(\mathfrak{a}))))$$

where $R_f(\mathfrak{X}')$ is the united extension of f over the classical interval $P_{\mathbb{R}}(\mathfrak{X}')$.

Proof See Lemma 3.1.1. ∎

The optimal modal interval extensions of a function f are the semantic ∗- and ∗∗-extensions, denoted by $f^*_{\mathbb{M}(t,\infty)}$ and $f^{**}_{\mathbb{M}(t,\infty)}$ and defined as follows.

Definition 9.4.3 (∗-semantic extension) The *∗-semantic extension* of a function of marks $f_{\mathbb{M}(t,n)} : \mathbb{M}(t,n)^k \to \mathbb{M}(t,n)$, associated to the continuous function $f : \mathbb{R}^k \to \mathbb{R}$, for the interval of marks argument $\mathfrak{X} \in I^*(\mathbb{M}(t,n)^k)$ is a function

$$f^*_{\mathbb{M}(t,\infty)} : I^*(\mathbb{M}(t,n)^k) \subseteq I^*(\mathbb{M}(t,\infty)^k) \longrightarrow I^*(\mathbb{M}(t,\infty))$$

defined by

$$f^*_{\mathbb{M}(t,\infty)}(\mathfrak{X}) = \langle f^*(P_{\mathbb{R}}(\mathfrak{X})), t, g_f(\mathfrak{X}), \infty, b \rangle,$$

9.4 Interval Extensions of Functions of Marks

where $P_\mathbb{R}(\mathcal{X})$ is the interval projection of $\mathcal{X}$, f^* is the *-semantic extension of f to the interval $P_\mathbb{R}(\mathcal{X})$, and $g_f(\mathcal{X})$ is the maximum of the granularities which would be obtained in the computation of the bounds of $f^*(P_\mathbb{R}(\mathcal{X}))$ in a theoretical way, i.e., with $n = \infty$.

In a dual way,

Definition 9.4.4 (-semantic extension)** The **-*semantic extension* of a function of marks $f_{\mathbb{M}(t,n)} : \mathbb{M}(t,n)^k \to \mathbb{M}(t,n)$, associated to the continuous function $f : \mathbb{R}^k \to \mathbb{R}$, for the interval of marks argument $\mathcal{X} \in I^*(\mathbb{M}(t,n)^k)$, is a function

$$f^{**}_{\mathbb{M}(t,\infty)} : I^*(\mathbb{M}(t,n)^k) \subseteq I^*(\mathbb{M}(t,\infty)^k) \longrightarrow I^*(\mathbb{M}(t,\infty))$$

defined by

$$f^{**}_{\mathbb{M}(t,\infty)}(\mathcal{X}) = \langle f^{**}(P_\mathbb{R}(\mathcal{X})), t, g_f(\mathcal{X}), \infty, b \rangle.$$

where $P_\mathbb{R}(\mathcal{X})$ is the interval projection of $\mathcal{X}$, f^{**} is the **-semantic extension of f to the interval $P_\mathbb{R}(\mathcal{X})$, and $g_f(\mathcal{X})$ is the maximum of the granularities which would be obtained in the computation of the bounds of $f^{**}(P_\mathbb{R}(\mathcal{X}))$ in a theoretical way, i.e., with $n - \infty$.

If $\mathcal{X}$ is unimodal proper, then

$$f^*_{\mathbb{M}(t,\infty)}(\mathcal{X}) = f^{**}_{\mathbb{M}(t,\infty)}(\mathcal{X}) = (R_{f_{\mathbb{M}(t,n)}}(\mathcal{X}'), \exists),$$

while if $\mathcal{X}$ is unimodal improper, then

$$f^*_{\mathbb{M}(t,\infty)}(\mathcal{X}) = f^{**}_{\mathbb{M}(t,\infty)}(\mathcal{X}) = (R_{f_{\mathbb{M}(t,n)}}(P_\mathbb{R}(\mathcal{X}')), \forall).$$

Example 9.4.1 For the real continuous $f(x_1, x_2) = x_1^2 + x_2^2$ the computation of the *-semantic and the **-semantic functions for $X = ([-1, 1], [1, -1])$ yields the following results:

$$\begin{aligned}
f^*([-1,1],[1,-1]) &= \bigvee_{x_1 \in [-1,1]'} \bigwedge_{x_2 \in [-1,1]'} [x_1^2 + x_2^2, x_1^2 + x_2^2] \\
&= \bigvee_{x_1 \in [-1,1]'} [x_1^2 + 1, x_1^2] = [1,1] \\
f^{**}([-1,1],[1,-1]) &= \bigwedge_{x_2 \in [-1,1]'} \bigvee_{x_1 \in [-1,1]'} [x_1^2 + x_2^2, x_1^2 + x_2^2] \\
&= \bigwedge_{x_2 \in [-1,1]'} [x_2^2, 1 + x_2^2] = [1,1].
\end{aligned}$$

Therefore the *- and **-semantic extensions for the corresponding mark function are

$$f^*_{M(t,\infty)}([\langle -1, 0.001\rangle, \langle 1, 0.001\rangle], [\langle 1, 0.001\rangle, \langle -1, 0.001\rangle])$$
$$= f^{**}_{M(t,\infty)}([\langle -1, 0.001\rangle, \langle 1, 0.001\rangle], [\langle 1, 0.001\rangle, \langle -1, 0.001\rangle])$$
$$= \langle [1, 1], t, 0.001, \infty, b\rangle,$$

because $g_f(\mathcal{X}) = 0.001$, by the computation of the granularities of the product and sum under the minimalist criterion. Both semantic extensions give the same result.

For the real continuous function $g(x_1, x_2) = (x_1 + x_2)^2$ and the interval $X = ([-1, 1], [1, -1])$

$$\begin{aligned}
g^*([-1, 1], [1, -1]) &= \bigvee_{x_1 \in [-1,1]'} \bigwedge_{x_2 \in [-1,1]'} [(x_1 + x_2)^2, (x_1 + x_2)^2] \\
&= \bigvee_{x_1 \in [-1,1]'} [\text{if } x_1 < 0 \text{ then } (x_1 - 1)^2 \text{ else } (x_1 + 1)^2, 0] \\
&= [1, 0]
\end{aligned}$$

$$\begin{aligned}
g^{**}([-1, 1], [1, -1]) &= \bigwedge_{x_2 \in [-1,1]'} \bigvee_{x_1 \in [-1,1]'} [(x_1 + x_2)^2, (x_1 + x_2)^2] \\
&= \bigwedge_{x_2 \in [-1,1]'} [0, \text{if } x_2 < 0 \text{ then } (x_2 - 1)^2 \text{ else } (x_2 + 1)^2] \\
&= [0, 1].
\end{aligned}$$

Therefore the *- and **-semantic extensions for the corresponding mark function are

$$g^*_{M(t,\infty)}([\langle -1, 0.001\rangle, \langle 1, 0.001\rangle], [\langle 1, 0.001\rangle, \langle -1, 0.001\rangle]) = \langle [1, 0], t, 0.001, \infty, b\rangle,$$
$$g^{**}_{M(t,\infty)}([\langle -1, 0.001\rangle, \langle 1, 0.001\rangle], [\langle 1, 0.001\rangle, \langle -1, 0.001\rangle]) = \langle [0, 1], t, 0.001, \infty, b\rangle.$$

because $g_g(\mathcal{X}) = 0.001$. For this function their semantic extensions give different results.

The equality between both extensions characterizes the following important concept.

Definition 9.4.5 (JM-commutativity) A function of marks associated to the continuous function $f: \mathbb{R}^k \to \mathbb{R}$, is *JM-commutable* in $\mathcal{X} \in I^*(\mathbf{M}^k)$ when

$$f^*_{M(t,\infty)}(\mathcal{X}) = f^{**}_{M(t,\infty)}(\mathcal{X}).$$

9.4.2 Properties of the *- and **-Semantics Functions

Interesting properties of the modal interval semantic extensions for a real continuous function are isotonicity

$$\mathcal{X} \subseteq \mathcal{Y} \implies f^*(\mathcal{X}) \subseteq f^*(\mathcal{Y}) \text{ and } f^{**}(\mathcal{X}) \subseteq f^{**}(\mathcal{Y})$$

9.4 Interval Extensions of Functions of Marks

and the inclusion

$$f^*(\mathfrak{X}) \subseteq f^{**}(\mathfrak{X})$$

Since the functions $f^*_{\mathbb{M}(t,\infty)}$ and $f^{**}_{\mathbb{M}(t,\infty)}$ are theoretical, that is to say, they are calculated in an exact way, the properties that they verify will coincide, roughly speaking, with the properties of the functions f^* and f^{**}. We will give as evident those properties that refer to this aspect.

Theorem 9.4.1 *If* $f_{\mathbb{M}(t,\infty)} : \mathbb{M}(t,\infty)^k \to \mathbb{M}(t,\infty)$ *is a function of marks, associated to the continuous function* $f : \mathbb{R}^k \to \mathbb{R}$, *and* $(\mathfrak{X}'_1, \mathfrak{X}'_2)$ *is any component splitting of* $\mathfrak{X}' \in I(\mathbb{M}^k)$, *then*

$$(\forall (\mathfrak{x}_1, \mathfrak{x}_2) \in (\mathfrak{X}'_1, \mathfrak{X}'_2)) \max_{\mathfrak{x}_1 \in \mathfrak{X}'_1} \min_{\mathfrak{x}_2 \in \mathfrak{X}'_2} f_{\mathbb{M}(t,\infty)}(\mathfrak{x}_1, \mathfrak{x}_2) \leq \min_{x_2 \in \mathfrak{X}'_2} \max_{\mathfrak{x}_1 \in \mathfrak{X}'_1} f_{\mathbb{M}(t,\infty)}(\mathfrak{x}_1, \mathfrak{x}_2).$$

Proof See the proof of Theorem 3.2.2. ∎

Lemma 9.4.2 *Let* $F_{1\mathbb{M}(t,n)}, F_{2\mathbb{M}(t,n)} : \mathbb{M}(t,n) \to I^*(\mathbb{M}(t,n))$. *Given* $\alpha \in]0,1]$ *and* $\mathfrak{X} \in I^*(\mathbb{M}(t,n))$, *for every* $\mathfrak{x} \in \mathfrak{X}'$

$$F_{1\mathbb{M}(t,n)}(\mathfrak{x}) \subseteq_\alpha F_{2\mathbb{M}(t,n)}(\mathfrak{x}) \Rightarrow \underset{(\mathfrak{x},\mathfrak{X})}{\Omega} F_{1\mathbb{M}(t,n)}(\mathfrak{x}) \subseteq_\alpha \underset{(\mathfrak{x},\mathfrak{X})}{\Omega} F_{2\mathbb{M}(t,n)}(\mathfrak{x}).$$

Proof See Lemma 3.2.1. ∎

Lemma 9.4.3 *Let* $F_{\mathbb{M}(t,n)} : \mathbb{M}(t,n) \to I^*(\mathbb{M}(t,n))$. *Given* $\mathfrak{X}_1, \mathfrak{X}_2 \in I^*(\mathbb{M}(t,n))$, *then*

$$\mathfrak{X}_1 \subseteq \mathfrak{X}_2 \Rightarrow \underset{(\mathfrak{x},\mathfrak{X}_1)}{\Omega} F_{\mathbb{M}(t,n)}(\mathfrak{x}) \subseteq \underset{(\mathfrak{x},\mathfrak{X}_2)}{\Omega} F_{\mathbb{M}(t,n)}(\mathfrak{x}).$$

Proof 1. If $\mathfrak{X}_1$ is proper, $\mathfrak{X}_2$ is proper, and $\mathfrak{X}'_1 \subseteq \mathfrak{X}'_2$, then

$$\underset{(\mathfrak{x},\mathfrak{X}_1)}{\Omega} F_{\mathbb{M}(t,n)}(\mathfrak{x}) = \bigvee_{\mathfrak{x} \in \mathfrak{X}'_1} F_{\mathbb{M}(t,n)}(\mathfrak{x})$$

$$= [\min_{\mathfrak{x} \in \mathfrak{X}'_1} \inf(F_{\mathbb{M}(t,n)}(\mathfrak{x})), \max_{\mathfrak{x} \in \mathfrak{X}'_1} \sup(F_{\mathbb{M}(t,n)}(\mathfrak{x}))]$$

$$\subseteq [\min_{\mathfrak{x} \in \mathfrak{X}'_2} \inf(F_{\mathbb{M}(t,n)}(\mathfrak{x})), \max_{\mathfrak{x} \in \mathfrak{X}'_2} \sup(F_{\mathbb{M}(t,n)}(\mathfrak{x}))]$$

$$= \bigvee_{\mathfrak{x} \in \mathfrak{X}'_2} F_{\mathbb{M}(t,n)}(\mathfrak{x}) = \underset{(\mathfrak{x},\mathfrak{X}_2)}{\Omega} F_{\mathbb{M}(t,n)}(\mathfrak{x}).$$

2. If $\mathfrak{X}_1$ is proper, $\mathfrak{X}_2$ is improper, and $\mathfrak{X}'_1 \subseteq \mathfrak{X}'_2$, then $\mathfrak{X}_1 = \mathfrak{X}_2 = [\mathfrak{a}, \mathfrak{a}]$ and $\underset{(\mathfrak{x},[\mathfrak{a},\mathfrak{a}])}{\Omega}$ is the identity operator.
3. If $\mathfrak{X}_1$ is improper,

$$\underset{(\mathfrak{x},\mathfrak{X}_1)}{\Omega} F_{\mathbb{M}(t,n)}(\mathfrak{x}) = \bigwedge_{\mathfrak{x} \in \mathfrak{X}'_1} F_{\mathbb{M}(t,n)}(\mathfrak{x})$$
$$= [\max_{\mathfrak{x} \in \mathfrak{X}'_1} \inf(F_{\mathbb{M}(t,n)}(\mathfrak{x})), \min_{\mathfrak{x} \in \mathfrak{X}'_1} \sup(F_{\mathbb{M}(t,n)}(\mathfrak{x}))].$$

▷ If $\mathfrak{X}_2$ is proper, the inclusion $\mathfrak{X}_1 \subseteq \mathfrak{X}_2$ is equivalent to $\mathfrak{X}'_1 \cap \mathfrak{X}'_2 \neq \emptyset$. Let $\hat{\mathfrak{x}} \in \mathfrak{X}'_1 \cap \mathfrak{X}'_2$

$$\max_{\mathfrak{x} \in \mathfrak{X}'_1} \inf(F_{\mathbb{M}(t,n)}(\mathfrak{x})) \geq \inf(F_{\mathbb{M}(t,n)}(\hat{\mathfrak{x}})) \geq \min_{\mathfrak{x} \in \mathfrak{X}'_2} \inf(F_{\mathbb{M}(t,n)}(\mathfrak{x})),$$
$$\min_{\mathfrak{x} \in \mathfrak{X}'_1} \sup(F_{\mathbb{M}(t,n)}(\mathfrak{x})) \leq \sup(F_{\mathbb{M}(t,n)}(\hat{X})) \leq \max_{\mathfrak{x} \in \mathfrak{X}'_2} \sup(F_{\mathbb{M}(t,n)}(\mathfrak{x})).$$

▷ If $\mathfrak{X}_2$ is improper, the inclusion $\mathfrak{X}_1 \subseteq \mathfrak{X}_2$ is equivalent to $\mathfrak{X}'_2 \subseteq \mathfrak{X}'_1$, and therefore

$$\max_{\mathfrak{x} \in \mathfrak{X}'_1} \inf(F_{\mathbb{M}(t,n)}(\mathfrak{x})) \geq \inf(F_{\mathbb{M}(t,n)}(\hat{\mathfrak{x}})) \geq \min_{\mathfrak{x} \in \mathfrak{X}'_2} \inf(F_{\mathbb{M}(t,n)}(\mathfrak{x})),$$
$$\min_{\mathfrak{x} \in \mathfrak{X}'_1} \sup(F_{\mathbb{M}(t,n)}(\mathfrak{x})) \leq \sup(F_{\mathbb{M}(t,n)}(\hat{\mathfrak{x}})) \leq \max_{\mathfrak{x} \in \mathfrak{X}'_2} \sup(F_{\mathbb{M}(t,n)}(\mathfrak{x})). \blacksquare$$

Theorem 9.4.2 *Let* $F_{1\mathbb{M}(t,n)}, F_{2\mathbb{M}(t,n)} : \mathbb{M}(t,n) \to I^*(\mathbb{M}(t,n))$. *Given* $\mathfrak{X}_1, \mathfrak{X}_2 \in I^*(\mathbb{M}(t,n))$ *and* $\alpha \in]0,1]$, *then*

$$(\mathfrak{X}_1 \subseteq \mathfrak{X}_2, F_{1\mathbb{M}(t,n)}(\mathfrak{x}) \subseteq_\alpha F_{2\mathbb{M}(t,n)}(\mathfrak{x})) \Rightarrow \underset{(\mathfrak{x},\mathfrak{X}_1)}{\Omega} F_{1\mathbb{M}(t,n)}(\mathfrak{x}) \subseteq_\alpha \underset{(\mathfrak{x},\mathfrak{X}_2)}{\Omega} F_{2\mathbb{M}(t,n)}(\mathfrak{x}),$$

whenever there exists compatibility of $\underset{(\mathfrak{x},\mathfrak{X}_1)}{\Omega} F_{1\mathbb{M}(t,n)}(\mathfrak{x})$ *and* $\underset{(\mathfrak{x},\mathfrak{X}_2)}{\Omega} F_{2\mathbb{M}(t,n)}(\mathfrak{x})$ *with* αt.

Proof Applying Lemma 9.4.3

$$\mathfrak{X}_1 \subseteq \mathfrak{X}_2 \Rightarrow \underset{(\mathfrak{x},\mathfrak{X}_1)}{\Omega} F_{1\mathbb{M}(t,n)}(\mathfrak{x}) \subseteq \underset{(\mathfrak{x},\mathfrak{X}_2)}{\Omega} F_{1\mathbb{M}(t,n)}(\mathfrak{x})$$

and using now Lemma 9.4.2

$$(\forall \mathfrak{x} \in \mathfrak{X}'_2) \, F_{1\mathbb{M}(t,n)}(\mathfrak{x}) \subseteq_\alpha F_{2\mathbb{M}(t,n)}(\mathfrak{x}) \Rightarrow \underset{(\mathfrak{x},\mathfrak{X}_2)}{\Omega} F_{1\mathbb{M}(t,n)}(\mathfrak{x}) \subseteq_\alpha \underset{(\mathfrak{x},\mathfrak{X}_2)}{\Omega} F_{2\mathbb{M}(t,n)}(\mathfrak{x}),$$

then

$$\underset{(\mathfrak{x},\mathfrak{X}_1)}{\Omega} F_{1\mathbb{M}(t,n)}(\mathfrak{x}) \subseteq \underset{(\mathfrak{x},\mathfrak{X}_2)}{\Omega} F_{1\mathbb{M}(t,n)}(\mathfrak{x}) \subseteq_\alpha \underset{(\mathfrak{x},\mathfrak{X}_2)}{\Omega} F_{2\mathbb{M}(t,n)}(\mathfrak{x}).$$

9.4 Interval Extensions of Functions of Marks

By transitivity

$$\bigcap_{(\mathfrak{x},\mathcal{X}_1)} F_{1\mathrm{M}(t,n)}(\mathfrak{x}) \subseteq_\alpha \bigcap_{(\mathfrak{x},\mathcal{X}_2)} F_{2\mathrm{M}(t,n)}(\mathfrak{x}).$$

■

Definition 9.4.6 (Saddle points and saddle value) Let $(\mathcal{X}'_1, \mathcal{X}'_2) = \mathcal{X}'$ be a component splitting of $I(\mathbb{M}(t,n)^k)$. The *saddle point set* in $(\mathcal{X}'_1, \mathcal{X}'_2)$ of the functions of marks associated to $f : \mathbb{R}^k \to \mathbb{R}$, denoted by $\mathrm{SDP}(f_{\mathrm{M}(t,n)}, \mathcal{X}'_1, \mathcal{X}'_2)$, is defined by means of the set of saddle points of the function f in $P_{\mathbb{R}}(\mathcal{X}'_1, \mathcal{X}'_2)$

$$\mathrm{SDP}(f_{\mathrm{M}(t,n)}, \mathcal{X}'_1, \mathcal{X}'_2) = \{(\mathfrak{x}^m_1, \mathfrak{x}^M_2) = \{(\langle x^m_1, g_1 \rangle, \langle x^M_2, g_2 \rangle) \in (\mathcal{X}'_1, \mathcal{X}'_2) \mid$$
$$(x^m_1, x^M_2) \in \mathrm{SDP}(f, P_{\mathbb{R}}(\mathcal{X}'_1, \mathcal{X}'_2))\}$$

and the *saddle value* $\mathrm{SDV}(f_{\mathrm{M}(t,n)}, \mathcal{X}'_1, \mathcal{X}'_2)$ is

$$\mathrm{SDV}(f_{\mathrm{M}(t,n)}, \mathcal{X}'_1, \mathcal{X}'_2) = \begin{cases} f_{\mathrm{M}(t,n)}(\mathfrak{x}^m_1, \mathfrak{x}^M_2) & \text{if } (\mathfrak{x}^m_1, \mathfrak{x}^M_2) \in \mathrm{SDP}(f_{\mathrm{M}(t,n)}, \mathcal{X}'_1, \mathcal{X}'_2) \\ \text{not defined} & \text{if } \mathrm{SDP}(f_{\mathrm{M}}, \mathcal{X}'_1, \mathcal{X}'_2) = \emptyset. \end{cases}$$

Starting from the properties of the saddle values of the function f, it is possible to show that if $(\mathfrak{x}^m_1, \mathfrak{x}^M_2)$ is a saddle point of the functions f_{M} associated to f in $(\mathcal{X}'_1, \mathcal{X}'_2)$, then

$$\min_{\mathfrak{x}_1 \in \mathcal{X}'_1} \max_{\mathfrak{x}_2 \in \mathcal{X}'_2} f_{\mathrm{M}(t,\infty)}(\mathfrak{x}_1, \mathfrak{x}_2) = f_{\mathrm{M}(t,\infty)}(\mathfrak{x}^m_1, \mathfrak{x}^M_2) = \max_{\mathfrak{x}_2 \in \mathcal{X}'_2} \min_{\mathfrak{x}_1 \in \mathcal{X}'_1} f_{\mathrm{M}(t,\infty)}(\mathfrak{x}_1, \mathfrak{x}_2).$$

From the properties of the function of marks, and computing the granularity under a maximalist approach, and whenever the granularities are compatible with αt, then

$$\mathrm{SDV}(f_{\mathrm{M}(t,n)}, \mathcal{X}'_1, \mathcal{X}'_2) \approx_\alpha f_{\mathrm{M}(t,\infty)}(\mathfrak{x}^m_1, \mathfrak{x}^M_2).$$

The next concepts and results can be proved using the parallelism between $f^*_{\mathrm{M}(t,\infty)}$ and $f^{**}_{\mathrm{M}(t,\infty)}$ and the functions f^* and f^{**}:

Theorem 9.4.3 (JM-commutativity) *If $f_{\mathrm{M}} : \mathbb{M}(t,n)^k \to \mathbb{M}(t,n)$ is a function of marks associated to the continuous function $f : \mathbb{R}^k \to \mathbb{R}$, for a given $\mathcal{X} = (\mathcal{X}_p, \mathcal{X}_i) \in I^*(\mathbb{M}^k)$ split in its proper and improper components, then*

$$\left. \begin{array}{l} \mathrm{SDP}(f_{\mathrm{M}(t,n)}, \mathcal{X}'_p, \mathcal{X}'_i) \neq \emptyset \\ \mathrm{SDP}(f_{\mathrm{M}(t,n)}, \mathcal{X}'_i, \mathcal{X}'_p) \neq \emptyset \end{array} \right\} \Leftrightarrow f^*_{\mathrm{M}(t,\infty)}(\mathcal{X}) = f^{**}_{\mathrm{M}(t,\infty)}(\mathcal{X}).$$

Theorem 9.4.4 (Commutativity condition) *f_{M} is JM-commutable over $\mathcal{X} \in I^*(\mathbb{M}^k)$ iff*

$$f^*_{M(t,\infty)}(\mathfrak{X}) = f^{**}_{M(t,\infty)}(\mathfrak{X}) = \langle [SDV(f_{M(t,\infty)}, \mathfrak{X}'_p, \mathfrak{X}'_i), SDV(f_{M(t,\infty)}, \mathfrak{X}'_i, \mathfrak{X}'_p)], g_f \rangle.$$

Example 9.4.2 For the real continuous $f(x_1, x_2) = x_1^2 + x_2^2$,

$$f^*([-1,1],[2,0]) = \bigvee_{x_1 \in [-1,1]'} \bigwedge_{x_2 \in [0,2]'} [x_1^2 + x_2^2, x_1^2 + x_2^2]$$

$$= \bigvee_{x_1 \in [-1,1]'} [x_1^2 + 4, x_1^2] = [4, 1],$$

$$f^{**}([-1,1],[2,0]) = \bigwedge_{x_2 \in [0,2]'} \bigvee_{x_1 \in [-1,1]'} [x_1^2 + x_2^2, x_1^2 + x_2^2]$$

$$= \bigwedge_{x_2 \in [0,2]'} [x_2^2, x_2^2 + 1] = [4, 1],$$

and

$$SDP(f, [-1,1]', [0,2]') = \{(0,2)\}, \qquad SDV(f, [-1,1]', [0,2]') = 4,$$
$$SDP(f, [0,2]', [-1,1]') = \{(1,0), (-1,0)\}, \quad SDV(f, [0,2]', [-1,1]') = 1.$$

The *-semantic and **-semantic extensions to the interval of marks

$$\mathfrak{X} = ([\langle -1, 0.001 \rangle, \langle 1, 0.001 \rangle], [\langle 2, 0.001 \rangle, \langle 0, 0.001 \rangle])$$

are

$$f^*([\langle -1, 0.001 \rangle, \langle 1, 0.001 \rangle], [\langle 2, 0.001 \rangle, \langle 0, 0.001 \rangle])$$
$$= f^{**}([\langle -1, 0.001 \rangle, \langle 1, 0.001 \rangle], [\langle 2, 0.001 \rangle, \langle 0, 0.001 \rangle])$$
$$= [\langle 4, 0.001 \rangle, \langle 1, 0.001 \rangle].$$

Therefore,

$$SDP(f, [\langle -1, 0.001 \rangle, \langle 1, 0.001 \rangle]', [\langle 0, 0.001 \rangle, \langle 2, 0.001 \rangle]')$$
$$= \{(\langle 0, 0.001 \rangle, \langle 2, 0.001 \rangle)\},$$
$$SDV(f, [\langle -1, 0.001 \rangle, \langle 1, 0.001 \rangle]', [\langle 0, 0.001 \rangle, \langle 2, 0.001 \rangle]')$$
$$= \langle 4, 0.001 \rangle,$$
$$SDP(f, [\langle 0, 0.001 \rangle, \langle 2, 0.001 \rangle]', [\langle -1, 0.001 \rangle, \langle 1, 0.001 \rangle]')$$
$$= \{(\langle 1, 0.001 \rangle, \langle 0, 0.001 \rangle), (\langle -1, 0.001 \rangle, \langle 0, 0.001 \rangle)\},$$
$$SDV(f, [\langle 0, 0.001 \rangle, \langle 2, 0.001 \rangle]', [\langle -1, 0.001 \rangle, \langle 1, 0.001 \rangle]')$$
$$= \langle 1, 0.001 \rangle.$$

9.4 Interval Extensions of Functions of Marks   251

Similarly, for $(X_1, X_2) = ([-1, 1], [1, -1])$, $f_1 = x_1 x_2$, $f_2 = \text{Abs}(x_1 x_2)$ and $f_3 = x_1^2 + x_2^2$ their corresponding mark extensions are also JM-commutable, but not for $f_4 = \text{Abs}(x_1 + x_2)$ and $f_5 = (x_1 + x_2)^2$.

Some important examples of JM-commutable functions are the one-variable continuous functions and those two-variable continuous function $f(x, y)$ which are partially monotonic in a domain $(\mathcal{X}', \mathcal{Y}')$, such as the arithmetic operators $x + y$, $x - y$, $x * y$, x/y and others such as x^y, $\max(x, y)$ and $\min(x, y)$, whose modal semantic extensions can be computed by means of arithmetic operations with the interval mark bounds.

9.4.3 Semantic Theorems

Theorem 9.4.5 (Semantic theorem for $f^*_{\mathbb{M}(t,\infty)}$) *Let* $\mathfrak{A} \in I^*(\mathbb{M}(t, n)^k)$, $f^*_{\mathbb{M}(t,\infty)}$ *be the $*$-semantic function associated to* $f_{\mathbb{M}(t,n)} : \mathbb{M}(t, n)^k \to \mathbb{M}(t, n)$ *and* $\mathfrak{Z} \in I^*(\mathbb{M}(t, \infty))$. *If all involved marks are valid, the following statements are equivalent:*

1. $f^*_{\mathbb{M}(t,\infty)}(\mathfrak{A}) \subseteq \mathfrak{Z}$.
2. $(\forall \mathcal{X}' \in I(\mathbb{M}(t, n)^k))$
   $((\mathfrak{x} \in \mathcal{X}') \in \text{Pred}^*(\mathfrak{A}) \Rightarrow (\mathfrak{z} \in R_{f_{\mathbb{M}(t,n)}}(\mathcal{X}')) \in \text{Pred}^*(\mathfrak{Z}))$.
3. $(\forall \mathfrak{a}_p \in \mathfrak{A}'_p) \, Q(\mathfrak{z}, \mathfrak{Z}') \, (\exists \mathfrak{a}_i \in \mathfrak{A}'_i) \, \mathfrak{z} = f_{\mathbb{M}(t,\infty)}(\mathfrak{a}_p, \mathfrak{a}_i)$.

Proof See Theorem 3.3.1. ∎

Corollary 9.4.1 (Semantic theorem over $I^*(\mathbb{M}(t, n))$) *Let* $\mathfrak{A} \in I^*(\mathbb{M}(t, n)^k)$, $f^*_{\mathbb{M}(t,\infty)}$ *be the $*$-semantic function associated to* $f_{\mathbb{M}(t,n)} : \mathbb{M}(t, n)^k \to \mathbb{M}(t, n)$ *and* $\mathfrak{Z} \in I^*(\mathbb{M}(t, n))$. *For certain value* $\xi(\alpha) \in [0, 1]$, *under a maximalist approach of the calculus of the granularity and whenever there is compatibility with $\xi(\alpha)t$ when it is necessary, it is true that*

$$f^*_{\mathbb{M}(t,\infty)}(\mathfrak{A}) \subseteq_\alpha \mathfrak{Z} \Rightarrow (\forall \mathfrak{a}_p \in \mathfrak{A}'_p) \, Q(\mathfrak{z}, \mathfrak{Z}) \, (\exists \mathfrak{a}_i \in \mathfrak{A}'_i) \, \mathfrak{z} \approx_{\xi(\alpha)} f_{\mathbb{M}(t,n)}(\mathfrak{a}_p, \mathfrak{a}_i).$$

Proof From Theorem 9.4.5 applied to the inclusion $f^*_{\mathbb{M}(t,\infty)}(\mathfrak{A}) \subseteq f^*_{\mathbb{M}(t,\infty)}(\mathfrak{A})$,

$$(\forall \mathfrak{a}_p \in \mathfrak{A}'_p) \, Q(\mathfrak{y}, f^*_{\mathbb{M}(t,\infty)}(\mathfrak{A})) \, (\exists \mathfrak{a}_i \in \mathfrak{A}'_i) \, \mathfrak{y} = f_{\mathbb{M}(t,\infty)}(\mathfrak{a}_p, \mathfrak{a}_i).$$

Then

$$f_{\mathbb{M}(t,\infty)}(\mathfrak{a}_p, \mathfrak{a}_i) \approx_{\beta_1} f_{\mathbb{M}(t,\infty)}(DI_n(\mathfrak{a}_p, \mathfrak{a}_i)) \approx_{\beta_2} f_{\mathbb{M}(t,n)}(DI_n(\mathfrak{a}_p, \mathfrak{a}_i)),$$

and therefore

$$f_{\mathbb{M}(t,\infty)}(\mathfrak{a}_p, \mathfrak{a}_i) \approx_{\beta_1 + \beta_2} f_{\mathbb{M}(t,n)}(DI_n(\mathfrak{a}_p, \mathfrak{a}_i)),$$

supposing the marks are valid with respect to $\beta_1 t$ and to $\beta_2 t$, respectively.

From hypothesis, $f^*_{\mathbb{M}(t,\infty)}(\mathfrak{A}) \subseteq_\alpha \mathfrak{Z}$ and splitting the cases by the different possible modalities of $f^*_{\mathbb{M}(t,\infty)}(\mathfrak{A})$ and $\mathfrak{Z}$:

1. When $f^*_{\mathbb{M}(t,\infty)}(\mathfrak{A})$ and $\mathfrak{Z}$ are proper,

$$f^*_{\mathbb{M}(t,\infty)}(\mathfrak{A}) \subseteq_\alpha \mathfrak{Z} \Leftrightarrow (f^*_{\mathbb{M}(t,\infty)}(\mathfrak{A}))' \subseteq_\alpha \mathfrak{Z}'.$$

In this case, and whenever the marks are compatible with αt, it is true that

$$(\forall \mathfrak{y} \in (f^*_{\mathbb{M}(t,\infty)}(\mathfrak{A}))')\ (\exists \mathfrak{z} \in \mathfrak{Z}')\ \mathfrak{y} \approx_\alpha \mathfrak{z},$$

Combining this weak equality with

$$(\forall \mathfrak{a}_p \in \mathfrak{A}'_p)\ (\exists \mathfrak{y} \in f^*_{\mathbb{M}(t,\infty)}(\mathfrak{A}))\ (\exists \mathfrak{a}_i \in \mathfrak{A}'_i)$$
$$\mathfrak{y} = f_{\mathbb{M}(t,\infty)}(\mathfrak{a}_p, \mathfrak{a}_i) \approx_{\beta_1+\beta_2} f_{\mathbb{M}(t,n)}(DI_n(\mathfrak{a}_p, \mathfrak{a}_i)),$$

then

$$(\forall \mathfrak{a}_p \in \mathfrak{A}'_p)\ (\exists \mathfrak{y} \in f^*_{\mathbb{M}(t,\infty)}(\mathfrak{A}))\ (\exists \mathfrak{z} \in \mathfrak{Z}')\ (\exists \mathfrak{a}_i \in \mathfrak{A}'_i)$$
$$\mathfrak{z} \approx_\alpha \mathfrak{y} = f_{\mathbb{M}(t,\infty)}(\mathfrak{a}_p, \mathfrak{a}_i) \approx_{\beta_1+\beta_2} f_{\mathbb{M}(t,n)}(DI_n(\mathfrak{a}_p, \mathfrak{a}_i)).$$

Applying the $(\alpha + \beta)$-transitivity of the weak equality between marks,

$$(\forall \mathfrak{a}_p \in \mathfrak{A}'_p)\ (\exists \mathfrak{z} \in \mathfrak{Z}')\ (\exists \mathfrak{a}_i \in \mathfrak{A}'_i)\ \mathfrak{z} \approx_{\alpha+\beta_1+\beta_2} f_{\mathbb{M}(t,n)}(DI_n(\mathfrak{a}_p, \mathfrak{a}_i)).$$

2. When $f^*_{\mathbb{M}(t,\infty)}(\mathfrak{A})$ and $\mathfrak{Z}$ are improper,

$$f^*_{\mathbb{M}(t,\infty)}(\mathfrak{A}) \subseteq_\alpha \mathfrak{Z} \Leftrightarrow \mathfrak{Z}' \subseteq_\alpha (f^*_{\mathbb{M}(t,\infty)}(\mathfrak{A}))'.$$

In this case, and whenever the marks are compatible with αt, it is true that

$$(\forall \mathfrak{z} \in \mathfrak{Z}')\ (\exists \mathfrak{y} \in (f^*_{\mathbb{M}(t,\infty)}(\mathfrak{A}))')\ \mathfrak{y} \approx_\alpha \mathfrak{z}.$$

Combining this weak equality with

$$(\forall \mathfrak{a}_p \in \mathfrak{A}'_p)\ (\forall \mathfrak{y} \in f^*_{\mathbb{M}(t,\infty)}(\mathfrak{A}))\ (\exists \mathfrak{a}_i \in \mathfrak{A}'_i)$$
$$\mathfrak{y} = f_{\mathbb{M}(t,\infty)}(\mathfrak{a}_p, \mathfrak{a}_i) \approx_{\beta_1+\beta_2} f_{\mathbb{M}(t,n)}(DI_n(\mathfrak{a}_p, \mathfrak{a}_i)),$$

then

$$(\forall \mathfrak{z} \in \mathfrak{Z}')\ (\exists \mathfrak{y} \in (f^*_{\mathbb{M}(t,\infty)}(\mathfrak{A}))')\ (\forall \mathfrak{a}_p \in \mathfrak{A}'_p)\ (\forall \mathfrak{y} \in f^*_{\mathbb{M}(t,\infty)}(\mathfrak{A}))\ (\exists \mathfrak{a}_i \in \mathfrak{A}'_i)$$
$$\mathfrak{z} \approx_\alpha \mathfrak{y} = f_{\mathbb{M}(t,\infty)}(\mathfrak{a}_p, \mathfrak{a}_i) \approx_{\beta_1+\beta_2} f_{\mathbb{M}(t,n)}(DI_n(\mathfrak{a}_p, \mathfrak{a}_i)),$$

Applying the $(\alpha + \beta)$-transitivity of the weak equality between marks,

9.4 Interval Extensions of Functions of Marks

$$(\forall \mathfrak{a}_p \in \mathfrak{A}'_p)\ (\forall \mathfrak{z} \in \mathfrak{Z}')\ (\exists \mathfrak{a}_i \in \mathfrak{A}'_i)\ \mathfrak{z} \approx_{\alpha+\beta_1+\beta_2} f_{\mathbb{M}(t,n)}(DI_n(\mathfrak{a}_p, \mathfrak{a}_i)).$$

3. $f^*_{\mathbb{M}(t,\infty)}(\mathfrak{A})$ improper and $\mathfrak{Z}$ proper,

$$f^*_{\mathbb{M}(t,\infty)}(\mathfrak{A}) \subseteq_\alpha \mathfrak{Z} \Leftrightarrow (\exists \mathfrak{y} \in (f^*_{\mathbb{M}(t,\infty)}(\mathfrak{A}))')\ (\exists \mathfrak{z} \in \mathfrak{Z}')\ \mathfrak{y} \approx_\alpha \mathfrak{z}.$$

As

$$(\forall \mathfrak{a}_p \in \mathfrak{A}'_p)\ (\forall \mathfrak{y} \in f^*_{\mathbb{M}(t,\infty)}(\mathfrak{A}))\ (\exists \mathfrak{a}_i \in \mathfrak{A}'_i)$$
$$\mathfrak{y} = f_{\mathbb{M}(t,\infty)}(\mathfrak{a}_p, \mathfrak{a}_i) \approx_{\beta_1+\beta_2} f_{\mathbb{M}(t,n)}(DI_n(\mathfrak{a}_p, \mathfrak{a}_i)),$$

then

$$(\forall \mathfrak{a}_p \in \mathfrak{A}'_p)\ (\forall \mathfrak{y} \in f^*_{\mathbb{M}(t,\infty)}(\mathfrak{A})')\ (\exists \mathfrak{a}_i \in \mathfrak{A}'_i)\ (\exists \mathfrak{y} \in (f^*_{\mathbb{M}(t,\infty)}(\mathfrak{A}))')\ (\exists \mathfrak{z} \in \mathfrak{Z}')$$
$$\mathfrak{z} \approx_\alpha \mathfrak{y} = f_{\mathbb{M}(t,\infty)}(\mathfrak{a}_p, \mathfrak{a}_i) \approx_{\beta_1+\beta_2} f_{\mathbb{M}(t,n)}(DI_n(\mathfrak{a}_p, \mathfrak{a}_i)),$$

and therefore,

$$(\forall \mathfrak{a}_p \in \mathfrak{A}'_p)\ \exists(\mathfrak{z} \in \mathfrak{Z}')\ (\exists \mathfrak{a}_i \in \mathfrak{A}'_i)\ \mathfrak{z} \approx_{\alpha+\beta_1+\beta_2} f_{\mathbb{M}(t,n)}(DI_n(\mathfrak{a}_p, \mathfrak{a}_i)).$$

4. When $f^*_{\mathbb{M}(t,\infty)}(\mathfrak{A})$ is proper and $\mathfrak{Z}$ improper,

$$f^*_{\mathbb{M}(t,\infty)}(\mathfrak{A}) \subseteq_\alpha \mathfrak{Z} \Leftrightarrow (\forall \mathfrak{y} \in (f^*_{\mathbb{M}(t,\infty)}(\mathfrak{A}))')\ (\forall \mathfrak{z} \in \mathfrak{Z}')\ \mathfrak{z} \approx_\alpha \mathfrak{y}.$$

Combining it with

$$(\forall \mathfrak{a}_p \in \mathfrak{A}'_p)\ (\exists \mathfrak{y} \in (f^*_{\mathbb{M}(t,\infty)}(\mathfrak{A}))')\ (\exists \mathfrak{a}_i \in \mathfrak{A}'_i)$$
$$\mathfrak{y} = f_{\mathbb{M}(t,\infty)}(\mathfrak{a}_p, \mathfrak{a}_i) \approx_{\beta_1+\beta_2} f_{\mathbb{M}(t,n)}(DI_n(\mathfrak{a}_p, \mathfrak{a}_i)),$$

then

$$(\forall \mathfrak{a}_p \in \mathfrak{A}'_p)\ (\exists \mathfrak{y} \in (f^*_{\mathbb{M}(t,\infty)}(\mathfrak{A}))')\ (\exists \mathfrak{a}_i \in \mathfrak{A}'_i)\ (\forall \mathfrak{y} \in (f^*_{\mathbb{M}(t,\infty)}(\mathfrak{A}))')\ (\forall \mathfrak{z} \in \mathfrak{Z}')$$
$$\mathfrak{z} \approx_\alpha \mathfrak{y} = f_{\mathbb{M}(t,\infty)}(\mathfrak{a}_p, \mathfrak{a}_i) \approx_{\beta_1+\beta_2} f_{\mathbb{M}(t,n)}(DI_n(\mathfrak{a}_p, \mathfrak{a}_i))$$

and therefore,

$$(\forall \mathfrak{a}_p \in \mathfrak{A}'_p)\ (\forall \mathfrak{z} \in \mathfrak{Z}')\ (\exists \mathfrak{a}_i \in \mathfrak{A}'_i)\ \mathfrak{z} \approx_{\alpha+\beta_1+\beta_2} f_{\mathbb{M}(t,n)}(DI_n(\mathfrak{a}_p, \mathfrak{a}_i)).\ \blacksquare$$

For the $**$-semantic function, dual theorems are valid.

Theorem 9.4.6 (Semantic theorem for $f^{**}_{\mathbb{M}(t,\infty)}$) *Let $f^*_{\mathbb{M}(t,\infty)}$ be the $*$-semantic function associated to $f_{\mathbb{M}(t,n)} : \mathbb{M}(t,n)^k \to \mathbb{M}(t,n)$ and $\mathfrak{Z} \in I^*(\mathbb{M}(t,\infty))$. If all involved marks are valid, the following statements are equivalent:*

1. $f^{**}_{\mathbb{M}(t,\infty)}(\mathfrak{A}) \supseteq F_{\mathbb{M}(t,\infty)}(\mathfrak{A}).$

2. $(\forall \mathfrak{X}' \in I(\mathbb{M}(t,n)^k))$
   $((\mathfrak{x} \notin \mathfrak{X}') \in \text{Copred}^*(\mathfrak{A}) \Rightarrow (\mathfrak{z} \notin R_{f_{\mathbb{M}(t,n)}}(\mathfrak{X}')) \in \text{Copred}^*(\mathfrak{Z}))$.
3. $(\forall \mathfrak{a}_i \in \mathfrak{A}'_i) \, Q(\mathfrak{z}, \mathfrak{Z}) \, (\exists \mathfrak{a}_p \in \mathfrak{A}'_p) \, \mathfrak{z} = f_{\mathbb{M}(t,\infty)}(\mathfrak{a}_p, \mathfrak{a}_i)$.

Corollary 9.4.2 (Semantic theorem over $I^*(\mathbb{M}(t,n))$) Let $f^*_{\mathbb{M}(t,\infty)}$ be the $*$-semantic function associated to $f_{\mathbb{M}(t,n)} : \mathbb{M}(t,n)^k \to \mathbb{M}(t,n)$ and $\mathfrak{Z} \in I^*(\mathbb{M}(t,n))$. For some value $\xi(\alpha) \in [0,1]$, under a maximalist approach of the calculus of the granularity and whenever there are compatibility with $\xi(\alpha)t$ when it is necessary, it is true that

$$f^{**}_{\mathbb{M}(t,\infty)}(\mathfrak{A}) \supseteq_\alpha \mathfrak{Z} \Rightarrow (\forall \mathfrak{a}_i \in \mathfrak{A}'_i) \, Q(\mathfrak{z}, \mathfrak{Z}) \, (\exists \mathfrak{a}_p \in \mathfrak{A}'_p) \, \mathfrak{z} \approx_{\xi(\alpha)} f_{\mathbb{M}(t,n)}(\mathfrak{a}_p, \mathfrak{a}_i).$$

Proof Dual of 9.4.1. ∎

Example 9.4.3 For the real continuous function $g(x_1, x_2) = (x_1 + x_2)^2$ and its interval of marks extensions to the interval

$$\mathfrak{X} = ([\langle -1, 0.001 \rangle, \langle 1, 0.001 \rangle], [\langle 1, 0.001 \rangle, \langle -1, 0.001 \rangle]),$$

as

$$g^*_{\mathbb{M}(t,\infty)}([\langle -1, 0.001 \rangle, \langle 1, 0.001 \rangle], [\langle 1, 0.001 \rangle, \langle -1, 0.001 \rangle]) = [\langle 1, 0.001 \rangle, \langle 0, 0.001 \rangle]$$

$$g^{**}_{\mathbb{M}(t,\infty)}([\langle -1, 0.001 \rangle, \langle 1, 0.001 \rangle], [\langle 1, 0.001 \rangle, \langle -1, 0.001 \rangle]) = [\langle 0, 0.001 \rangle, \langle 1, 0.001 \rangle],$$

the $*$-semantic theorem states that

$$(\forall \mathfrak{x}_1 \in [\langle -1, 0.001 \rangle, \langle 1, 0.001 \rangle]') \, (\forall \mathfrak{z} \in [\langle 0, 0.001 \rangle, \langle 1, 0.001 \rangle]')$$
$$(\exists \mathfrak{x}_2 \in [\langle -1, 0.001 \rangle, \langle 1, 0.001 \rangle]') \, \mathfrak{z} = (\mathfrak{x}_1 + \mathfrak{x}_2)^2$$

the $**$-semantic theorem states that

$$(\forall \mathfrak{x}_2 \in [\langle -1, 0.001 \rangle, \langle 1, 0.001 \rangle]') \, (\forall \mathfrak{z} \in [\langle 0, 0.001 \rangle, \langle 1, 0.001 \rangle]')$$
$$(\exists \mathfrak{x}_1 \in [\langle -1, 0.001 \rangle, \langle 1, 0.001 \rangle]') \, \mathfrak{z} = (\mathfrak{x}_1 + \mathfrak{x}_2)^2$$

9.5 Syntactic Extensions

In classical set-theoretical interval analysis, since the domain of values of a general continuous function is generally not computable, set-theoretical interval syntactic extensions $fR(X'_1, \ldots, X'_k)$ are defined like their corresponding real functions $f(x_1, \ldots, x_k)$ replacing real operands and operators by their corresponding interval operands and operators. The relation between both extensions is

9.5 Syntactic Extensions

$$R_f(X'_1, \ldots, X'_k) \subseteq fR(X'_1, \ldots, X'_k),$$

where $fR(X'_1, \ldots, X'_k)$, computable from the bounds of the intervals, usually represents an overestimation of $R_f(X'_1, \ldots, X'_k)$. Syntactic interval functions have the property of being "inclusive",

$$fR(A'_1, \ldots, A'_k) \subseteq fR(B'_1, \ldots, B'_k).$$

This classical interval syntactic extension of f satisfies only one kind of interval predicate compatible with the outer rounding of $f(X')$: if $Z' = fR(X'_1, \ldots, X'_k)$, the only valid semantic statement will be

$$(\forall x_1 \in X'_1) \cdots (\forall x_k \in X'_k)\,(\exists z \in \mathrm{Out}(fR(X'_1, \ldots, X'_k)))\, z = f(x_1, \ldots, x_k).$$

In modal interval analysis, when the continuous function f has a syntactic tree, there exist modal syntactic extensions which are obtained by using the computing program defined by the expression of the function: if f is a rational function from $\mathbb{R}^k$ to $\mathbb{R}$, its rational extension to the modal intervals $X_1, \ldots, X_k$, denoted by $fR(X_1, \ldots, X_k)$, is the function fR from $I^*(\mathbb{R}^k)$ to $I^*(\mathbb{R})$ defined by the computational program indicated by the syntax of f when the real operators, supposed JM-commutable functions, are transformed into their semantic extensions.

Moreover, modal interval analysis is a semantic system formed by three levels: the theoretic real level, the interval level, and the level of intervals with digital bounds. Similarly, three levels are involved in the intervals of marks theory: the theoretic level, the interval level, and the level of intervals with mark bounds. So, together with the united interval extension of a continuous function, it is necessary to define set-theoretical interval extensions to allow the specification of effective computations through their interval of marks operators.

The corresponding concepts for intervals of marks are described in the following definitions.

Definition 9.5.1 (Set-theoretical operator for intervals of marks) The function of marks $\Omega_{\mathbb{M}(t,n)} : \mathbb{M}(t,n)^2 \to \mathbb{M}(t,n)$ associated to the continuous real operator $w : \mathbb{R}^2 \to \mathbb{R}$, for the interval arguments $\mathcal{X}', \mathcal{Y}' \in I(\mathbb{M}(t,n))$, is defined by

$$\Omega_{\mathbb{M}(t,n)}(\mathcal{X}', \mathcal{Y}') = [\langle \underline{z}, g_{\underline{z}} \rangle, \langle \overline{z}, g_{\overline{z}} \rangle],$$

where

1) $\langle \underline{z}, g_{\underline{z}} \rangle$ and $\langle \overline{z}, g_{\overline{z}} \rangle$ are obtained from $\omega(P_\mathbb{R}(\mathcal{X}'), P_\mathbb{R}(\mathcal{Y}'))$, the interval extension of w, by substitution of each bound with the marks bounds and the operator with the corresponding mark operators,
2) the granularity g_z is the maximum of $g_{\underline{z}}$ and $g_{\overline{z}}$.

Example 9.5.1 For $\mathcal{X}' = [\langle \underline{x}, g_x \rangle, \langle \overline{x}, g_x \rangle]$ and $\mathcal{Y}' = [\langle \underline{y}, g_y \rangle, \langle \overline{y}, g_y \rangle]$,

1) $\mathcal{X}' + \mathcal{Y}' = [\langle \underline{x}, g_x \rangle + \langle \underline{y}, g_y \rangle, \langle \overline{x}, g_x \rangle + \langle \overline{y}, g_y \rangle] = [\langle \underline{x} + \underline{y}, g_z \rangle, \langle \overline{x} + \overline{y}, g_z \rangle]$.
2) $\mathcal{X}' - \mathcal{Y}' = [\langle \underline{x}, g_x \rangle - \langle \overline{y}, g_y \rangle, \langle \overline{x}, g_x \rangle - \langle \underline{y}, g_y \rangle] = [\langle \underline{x} - \overline{y}, g_z \rangle, \langle \overline{x} - \underline{y}, g_z \rangle]$.
3) $\mathcal{X}' * \mathcal{Y}' = \ldots = [\langle \min(\underline{x}\underline{y}, \underline{x}\overline{y}, \overline{x}\underline{y}, \overline{x}\overline{y}), g_z \rangle, \max(\underline{x}\underline{y}, \underline{x}\overline{y}, \overline{x}\underline{y}, \overline{x}\overline{y}), g_z \rangle]$.
4) If $0 \notin P_\mathbb{R}(\mathcal{Y})$, then

$$\mathcal{X}'/\mathcal{Y}' = \ldots = [\langle \min(\underline{x}/\underline{y}, \underline{x}/\overline{y}, \overline{x}/\underline{y}, \overline{x}/\overline{y}), g_z \rangle, \max(\underline{x}/\underline{y}, \underline{x}/\overline{y}, \overline{x}/\underline{y}, \overline{x}/\overline{y}), g_z \rangle].$$

5) $\min(\mathcal{X}', \mathcal{Y}') = \ldots = [\langle \min(\underline{x}, \underline{y})], g_z \rangle, \langle \min(\overline{x}, \overline{y}), g_z \rangle]$.
6) $\max(\mathcal{X}', \mathcal{Y}') = \ldots = [\langle \max(\underline{x}, \underline{y})], g_z \rangle, \langle \max(\overline{x}, \overline{y}), g_z \rangle]$.

The granularity g_z is the maximum of the granularities obtained in the computation of the mark bounds for the resulting intervals.

Definition 9.5.2 (Set-theoretical extension for intervals of marks) If $f_{\mathbb{M}(t,n)} : \mathbb{M}(t,n)^k \to \mathbb{M}(t,n)$ is the function of marks associated to the continuous function $f : \mathbb{R}^k \to \mathbb{R}$, of which operators in its syntactic tree belong to $\{+, -, *, /, \min, \max\}$, the *set-theoretical syntactic extension* of $f_{\mathbb{M}(t,n)}$ is a function $f_{\mathbb{M}(t,n)} R : \mathbb{M}(t,n)^k \to \mathbb{M}(t,n)$ obtained from the syntactic tree of f by substitution of

1) the operands by their corresponding intervals of marks,
2) the operators by the corresponding set-theoretical operators for intervals of marks,

and every incidence of the multi-incident variables is considered as independent variable.

This set-theoretical extension is inclusive.

Lemma 9.5.1 *For every $\mathcal{X}', \mathcal{Y}' \in I(\mathbb{M}(t,n)^k)$ if the greater granularity obtained, with the maximal approach, in the computations of $f_{\mathbb{M}(t,n)} R(\mathcal{X}')$ and $f_{\mathbb{M}(t,n)} R(\mathcal{Y}')$, g is compatible with αt, then*

$$\mathcal{X}' \subseteq \mathcal{Y}' \Rightarrow f_{\mathbb{M}(t,n)} R(\mathcal{X}') \subseteq_{2\alpha} f_{\mathbb{M}(t,n)} R(\mathcal{Y}').$$

Proof From the inclusivity of the interval operators and the properties of the inclusion of intervals of marks,

$$f_{\mathbb{M}(t,n)} R(\mathcal{X}') \approx_\alpha f_{\mathbb{M}(t,\infty)} R(\mathcal{X}') \subseteq f_{\mathbb{M}(t,\infty)} R(\mathcal{Y}') \approx_\alpha f_{\mathbb{M}(t,n)} R(\mathcal{Y}'). \blacksquare$$

The relation between the united and the set-theoretical extension is given by the following result.

Lemma 9.5.2 *If $R_{f_{\mathbb{M}(t,n)}}$ and $f_{\mathbb{M}(t,n)} R$ are the united and the interval extensions of $f_{\mathbb{M}(t,n)}$, respectively, over the interval of marks $\mathcal{X}' \in I(\mathbb{M}(t,n)^k)$ and g_z is the granularity associated to $f_{\mathbb{M}(t,n)} R(\mathcal{X}')$, computed with the maximal approach, then*

9.5 Syntactic Extensions

$$(\forall \alpha \in]g_z/t, 1]) \ R_{f_{\mathbb{M}(t,n)}}(\mathcal{X}') \subseteq_\alpha f_{\mathbb{M}(t,n)} R(\mathcal{X}').$$

Proof As $P_\mathbb{R}(f_{\mathbb{M}(t,\infty)} R(\mathcal{X}')) = fR(P_\mathbb{R}(\mathcal{X}'))$, one has

$$R_{f_{\mathbb{M}(t,n)}}(\mathcal{X}') \subseteq f_{\mathbb{M}(t,\infty)} R(\mathcal{X}') \approx_\alpha f_{\mathbb{M}(t,n)} R(\mathcal{X}').$$
∎

Theorem 9.5.1 (Semantic for $fR_{\mathbb{M}(t,n)} R$) *If $f_{\mathbb{M}(t,n)} R$ is the rational set extension of the function $f_{\mathbb{M}(t,n)}$ and $[\underline{\mathfrak{z}}, \overline{\mathfrak{z}}] = f_{\mathbb{M}(t,n)} R(\mathcal{X}')$, then*

$$(\forall \underline{z} \in Iv'(\underline{\tilde{\mathfrak{z}}})) \ (\forall \overline{z} \in Iv'(\overline{\tilde{\mathfrak{z}}})) \ (\exists \underline{x} \in Iv'(\underline{X})) \ (\exists \overline{x} \in Iv'(\overline{X})) \ [\underline{z}, \overline{z}] = fR([\underline{x}, \overline{x}]).$$

Proof If $[\underline{\mathfrak{z}}, \overline{\mathfrak{z}}] = f_{\mathbb{M}(t,n)} R(\mathcal{X}')$, then

$$\underline{\mathfrak{z}} = f_{1\mathbb{M}(t,n)}(\underline{X}, \overline{X})$$
$$\overline{\mathfrak{z}} = f_{2\mathbb{M}(t,n)}(\underline{X}, \overline{X}).$$

Using the semantics of the functions of marks

$$(\forall \underline{z} \in Iv'(\underline{\tilde{\mathfrak{z}}})) \ (\exists x_1 \in Iv'(\underline{X}, \overline{X})) \ \underline{z} = f_1(x_1)$$
$$(\forall \overline{z} \in Iv'(\overline{\tilde{\mathfrak{z}}})) \ (\exists x_2 \in Iv'(\underline{X}, \overline{X})) \ \overline{z} = f_2(x_2)$$

where f_1 and f_2 are the functions that provide the infimum and supremum of fR. Taking $\underline{x} = x_1$ and $\overline{x} = x_2$ the proposition is proved. ∎

The generalization of these concepts for modal interval of marks is very similar to the generalization in the case of modal intervals which was treated in Chap. 3. It would be superfluous to insist on many details. The main features of this generalization are contained in the following results.

Definition 9.5.3 (Modal syntactic operator) A *modal syntactic operator* over $\mathcal{X} \in I^*(\mathbb{M}^k)$ is a function $f_\mathbb{M} : \mathbb{M}^k \to \mathbb{M}$ JM-commutable over $\mathcal{X}$.

For a modal syntactic operator, and under a maximalist approach for assignation of granularities,

$$f_\mathbb{M}^*(\mathcal{X}) = f_\mathbb{M}^{**}(\mathcal{X}) \approx_\alpha \langle [\text{SDV}(f_{\mathbb{M}(t,n)}, \mathcal{X}'_p, \mathcal{X}'_i), \text{SDV}(f_{\mathbb{M}(t,n)}, \mathcal{X}'_i, \mathcal{X}'_p)], g_f \rangle$$

whenever the marks are compatible with αt.

Definition 9.5.4 (Computed operator) If $f_\mathbb{M} : \mathbb{M}^k \to \mathbb{M}$ is a modal syntactic operator over mark intervals, the *computed operator* of $f_\mathbb{M}$ in $\mathbb{M}(t, n)$ over $\mathcal{X} \in I^*(\mathbb{M}(t, n))$, denoted by $F_{\mathbb{M}(t,n)}(\mathcal{X})$, is defined by

$$F_{\mathbb{M}(t,n)}(\mathcal{X}) = \langle [\text{SDV}(f_{\mathbb{M}(t,n)}, \mathcal{X}'_p, \mathcal{X}'_i), \text{SDV}(f_{\mathbb{M}(t,n)}, \mathcal{X}'_i, \mathcal{X}'_p)], g_f \rangle.$$

Definition 9.5.5 (Modal syntactic extension) When all the operators of the syntax tree of $f_\mathbb{M}$ are modal syntactic, the function, which results from the replacement of each operator by its $*$-semantic extension is called the *modal syntactic function* of $f_\mathbb{M}$ and denoted by $fR_{\mathbb{M}(t,\infty)}$.

Definition 9.5.6 (Computed modal syntactic extension) When all the operators of the syntax tree of $f_\mathbb{M}$ are modal syntactic, the function which results from the replacement of each operator by its computed operator is called the *modal syntactic function* of $f_\mathbb{M}$ and denoted by $FR_{\mathbb{M}(t,n)}$.

Theorem 9.5.2 (Relation between the modal syntactic extensions) *If* $f_\mathbb{M} : \mathbb{M}^k \to \mathbb{M}$ *is a function of marks associated to the continuous function* $f : \mathbb{R}^k \to \mathbb{R}$ *and there exists the modal syntactic function* $fR_{\mathbb{M}(t,\infty)}$ *and the computed modal syntactic function* $FR_{\mathbb{M}(t,n)}$ *over the mark interval* $\mathfrak{X} \in I^*(\mathbb{M}(t,n))$, *whenever the resulting granularity of the calculus* $FR_{\mathbb{M}(t,n)}(\mathfrak{X})$ *are compatible with* αt, *and have been calculated under a maximalist approach, one has*

$$fR_{\mathbb{M}(t,\infty)}(\mathfrak{X}) \approx_\alpha FR_{\mathbb{M}(t,n)}(\mathfrak{X}).$$

Proof All the computed operators of the syntax tree fulfill this relationship and the increase in the granularity at each step implies that we can apply induction to this process. ∎

Theorem 9.5.3 (Semantics for a computed modal syntactic functions) *Let* $FR_{\mathbb{M}(t,n)} : I^*(\mathbb{M}(t,n)^k) \to I^*(\mathbb{M}(t,n))$ *be the computed modal syntactic function and let* $fR : I^*(\mathbb{R}^k) \to I^*(\mathbb{R})$ *be the modal syntactic function, both associated to* $f : \mathbb{R}^k \to \mathbb{R}$. *If* $\mathfrak{X}_1 = [\underline{\mathfrak{x}}_1, \overline{\mathfrak{x}}_1], \ldots, \mathfrak{X}_k = [\underline{\mathfrak{x}}_k, \overline{\mathfrak{x}}_k] \in I^*(\mathbb{M}(t,n))$, *and* $[\underline{\mathfrak{z}}, \overline{\mathfrak{z}}] = FR_{\mathbb{M}(t,n)}(\mathfrak{X}_1, \ldots, \mathfrak{X}_k)$, *then*

$$(\forall \underline{z} \in Iv'(\underline{\tilde{\mathfrak{z}}}))\, (\forall \overline{z} \in Iv'(\overline{\tilde{\mathfrak{z}}}))$$
$$(\exists \underline{x}_1 \in Iv'(\underline{\mathfrak{x}}_1))\, (\exists \overline{x}_1 \in Iv'(\overline{\mathfrak{x}}_1)) \cdots (\exists \underline{x}_k \in Iv'(\underline{\mathfrak{x}}_k))\, (\exists \overline{x}_k \in Iv'(\overline{\mathfrak{x}}_k))$$
$$[\underline{z}, \overline{z}] = fR([\underline{x}_1, \overline{x}_1], \ldots, [\underline{x}_k, \overline{x}_k]).$$

Proof Start from the semantics of the functions of marks and take into account that the functions and bounds involved in $F_{\mathbb{M}(t,n)}R$ and fR are the same. ∎

Remark 9.5.1 In the construction of modal intervals and their modal syntactic extensions, when it comes to interpretability and optimality, the main problem was the multi-incidence of the variables in the syntactic tree. In modal syntactic extensions to intervals of marks, multi-incidences are not a problem because they only affect the computation of fR.

9.5 Syntactic Extensions

9.5.1 Arithmetic Operations for Intervals of Marks

As the construction of the intervals of marks extensions for a real continuous function has been made in a parallel way to the interval extension developed in MIA, the extension of the arithmetic operators (sum, difference, product and quotient) to mark intervals is analogous to the extension to modal intervals with similar proofs. In this section the computational results will be shown.

Let us suppose given two intervals of marks $\mathcal{X} = [\underline{\mathfrak{x}}, \overline{\mathfrak{x}}] \in I^*(\mathbb{M})$ and $\mathcal{Y} = [\underline{\mathfrak{y}}, \overline{\mathfrak{y}}] \in I^*(\mathbb{M})$, where

$$\underline{\mathfrak{x}} = \langle \underline{x}, g_x \rangle \in \mathbb{M}(t, n) \qquad \overline{\mathfrak{x}} = \langle \overline{x}, g_x \rangle \in \mathbb{M}(t, n)$$

and

$$\underline{\mathfrak{y}} = \langle \underline{y}, g_y \rangle \in \mathbb{M}(t, n) \qquad \overline{\mathfrak{y}} = \langle \overline{y}, g_y \rangle \in \mathbb{M}(t, n).$$

Sum of intervals of marks. The sum of $\mathcal{X}$ and $\mathcal{Y}$ is denoted by $\mathcal{X} + \mathcal{Y}$ and it turns out to be

$$\mathcal{X} + \mathcal{Y} = [\langle \underline{x}, g_x \rangle + \langle \underline{y}, g_y \rangle, \langle \overline{x}, g_x \rangle + \langle \overline{y}, g_y \rangle]$$

with perhaps the necessary coercion of one of the bounds to the greater of the granularities.

Difference of intervals of marks. The difference of $\mathcal{X}$ and $\mathcal{Y}$ is denoted by $\mathcal{X} - \mathcal{Y}$ and it turns out to be

$$\mathcal{X} - \mathcal{Y} = [\langle \underline{x}, g_x \rangle + \langle -\overline{y}, g_y \rangle, \langle \overline{x}, g_x \rangle + \langle -\overline{y}, g_y \rangle]$$

with perhaps the necessary coercion of one of the bounds to the greater of the granularities.

Product of a mark and an interval of marks. The product of $\mathcal{X}$ with a mark $\mathfrak{r} = \langle r, g_r \rangle \in \mathbb{M}(t, n)$ is denoted by $\mathfrak{r} * \mathcal{X}$ and turns out to be

$$\mathfrak{r} * \mathcal{X} = \begin{cases} [\langle r, g_r \rangle * \langle \underline{x}, g_x \rangle, \langle r, g_r \rangle * \langle \overline{x}, g_x \rangle] & \text{if } \langle r, g_r \rangle \geq \langle 0, b^{-n} \rangle \\ [\langle r, g_r \rangle * \langle \overline{x}, g_x \rangle, \langle r, g_r \rangle * \langle \underline{x}, g_x \rangle] & \text{if } \langle r, g_r \rangle \leq \langle 0, b^{-n} \rangle \end{cases}$$

Product of intervals of marks. The product of $\mathcal{X}$ and $\mathcal{Y}$ is denoted by $\mathcal{X} * \mathcal{Y}$ and turns out to be

$\mathfrak{X} * \mathfrak{Y} =$ if $\underline{\mathfrak{x}} \geq 0, \overline{\mathfrak{x}} \geq 0, \underline{\mathfrak{y}} \geq 0, \overline{\mathfrak{y}} \geq 0$, then $[\underline{\mathfrak{x}} * \underline{\mathfrak{y}}, \overline{\mathfrak{x}} * \overline{\mathfrak{y}}]$

if $\underline{\mathfrak{x}} \geq 0, \overline{\mathfrak{x}} \geq 0, \underline{\mathfrak{y}} \geq 0, \overline{\mathfrak{y}} < 0$, then $[\underline{\mathfrak{x}} * \underline{\mathfrak{y}}, \underline{\mathfrak{x}} * \overline{\mathfrak{y}}]$

if $\underline{\mathfrak{x}} \geq 0, \overline{\mathfrak{x}} \geq 0, \underline{\mathfrak{y}} < 0, \overline{\mathfrak{y}} \geq 0$, then $[\overline{\mathfrak{x}} * \underline{\mathfrak{y}}, \overline{\mathfrak{x}} * \overline{\mathfrak{y}}]$

if $\underline{\mathfrak{x}} \geq 0, \overline{\mathfrak{x}} \geq 0, \underline{\mathfrak{y}} < 0, \overline{\mathfrak{y}} < 0$, then $[\overline{\mathfrak{x}} * \underline{\mathfrak{y}}, \underline{\mathfrak{x}} * \overline{\mathfrak{y}}]$

if $\underline{\mathfrak{x}} \geq 0, \overline{\mathfrak{x}} < 0, \underline{\mathfrak{y}} \geq 0, \overline{\mathfrak{y}} \geq 0$, then $[\underline{\mathfrak{x}} * \underline{\mathfrak{y}}, \overline{\mathfrak{x}} * \underline{\mathfrak{y}}]$

if $\underline{\mathfrak{x}} \geq 0, \overline{\mathfrak{x}} < 0, \underline{\mathfrak{y}} \geq 0, \overline{\mathfrak{y}} < 0$, then $[\max(\underline{\mathfrak{x}} * \underline{\mathfrak{y}}, \overline{\mathfrak{x}} * \overline{\mathfrak{y}}), \min(\underline{\mathfrak{x}} * \overline{\mathfrak{y}}, \overline{\mathfrak{x}} * \underline{\mathfrak{y}})]$

if $\underline{\mathfrak{x}} \geq 0, \overline{\mathfrak{x}} < 0, \underline{\mathfrak{y}} < 0, \overline{\mathfrak{y}} \geq 0$, then $[\langle 0, g_{x,y}^M \rangle, \langle 0, g_{x,y}^M \rangle]$

if $\underline{\mathfrak{x}} \geq 0, \overline{\mathfrak{x}} < 0, \underline{\mathfrak{y}} < 0, \overline{\mathfrak{y}} < 0$, then $[\overline{\mathfrak{x}} * \overline{\mathfrak{y}}, \underline{\mathfrak{x}} * \overline{\mathfrak{y}}]$

if $\underline{\mathfrak{x}} < 0, \overline{\mathfrak{x}} \geq 0, \underline{\mathfrak{y}} \geq 0, \overline{\mathfrak{y}} \geq 0$, then $[\underline{\mathfrak{x}} * \overline{\mathfrak{y}}, \overline{\mathfrak{x}} * \overline{\mathfrak{y}}]$

if $\underline{\mathfrak{x}} < 0, \overline{\mathfrak{x}} \geq 0, \underline{\mathfrak{y}} \geq 0, \overline{\mathfrak{y}} < 0$, then $[\langle 0, g_{x,y}^M \rangle, \langle 0, g_{x,y}^M \rangle]$

if $\underline{\mathfrak{x}} < 0, \overline{\mathfrak{x}} \geq 0, \underline{\mathfrak{y}} < 0, \overline{\mathfrak{y}} \geq 0$, then $[\min(\underline{\mathfrak{x}} * \overline{\mathfrak{y}}, \overline{\mathfrak{x}} * \underline{\mathfrak{y}}), \max(\underline{\mathfrak{x}} * \underline{\mathfrak{y}}, \overline{\mathfrak{x}} * \overline{\mathfrak{y}})]$

if $\underline{\mathfrak{x}} < 0, \overline{\mathfrak{x}} \geq 0, \underline{\mathfrak{y}} < 0, \overline{\mathfrak{y}} < 0$, then $[\overline{\mathfrak{x}} * \underline{\mathfrak{y}}, \underline{\mathfrak{x}} * \underline{\mathfrak{y}}]$

if $\underline{\mathfrak{x}} < 0, \overline{\mathfrak{x}} < 0, \underline{\mathfrak{y}} \geq 0, \overline{\mathfrak{y}} \geq 0$, then $[\underline{\mathfrak{x}} * \overline{\mathfrak{y}}, \overline{\mathfrak{x}} * \underline{\mathfrak{y}}]$

if $\underline{\mathfrak{x}} < 0, \overline{\mathfrak{x}} < 0, \underline{\mathfrak{y}} \geq 0, \overline{\mathfrak{y}} < 0$, then $[\overline{\mathfrak{x}} * \overline{\mathfrak{y}}, \overline{\mathfrak{x}} * \underline{\mathfrak{y}}]$

if $\underline{\mathfrak{x}} < 0, \overline{\mathfrak{x}} < 0, \underline{\mathfrak{y}} < 0, \overline{\mathfrak{y}} \geq 0$, then $[\underline{\mathfrak{x}} * \overline{\mathfrak{y}}, \underline{\mathfrak{x}} * \underline{\mathfrak{y}}]$

if $\underline{\mathfrak{x}} < 0, \overline{\mathfrak{x}} < 0, \underline{\mathfrak{y}} < 0, \overline{\mathfrak{y}} < 0$, then $[\overline{\mathfrak{x}} * \overline{\mathfrak{y}}, \underline{\mathfrak{x}} * \underline{\mathfrak{y}}]$

Quotient of intervals of marks. The quotient of $\mathfrak{X}$ and $\mathfrak{Y}$ is denoted by $\mathfrak{X}/\mathfrak{Y}$ and it turns out to be

$\mathfrak{X}/\mathfrak{Y} =$ if $\underline{\mathfrak{x}} \geq 0, \overline{\mathfrak{x}} \geq 0, \underline{\mathfrak{y}} > 0, \overline{\mathfrak{y}} > 0$, then $[\underline{\mathfrak{x}}/\overline{\mathfrak{y}}, \overline{\mathfrak{x}}/\underline{\mathfrak{y}}]$

if $\underline{\mathfrak{x}} \geq 0, \overline{\mathfrak{x}} \geq 0, \underline{\mathfrak{y}} > 0, \overline{\mathfrak{y}} > 0$, then $[\underline{\mathfrak{x}}/\overline{\mathfrak{y}}, \overline{\mathfrak{x}}/\underline{\mathfrak{y}}]$

if $\underline{\mathfrak{x}} \geq 0, \overline{\mathfrak{x}} \geq 0, \underline{\mathfrak{y}} < 0, \overline{\mathfrak{y}} < 0$, then $[\overline{\mathfrak{x}}/\overline{\mathfrak{y}}, \underline{\mathfrak{x}}/\underline{\mathfrak{y}}]$

if $\underline{\mathfrak{x}} \geq 0, \overline{\mathfrak{x}} < 0, \underline{\mathfrak{y}} > 0, \overline{\mathfrak{y}} > 0$, then $[\underline{\mathfrak{x}}/\underline{\mathfrak{y}}, \overline{\mathfrak{x}}/\overline{\mathfrak{y}}]$

if $\underline{\mathfrak{x}} \geq 0, \overline{\mathfrak{x}} < 0, \underline{\mathfrak{y}} < 0, \overline{\mathfrak{y}} < 0$, then $[\overline{\mathfrak{x}}/\underline{\mathfrak{y}}, \underline{\mathfrak{x}}/\overline{\mathfrak{y}}]$

if $\underline{\mathfrak{x}} < 0, \overline{\mathfrak{x}} \geq 0, \underline{\mathfrak{y}} > 0, \overline{\mathfrak{y}} > 0$, then $[\underline{\mathfrak{x}}/\underline{\mathfrak{y}}, \overline{\mathfrak{x}}/\underline{\mathfrak{y}}]$

if $\underline{\mathfrak{x}} < 0, \overline{\mathfrak{x}} \geq 0, \underline{\mathfrak{y}} < 0, \overline{\mathfrak{y}} < 0$, then $[\overline{\mathfrak{x}}/\overline{\mathfrak{y}}, \underline{\mathfrak{x}}/\overline{\mathfrak{y}}]$

if $\underline{\mathfrak{x}} < 0, \overline{\mathfrak{x}} < 0, \underline{\mathfrak{y}} > 0, \overline{\mathfrak{y}} > 0$, then $[\underline{\mathfrak{x}}/\underline{\mathfrak{y}}, \overline{\mathfrak{x}}/\overline{\mathfrak{y}}]$

if $\underline{\mathfrak{x}} < 0, \overline{\mathfrak{x}} < 0, \underline{\mathfrak{y}} < 0, \overline{\mathfrak{y}} < 0$, then $[\overline{\mathfrak{x}}/\underline{\mathfrak{y}}, \underline{\mathfrak{x}}/\overline{\mathfrak{y}}]$

supposing $0 \notin \mathfrak{Y}'$.

9.5 Syntactic Extensions

Example 9.5.2 For the function $f(x_1, x_2, x_3, x_4) = (x_1 + x_2)(x_3 + x_4)$ its interval of marks syntactic extension in a scale DI_4 to the intervals

$$\mathfrak{X}_1 = [\langle -2, 0.001 \rangle, \langle 2, 0.001 \rangle]$$
$$\mathfrak{X}_2 = [\langle 1, 0.003 \rangle, \langle -1, 0.003 \rangle]$$
$$\mathfrak{X}_3 = [\langle -3, 0.009 \rangle, \langle 3, 0.009 \rangle]$$
$$\mathfrak{X}_4 = [\langle 2, 0.007 \rangle, \langle -2, 0.007 \rangle]$$

is

$$\begin{aligned}
FR_{M,4} &= (\mathfrak{X}_1 + \mathfrak{X}_2) * (\mathfrak{X}_3 + \mathfrak{X}_4) \\
&= ([\langle -2, 0.001 \rangle, \langle 2, 0.001 \rangle] + [\langle 1, 0.003 \rangle, \langle -1, 0.003 \rangle]) \\
&\quad *([\langle -3, 0.009 \rangle, \langle 3, 0.009 \rangle] + [\langle 2, 0.007 \rangle, \langle -2, 0.007 \rangle]) \\
&= [\langle -1, 0.003 \rangle, \langle 1, 0.003 \rangle] * ([\langle -1, 0.027 \rangle, \langle 1, 0.027 \rangle]) \\
&= [\min(\langle -1, 0.003 \rangle * \langle 1, 0.027 \rangle, \langle 1, 0.003 \rangle * \langle -1, 0.027 \rangle), \\
&\quad \max(\langle -1, 0.003 \rangle * \langle -1, 0.027 \rangle, \langle 1, 0.003 \rangle * \langle 1, 0.027 \rangle)] \\
&= [\langle -1, 0.027 \rangle, \langle 1, 0.027 \rangle]
\end{aligned}$$

For a tolerance $t = 0.05$, the associated intervals to the mark bounds of these intervals of marks are

$Iv(\underline{\mathfrak{X}_1}) = [-1.9, -2.1]$ $Iv(\overline{\mathfrak{X}_1}) = [2.1, 1.9]$
$Iv(\underline{\mathfrak{X}_2}) = [1.05, 0.95]$ $Iv(\overline{\mathfrak{X}_2}) = [-0.95, -1.05]$
$Iv(\underline{\mathfrak{X}_3}) = [-2.85, -3.15]$ $Iv(\overline{\mathfrak{X}_3}) = [3.15, 2.85]$
$Iv(\underline{\mathfrak{X}_4}) = [2.1, 1.9]$ $Iv(\overline{\mathfrak{X}_4}) = [-1.9, -2.1]$
$Iv(\underline{\tilde{3}}) = [-0.95, -1.05]$ $Iv(\overline{\tilde{3}}) = [1.05, 0.95]$.

Therefore, from Theorem 9.5.1, the logical formula

$$(\forall \underline{z} \in [-1.05, -0.95]')\, (\forall \overline{z} \in [0.95, 1.05]')$$
$$(\exists \underline{x_1} \in [-2.1, -1.9]')\, (\exists \overline{x_1} \in [1.9, 2.1]')$$
$$(\exists \underline{x_2} \in [0.95, 1.05]')\, (\exists \overline{x_2} \in [-1.05, -0.95]')$$
$$(\exists \underline{x_3} \in [-3.15, -2.85]')\, (\exists \overline{x_3} \in [2.85, 3.15]')$$
$$(\exists \underline{x_4} \in [1.9, 2.1]')\, (\exists \overline{x_4} \in [-2.1, -1.9]')$$
$$[\underline{z}, \overline{z}] = ([\underline{x_1}, \overline{x_1}] + [\underline{x_2}, \overline{x_2}]) * ([\underline{x_3}, \overline{x_3}] + [\underline{x_4}, \overline{x_4}])$$

is valid.

Fig. 9.1 Physical system

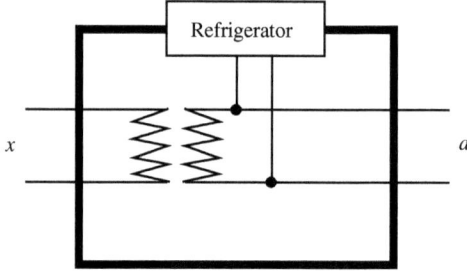

Example 9.5.3 Coming back to Example 8.2.1 of Chap. 8 (see Fig. 9.1), let us consider the physical system [46] of a transformer with a refrigerator. Let x be the input power, a be the output power, y be the power consumption of the refrigerator, b be the heat balance in the system, σ be the fraction of input power converted in heat by the transformer, and ρ be the fraction of refrigerator power transformed into heat.

The mathematical model is

$$\begin{cases} (1-\sigma)x - y = a \\ \sigma x - (1-2\rho)y = b, \end{cases}$$

where the first equation expresses the electric balance, and the second is the thermal balance equation. The problem is to design the refrigerator, i.e., to calculate intervals of marks $\mathcal{X}$ and $\mathcal{Y}$ for given $\mathfrak{A}$ and $\mathfrak{B}$ in such a way that

$$(\forall a \in \mathfrak{A}')\,(\forall \mathfrak{y} \in \mathcal{Y}')\,(\exists \mathfrak{x} \in \mathcal{X}')\, a = (1-\sigma)\mathfrak{x} - \mathfrak{y}$$

for the electrical balance and

$$(\forall \mathfrak{x} \in \mathcal{X}')\,(\exists \mathfrak{b} \in \mathfrak{B}')\,(\exists \mathfrak{y} \in \mathcal{Y}')\, \mathfrak{b} = \sigma \mathfrak{x} - (1-2\rho)\mathfrak{y}$$

for the thermal balance. In accordance with the *-semantic theorem, it is necessary to solve the following interval linear system of marks

$$\begin{cases} (1-\sigma) * \mathcal{X} - \text{Dual}(\mathcal{Y}) \subseteq_\alpha \mathfrak{A} \\ \sigma * \text{Dual}(\mathcal{X}) - (1-2\rho) * \mathcal{Y} \subseteq_\alpha \mathfrak{B} \end{cases} \quad (9.1)$$

with $\mathfrak{A}$, $\mathcal{X}$ and $\mathcal{Y}$ improper intervals and $\mathfrak{B}$ a proper interval. Let us consider a system of marks of a digital scale DI_3 and a tolerance $t = 0.05$. For the values $\sigma = \langle 1.111e-1, 0.01 \rangle$, $\rho = \langle 1.111e-1, 0.01 \rangle$, $A = [\langle 16, 0.005 \rangle, \langle 8, 0.005 \rangle]\,W$ and $b = [\langle 0.2, 0.005 \rangle, \langle 0.4, 0.005 \rangle]\,W$ let us consider the system

9.5 Syntactic Extensions

$$\begin{cases} \langle 8.889e-1, 0.01 \rangle * [\underline{x}, \overline{x}] - [\overline{y}, \underline{y}] & \approx_\alpha \\ \qquad\qquad [\langle 1.600e+1, 0.005 \rangle, \langle 8.000e0, 0.005 \rangle] \\ \langle 1.111e-1, 0.01 \rangle * [\overline{x}, \underline{x}] - \langle 7.778e-1, 0.01 \rangle * [\underline{y}, \overline{y}] \approx_\alpha \\ \qquad\qquad [\langle 2.000e-1, 0.005 \rangle, \langle 4.000e-1, 0.005 \rangle] \end{cases}$$

which can be split into two systems of marks

$$\begin{cases} \langle 8.889e-1, 0.01 \rangle * \underline{x} - \underline{y} & \approx_\alpha \langle 1.600e+1, 0.005 \rangle \\ \langle 1.111e-1, 0.01 \rangle * \underline{x} - \langle 7.778e-1, 0.01 \rangle * \underline{y} \approx_\alpha \langle 4.000e-1, 0.005 \rangle \end{cases}$$

and

$$\begin{cases} \langle 8.889e-1, 0.01 \rangle * \overline{x} - \overline{y} & \approx_\alpha \langle 8.000e+1, 0.005 \rangle \\ \langle 1.111e-1, 0.01 \rangle * \overline{x} - \langle 7.778e-1, 0.01 \rangle * \overline{y} \approx_\alpha \langle 2.000e-1, 0.005 \rangle \end{cases}$$

Their solutions are

$$\underline{x} = \langle 2.076e+1, 0.0123 \rangle \qquad \underline{y} = \langle 2.451e0, 0.0926 \rangle$$

and

$$\overline{x} = \langle 1.038e+1, 0.0122 \rangle \qquad \overline{y} = \langle 1.226e0, 0.0952 \rangle,$$

therefore, the solution of the interval of marks system is

$$\mathcal{X} = [\langle 2.076e+1, 0.0123 \rangle, \langle 1.038e+1, 0.0122 \rangle] \, W$$

and

$$\mathcal{Y} = [\langle 2.451e0, 0.0926 \rangle, \langle 1.226e0, 0.0952 \rangle] \, W$$

which is the optimal solution of the interval system (1). In accordance with the *-semantic theorem, this solution means

$$(\forall \mathfrak{a} \in [\langle 8.000e-1, 0.005 \rangle, \langle 1.600e+1, 0.005 \rangle]')$$
$$(\forall \mathfrak{y} \in [\langle 1.226e0, 0.0952 \rangle, \langle 2.451e0, 0.0926 \rangle]')$$
$$(\exists \mathfrak{x} \in [\langle 1.038e+1, 0.0122 \rangle, \langle 2.076e+1, 0.0123 \rangle]') \, \mathfrak{a} = (1-\sigma)\mathfrak{x} - \mathfrak{y}$$

for the electrical balance and

$$(\forall \mathfrak{x} \in [\langle 1.038e+1, 0.0122\rangle, \langle 2.076e+1, 0.0123\rangle]')$$
$$(\exists \mathfrak{b} \in [\langle 2.000e-1, 0.005\rangle, \langle 4.000e-1, 0.005\rangle]')$$
$$(\exists \mathfrak{y} \in [\langle 1.226e0, 0.0952\rangle, \langle 2.451e0, 0.0926\rangle]')\ \mathfrak{b} = \sigma\mathfrak{a} - (1-2\sigma)\mathfrak{b}$$

for the thermal balance. So, a problem theoretically solvable without computational solution in the context of real intervals, has a solution in the context of intervals of marks.

Chapter 10
Some Related Problems

10.1 Introduction

This chapter presents some applications of modal intervals to practical problems in different fields.

First, the minimax problem, tackled from the definitions of the modal *- and **-semantic extensions of a continuous function. Many real life problems of practical importance can be modelled as continuous minimax optimization problems. Well known applications to engineering, finance, optics and other fields demonstrate the importance of having reliable methods to solve continuous minimax problems and some algorithms have been developed for the continuous case [2, 15–17, 53, 70]. Classic examples of this are the Chebyshev approximation problem, finding optimal strategies in Game Theory, minimizing the effect of tolerances in engineering design, and the satisfaction problem for first-order logic formulas, [71,98]. Applications to optics [17], control [41,44], finance [76] and industrial engineering [92] are also well known. As an example, the Chebyshev approximation will be introduced in order to illustrate how it can be cast as a minimax optimization problem. The objective in the Chebyshev approximation problem is to approximate, as closely as possible, an given function f using operations that can be performed on the computer or calculator, typically with an accuracy close to that of the underlying computer's floating point arithmetic. In its classic form, this is accomplished by using a polynomial P of high degree, and/or narrowing the domain over which the polynomial has to approximate the function. Once the domain and degree of the polynomial are chosen, the polynomial itself is chosen in such a way as to minimize the maximum value of $|P(x) - f(x)|$, where $P(x)$ is the approximating polynomial and $f(x)$ is the actual function. In a more general form, given a continuous function f from $Y \subseteq \mathbb{R}^m$ to $\mathbb{R}$ and a set of approximating continuous functions p_x from $\mathbb{R}^m$ to $\mathbb{R}$ parameterized by $x \in \mathbb{R}^k$ belonging to a function space P_k, the Chebyshev approximation is to find p_x solving the minimax problem

$$\min_{x} \max_{y \in Y} |p_x(y) - f(y)|.$$

While global optimization has received much attention from the interval community [5,6,34,67] and interval methods are now known to be a very powerful approach to this problem, only a few studies have addressed minimax problems over continuous domains with interval techniques [41,44,98].

The second question is the characterization of solution sets of equations and systems when the unknowns and coefficients are within some given intervals and to tackle a set-theoretical problem: to find outer or inner interval estimates for the solutions of a real linear system $Ax = b$ when the coefficients are known to be within certain intervals $a_{ij} \in A'_{ij}$ and similarly for the right-hand sides: $b_i \in B'_i$. This problem appears in several areas of the design of control in physical systems and it has been treated in the context of classical intervals [56,57,59,79–84]. If $Ax = b$ is a real-valued system of linear equations, considering intervals A'_{ij} of variation for the coefficients and intervals B'_i of variation for the right-hand sides, a kind of *solution set* appears: the so called *AE-solution sets* [80], i.e., the set of solutions for the system with logical specifications for the selection of the coefficients

$$\Xi_{\alpha\beta} = \{x \in \mathbb{R}^n \mid (\forall a_{i_1 j_1} \in A'_{i_1 j_1}) \ldots (\forall a_{i_p j_p} \in A'_{i_p j_p})(\forall b_{k_1} \in B'_{k_1}) \ldots (\forall b_{k_q}, B'_{k_q})$$
$$(\exists a_{i_{p+1} j_{p+1}} \in A'_{i_{p+1} j_{p+1}}) \ldots (\exists a_{i_n j_n} \in A'_{i_n j_n})$$
$$(\exists b_{k_{q+1}} \in B'_{k_{q+1}}) \ldots (\exists b_{k_n} \in B'_{k_n}) \, Ax = b\}.$$

This $\Xi_{\alpha\beta}$-solution set is a particular case of sets such as the AE-solution sets

$$\Xi = \{x \in \mathbb{R}^n \mid (\forall v \in V')(\exists w \in W') f(v, w, x) = 0\}, \quad (10.1)$$

where f is a vectorial continuous function, whose characteristics are studied in detail together with specialization to the linear case of AE-solution sets. Such a characterization can be obtained in different ways: *pavings*, set of intervals contained and covering Ξ, *inner interval estimates*, any interval X such that $X \subseteq \Xi$, *outer interval estimates*, any interval X such that $X \supseteq \Xi$, or other estimates such as *weak inner estimates* or the *hull*.

The third part is an introduction to control problems from a semantic point of view and modal intervals as a tool to treat them. Feedback is a necessary technique to deal with uncertainty in the control of systems. A usual way to represent the behaviour of a dynamic system is by means of a discrete-time transfer functions such as

$$G(z^{-1}) = \frac{B(z^{-1})}{A(z^{-1})} = \frac{\sum_{i=1}^{m} b_i z^{-i}}{1 - \sum_{j=1}^{n} a_j z^{-j}}. \quad (10.2)$$

10.1 Introduction

For instance, a first-order system can be represented by

$$G(z^{-1}) = \frac{B(z^{-1})}{A(z^{-1})} = \frac{b_1 z^{-1}}{1 - a_1 z^{-1}} \tag{10.3}$$

which is equivalent to the difference equation

$$y((n+1)\Delta t) = a_1 y(n\Delta t) + b_1 u(n\Delta t), \tag{10.4}$$

in which $u(n\Delta t)$ and $y(n\Delta t)$ are, respectively, the input and the output of the system at time $n\Delta t$; Δt is the sampling time; and n is an integer indicating the step in the simulation process. With a more comfortable notation, the starting difference equation for one simulation step can be written

$$y(k+1) = ay(k) + bu(k), \tag{10.5}$$

where $k + i = (n+i)\Delta t$.

The main goal of control is to keep the output of the system at a desired value (the *setpoint*) at any step

$$y(k) = y_{sp}(k). \tag{10.6}$$

This problem is usually approached by a mathematical function or an algorithm, the *controller*, which uses the measurement of the system's output to compute the necessary *control variable (feedback)*. In the case of a first-order system, this computation can be performed easily using the system's model:

$$u(k) = \frac{y_{sp}(k+1) - ay(k)}{b}. \tag{10.7}$$

For instance, if $a = 0.3$, $b = 0.5$, $y_{sp}(k+1) = 7$, and $y(k) = 2$, the necessary control variable is $u(k) = 12.8$.

In the absence of uncertainty, the use of formulas or tables for open-loop control would be enough. However, there is always uncertainty: in the model of the system, in the measurements, in the actuators, in the perturbations, etc. An additional way to deal with uncertainty is taking it into account in the whole control procedure by means of intervals: the uncertainty in the parameters of the model can be represented by means of interval parameters, the uncertainty in the measurements can be represented by interval measurements, etc. An interval is a set of real numbers with different meanings: in some cases it means that one or several, but unknown, values belonging to the interval have a property and in other cases it means that all the values have the property. By combining these meanings, different control problems can be stated as logical problems.

10.2 Minimax

A new approach, based on modal interval analysis, to deal with continuous minimax problems over the reals is presented in this section. Continuous minimax, and global optimization as a particular case, is reduced to the computation of some semantic extensions. Modal intervals allow of computing these extensions efficiently and obtaining guaranteed results. In some simple cases, the results are obtained by means of simple interval arithmetic computations. Nevertheless, in many cases, a branch-and-bound algorithm will be required and several examples will illustrate its behaviour. This algorithm can be applied, with minor and obvious changes, when the variables take values belonging to discrete sets instead of intervals.

This new approach to the solution of minimax problems can be seen as a collateral result of the implementation of the f^* algorithm presented in Chap. 7. Two versions of an algorithm for solving continuous minimax optimization are presented: one devoted to solving unconstrained minimax optimization, and another to solve constrained minimax optimization.

The continuous minimax problem [15, 98] is defined as follows. If f is a $\mathbb{R}^k$ to $\mathbb{R}$ continuous function $z = f(x_1, \ldots, x_k)$ defined on an k-dimensional interval domain $X' = U' \times V'$,

- the *unconstrained minimax problem* is to find a minmax point

$$x^*_{minimax} = (u^*, v^*) \in U' \times V'$$

such that

$$f(x^*_{minimax}) = \min_{u \in U'} \max_{v \in V'} f(u, v),$$

together with the minimax value $f(x^*_{minimax})$.
- The *constrained minimax problem* is to find $x^*_{minimax}$ and $f(x^*_{minimax})$ such that

$$f(x^*_{minimax}) = \min_{u \in U'} \max_{v \in V'} f(u, v),$$

subject to some constraints

$$g_r(u, v) \leq 0 \qquad (r = 1, \ldots, m),$$

where the g_r are continuous functions defined on X'.

Remark 10.2.1 Constraints of the type $x_i \lesseqgtr k$ will be considered to have been previously removed by modifying the initial interval domain X'.

Example 10.2.1 For the minimax optimization

$$\min_{u \in [-1,1]'} \max_{v \in [-1,1]'} (v - 1)^2 + u^2,$$

10.2 Minimax

subject to the constraint $u^2 + v^2 - 1 \leq 0$, defining

$$h(u) = \max_{v \in [-1,1]'} (v-1)^2 + u^2,$$

the problem is to find

$$\min_{u \in [-1,1]'} h(u).$$

Isolating v from the constraint

$$v = -\sqrt{1 - u^2},$$

then

$$h(u) = (-\sqrt{1 - u^2} - 1)^2 + u^2$$

and

$$\min_{u \in [-1,1]'} (-\sqrt{1 - u^2} - 1)^2 + u^2.$$

Thus, we have $u^* = 1$ or $u^* = -1$, which yields

$$v^* = -\sqrt{1 - u^2} = 0$$

and consequently

$$f(u^*, v^*) = (-1)^2 + 1^2 = 2.$$

Obviously, not all minimax optimization problems can be analytically solved as easily as this one.

10.2.1 Solution of Unconstrained Problems

The solution to an unconstrained minimax problem for a continuous function f using modal intervals is closely related to the semantic extension f^*. Specifically, in accordance with Definition 3.2.1, the minimax value of f in X' is the infimum of the interval $f^*(U, V)$,

$$\min_{u \in U'} \max_{v \in V'} f(u, v) = \text{Inf}(f^*(U, V))$$

with U a proper interval and V an improper one.

Nevertheless, the computation of $f^*(U,V)$ depends on the monotonicity properties of f in (U',V'). Two cases must therefore be considered.

Case 1: f is optimal in X. When f is monotonic with respect to all its variables and their incidences and it has an optimal rational extension fR, then it is possible to compute $f^*(U,V)$ using interval arithmetic. By the Coercion to Optimality Theorem 4.2.15

$$f^*(U,V) = fR(UD,VD) = f^{**}(U,V),$$

and so the minimax problem has been reduced to computing $fR(UD,VD)$. Moreover, the minimax points can be obtained by applying the following rule,

$$u_i^* = \begin{cases} \text{Inf}(U_i') & \text{if } \partial f/\partial u_i \geq 0, \\ \text{Sup}(U_i') & \text{if } \partial f/\partial u_i \leq 0, \end{cases} \quad (10.8)$$

and

$$v_i^* = \begin{cases} \text{Sup}(V_i') & \text{if } \partial f/\partial v_i \geq 0, \\ \text{Inf}(V_i') & \text{if } \partial f/\partial v_i \leq 0, \end{cases} \quad (10.9)$$

where i is the variable subindex inside its corresponding vector u or v.

Example 10.2.2 Given the minimax problem

$$\min_{u \in U'} \max_{v \in V'} f(u,v),$$

where f is the continuous function

$$f(u,v) = u^2 + v^2 + 2uv - 20u - 20v + 100$$

and $U' = [0,2]'$, $V' = [2,8]'$. This function can be written as

$$f(u,v) = u_1^2 + v_1^2 + 2u_2v_2 - 20u_3 - 20v_3 + 100,$$

where the subindices represent the different incidences of each variable. First of all, the monotonicity for each variable and for each of its incidences, considered as different variables, has to be computed by means of the computation of the ranges of the partial derivatives with respect to each variable and its incidences.

$$\partial f(u,v)/\partial u = 2u + 2v - 20 \in 2*[0,2] + 2*[2,8] - 20$$
$$= [-16, 0] \leq 0,$$
$$\partial f(u,v)/\partial u_1 = 2u \in 2*[0,2] = [0,4] \geq 0,$$
$$\partial f(u,v)/\partial u_2 = 2v \in 2*[2,8] = [4,16] \geq 0,$$

10.2 Minimax

$$\partial f(u,v)/\partial u_3 = -20 \le 0,$$
$$\partial f(u,v)/\partial v = 2v + 2u - 20 \in 2*[2,8] + 2*[0,2] - 20$$
$$= [-16,0] \le 0,$$
$$\partial f(u,v)/\partial v_1 = 2v \in [4,16] \ge 0,$$
$$\partial f(u,v)/\partial v_2 = 2u \in [0,4] \ge 0,$$
$$\partial f(u,v)/\partial v_3 = -20 \le 0.$$

As f is totally monotonic with respect to the multi-incident variables (u,v), by Theorem 4.2.15, we have

$$fR(UD, VD) = \text{Dual}(U_1)^2 + \text{Dual}(V_1)^2$$
$$+ 2 * \text{Dual}(U_2) * \text{Dual}(V_2) - 20 * U_3 - 20 * V_3 + 100$$
$$= [2,0]^2 + [2,8]^2 + 2*[2,0]*[2,8] - 20*[0,2] - 20*[8,2] + 100$$
$$= [36,4],$$

which is an optimal computation for $f^*(U,V)$. Therefore,

$$f(u^*, v^*) = \min_{u \in [0,2]'} \max_{v \in [2,8]'} f(u,v) = \text{Inf}(f^*(U,V)) = 36.$$

As the function is totally monotonic with respect to all its variables, the minimax value is achieved at certain bounds of the respective intervals. Taking into account the signs of the derivatives, the *minimax point* is $(u^*, v^*) = (2,2)$.

Case 2: f is not optimal in X. This case occurs when f^* and f^{**} are different or when f is not monotonic with respect to all its variables and their incidences. Hence, it is not possible to compute $f^*(X)$ using simple arithmetic computations, and only interpretability theorems can be applied. By Theorem 4.2.10,

$$f^*(U,V) \subseteq fR(UD, VDt^*)$$

or, using Theorem 4.2.12,

$$f^*(U,V) \subseteq fR(UD, VDT^*).$$

In any case, fR will provide only an approximation to f^*. Therefore, only a lower bound for the minimax value can be obtained.

Example 10.2.3 Let consider the previous Example 10.2.2 but changing the interval U to $[0,6]$. In this case, none of the variables are monotonic and Theorem 4.2.15 can not be applied. Using Theorem 4.2.10,

$$\begin{aligned}
fR(UD, VDt^*) &= U_1^2 + [\text{mid}(V)]^2 + 2 * U_2 * [\text{mid}(V)] \\
&\quad -20 * U_3 - 20 * [\text{mid}(V)] + 100 \\
&= [0, 6]^2 + [5, 5]^2 + 2 * [0, 6] * [5, 5] - 20 * [0, 6] - 20 * [5, 5] + 100 \\
&= [-95, 121],
\end{aligned}$$

where $[\text{mid}(V)]$ is the point-wise interval corresponding to the midpoint of (V). Therefore,

$$f(u^*, v^*) = \min_{u \in [0,6]'} \max_{v \in [2,8]'} f(u, v) \geq \text{Inf}(f^*(U, V)) = -95.$$

Or applying Theorem 4.2.12,

$$\begin{aligned}
fR(UD, VDT^*) &= U_1^2 + \text{Dual}(V_1)^2 + 2 * U_2 * \text{Dual}(V_2) - 20 * U_3 - 20 * V_3 + 100 \\
&= [0, 6]^2 + [2, 8]^2 + 2 * [0, 6] * [2, 8] \\
&\quad -20 * [0, 6] - 20 * [8, 2] + 100 \\
&= [-56, 136].
\end{aligned}$$

Therefore,

$$f(u^*, v^*) = \min_{u \in [0,6]'} \max_{v \in [2,8]'} f(u, v) \geq \text{Inf}(f^*(U, V)) = -56.$$

The results obtained are approximations to the minimax value, which actually is $f(u^*, v^*) = 9$.

Summarizing, the computation of the minimax value of a continuous function can be done through the computation of its f^* extension. However, a good approximation to the minimax value can only be achieved under certain conditions of monotonicity. When these monotonicity conditions are not satisfied, to reduce the overestimation effect, a branch-and-bound algorithm is presented below.

10.2.2 Minimax Algorithm

Computing the minimax value of a continuous function f can be seen as computing (or approximating) its f^* extension. This section describes a particularization of the f^* algorithm presented in Chap. 7 which allows of approximating the minimax value of a continuous function f. Moreover, a minimax optimization problem normally requires the minimax point, therefore the proposed algorithm also returns a list of boxes (interval vectors) which are candidates for containing

10.2 Minimax

the minimax point(s). First the unconstrained version of the minimax algorithm is presented, and then the constrained version is explained.

Let $X = (U, V)$ be a modal interval vector split into proper U and improper V components. Let $\{U_1, \ldots, U_r\}$ be a partition of U and, for every $j = 1, \ldots, r$, let $\{V_{1j}, \ldots, V_{s_j}\}$ be a partition of V. Each interval $U_j \times V_{k_j}$ is called a *Cell*, each V_{*j}-partition is called a *Strip*, and the U-partition is called the *Strips' List*. The presented algorithm is based on the following theorem.

Theorem 10.2.1 *Given f, a real continuous function from $\mathbb{R}^k$ to $\mathbb{R}$, then*

$$\min_{u \in U'} \max_{v \in V'} f(u, v)$$

belongs to the interval

$$[\text{Inf}(\bigvee_{j \in \{1, \ldots, r\}} \bigwedge_{k_j \in \{1_j, \ldots, s_j\}} \text{Out}(fR(U_j, \check{v}_{k_j}))),$$

$$\text{Inf}(\bigvee_{j \in \{1, \ldots, r\}} \bigwedge_{k_j \in \{1_j, \ldots, s_j\}} \text{Inn}(fR(\check{u}_j, V_{k_j})))],$$

where $\check{u}_j$ is any point of U'_j ($j = 1, \ldots, r$) and $\check{v}_{k_j}$ is any point of V'_{k_j} ($k_j = 1_j, \ldots, s_j$) (for example the midpoints of the intervals or their bounds).

Proof Directly from Theorem 7.5.1. ∎

Remark 10.2.2 Taking advantage of the possible monotonicities of f, in accordance with Theorem 7.5.2 (see also (7.16) and (7.17)), inner and outer approximation for each cell can be

$$\text{Inn}(Cell) = \text{Inn}(fR(\check{u}D, VD)) \qquad (10.10)$$

and

$$\text{Out}(Cell) = \text{Out}(fR(UD, VDt^*)). \qquad (10.11)$$

The unconstrained version of the minimax algorithm is similar to the f^* algorithm presented in Chap. 7. However, as only the minimax bound of the f^* computation is required, some slight modifications are introduced to focus on the computation of this bound. These modifications consist of the following criteria.

10.2.2.1 Stopping Criteria

The stopping condition is replaced by

While $\{Minimax(\text{Inn}, \text{Out}) < \varphi\}$,

where φ is the desired precision for the output and *Minimax* is the function defined by

$$\text{Minimax}(\text{Inn}, \text{Out}) = |\text{Inf}(\text{Out}) - \text{Inf}(\text{Inn})|.$$

Remark 10.2.3 The stopping condition concerning the satisfaction of the logical formula is nonsensical in the minimax algorithm.

10.2.2.2 Bounding Criteria

The bounding criteria is replaced by

- a *Cell* is not bisected Inf(Inn(*Cell*)) $\leq$ Inf(Out(*Strip*)), because no division of any improper component V will improve the minimax approximation.

Similarly,

- a *Strip* is not bisected when Inf(Out(*Strip*)) $\geq$ Inf(Inn), because no division through any proper component U will improve the minimax approximation. Moreover, this *Strip* can be eliminated from the *StripSet*.

10.2.2.3 Selection Strategy

The selection strategy is modified and consists of selecting the *Strip* and the *Cell* with the biggest *Minimax*(Inn(.), Out(.)), and whose left bound approximations Inf(Inn(.)) and Inf(Out(.))) match at least one of these bounds with one of the left bounds of the global approximation Inf(Inn) or Inf(Out).

10.2.2.4 Return Value

Instead of returning the inner and outer approximations of f^*, the minimax optimization algorithm returns *Minimax*(Inn, Out).

10.2.2.5 Minimax Point Selection

As mentioned before, a minimax optimization problem normally requires the minimax point.

Algorithm 2 takes the *StripSet* resulting from applying the minimax algorithm and returns a list of boxes which are candidates for containing the minimax point(s).

The result of the algorithm implementation is to have inner and outer estimates of the minimax, together with a cell or a list of cells which are candidates for enclosing the minimax points, with a maximum width smaller than ϵ, and the union of these boxes for each point.

10.2 Minimax

Algorithm 2 SelectMinimaxBoxes

Require: Minimax approximation (*Minimax*(Inn, Out)) and *StripSet*.
Ensure: List of candidate boxes to contain the minimax point (*MinimaxList*).
1: **for** *Cell* in *Strip* in *StripSet* **do**
2:     Compute *Cell* approximations

$$\text{Inn}(Cell) = \text{Inn}(\text{Inn}(fR(\check{u}D, VD))),$$
$$\text{Out}(Cell) = \text{Out}(fR(UD, VDt^*)).$$

3:     **if** *Minimax*(*Cell*) > *Minimax*(Inn, Out) **then**
4:        Eliminate *Cell*.
5:     **else**
6:        Enqueue *Cell* to *MinimaxList*.
7:     **end if**
8: **end for** 9: **return** *MinimaxList*.

Example 10.2.4 Following the previous examples, for the same continuous function

$$f(u, v) = u^2 + v^2 + 2uv - 20u - 20v + 100$$

and the intervals $U' = [0, 6]'$, $V' = [2, 8]'$, this problem can be solved analytically and the result is a minimax value of 9 and two minimax points (5,2) and (5,8). Using the algorithm with an $\epsilon = 10^{-4}$ and a tolerance of 10^{-2}, the following result is obtained in 0.5 s on a Pentium IVM 1.5 GHz,

$$\min_{u \in [0,6]'} \max_{v \in [2,8]'} f(u, v) \in [8.999999, 9.000001]'$$

$$x^*_{minimax} \in \{([4.999999, 5.000001]', [1.999999, 2.000001]'),$$
$$([4.999999, 5.000001]', [7.999999, 8.000001]')\}.$$

Example 10.2.5 Given the minimax optimization problem [15]

$$\min_{z \in [-\pi,\pi]'} \max_{y \in [-\pi,\pi]'} f(y, z),$$

where

$$f(y, z) = (\cos y + \Sigma_{k=1}^{m} \cos((k+1)y + z_k))^2,$$

for $m = 1$, $\varphi = 10^{-3}$ and $\epsilon = 10^{-3}$, the following result is obtained in 0.6 s.

- *Minimax* : [3.098176, 3.100176]'.
- *MinimaxList* : {([-1.573066, -1.471875]', [0.785000, 0.490625]'),
   ([-1.573066, -1.471875]', [2.551250, 2.355000]'),
   ([1.471875, 1.573066]', [-0.490625, -0.785000]'),
   ([1.471875, 1.573066]', [-2.355000, -2.551250]')}.

Example 10.2.6 Given the minimax optimization problem [33]

$$\min_{x_1 \in [-1,2]'} \max_{(x_2,x_3) \in ([-1,1]',[-1,1]')} f(x_1, x_2, x_3),$$

where

$$f(x_1, x_2, x_3) = \sum_{k=1}^{10} (e^{-0.1kx_1} - e^{-0.1kx_2} - (e^{-0.1k} - e^{-k})x_3)^2,$$

for $\varphi = 10^{-3}$, $\epsilon = 10^{-3}$ and a computation time of 0.5 s, the following result is obtained.

- Minimax : [13.79551, 13.80204]'.
- MinimaxList : {([1.859375,2]',[0.999999,1.]'[-0.375000,-0.335937]'),
  ([−0.343750, −0.343017]', [0.999999, 1.0]', [−1.0, −0.999999]')}.

10.2.3 Solution of Constrained Problems

The problem of finding the minimax value when there exist general constraints defined by inequalities in the form

$$g_r(x, y) \le 0 \qquad (r = 1, \ldots, m)$$

cannot be solved with procedures based on simple interval computations because, even if all the functions involved were optimal, the feasible region need not an interval.

An approximate solution can be obtained by adapting the unconstrained minimax algorithm to the feasible region. Each cell of the partition must be situated with regard to the feasible region defined by the constraints: if the cell is out, it must be eliminated; if the cell is in, it must be considered as a member of the partition; while in other cases it must be kept for subsequent divisions.

Specifically, let Σ be the feasible region and let (U'_j, V'_{k_j}) be a sub-box of the initial interval domain $X' = (U', V')$. The following propositions must be tested

$$\forall (x \in U'_j) \; \forall (y \in V'_k) \; g_r(x, y) \le 0 \qquad (r = 1, \ldots, m)$$

by means of the following modal interval inclusions:

$$g_r R(U_j, V_k) \subseteq (-\infty, 0] \qquad (r = 1, \ldots, m),$$

with U_j and V_k proper intervals, because the rational extension $g_r R(U_j, V_k)$ is an outer approximation to the range $g_r(U'_j, V'_k)$.

10.2 Minimax

Fig. 10.1 Feasibility region and partitions

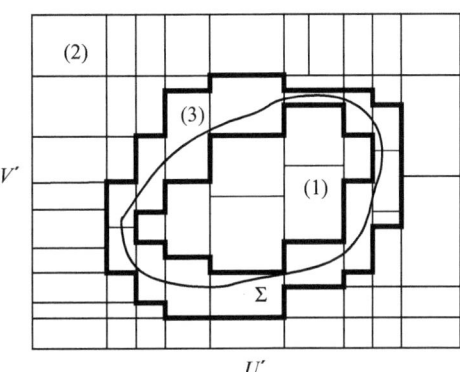

If all of the m inclusions are true, the cell (U_j, V_k) is inside the feasible region and it is included in a partition named $\Pi^{(1)}$. If any of the inclusions

$$g_r R(U_j, V_k) \subseteq\,]0, \infty) \qquad (r = 1, \ldots, m)$$

are true, the cell is outside of the feasible region, it is included in a partition named $\Pi^{(2)}$, and it must be eliminated. Otherwise, the cell intersects the feasible region and it is included in a partition named $\Pi^{(3)}$ and it must be kept for subsequent divisions. Figure 10.1 illustrates a feasible region in two dimensions.

For any partition $\Pi^{(l)}$ (with $l = 1, 3$), let $r^{(l)}$ be the number of strips and let $s_j^{(l)}$ be the number of cells of the strip j. By Theorem 10.2.1,

$$\min_{u \in U'} \max_{v \in V'} f(u, v) \leq \mathrm{Inf}(\bigvee_{j \in \{1^{(1)}, \ldots, r^{(1)}\}} \bigwedge_{k_j \in \{1_j^{(1)+(3)}, \ldots, s_j^{(1)+(3)}\}} \mathrm{Inn}(fR(\check{u}_j, V_{k_j})))$$

and

$$\min_{u \in U'} \max_{v \in V'} f(u, v) \geq \mathrm{Inf}((\bigvee_{j \in \{1^{(1)}, \ldots, r^{(1)}\}} \bigwedge_{k_j \in \{1_j^{(1)}, \ldots, s_j^{(1)}\}} \mathrm{Out}(fR(U_j, \check{v}_{k_j}))) \vee E),$$

where

$$E = \bigvee_{j \in \{1^{(3)}, \ldots, r^{(3)}\} - \{1^{(1)}, \ldots, r^{(1)}\}} \bigvee_{k_j \in \{1_j^{(3)}, \ldots, s_j^{(3)}\}} \mathrm{Out}(fR(U_j, \mathrm{Dual}(V_{k_j}))).$$

is a necessary allowance when the cell is in $\Pi^{(3)}$ intersecting, but not contained in, the feasible region.

By the previous reasoning, and taking advantage of the possible monotonicities of f (see Remark 10.2.2), the necessary additional (or substitutive) steps for solving the constrained version of the minimax optimization problem, in comparison to the unconstrained version, are summarized in the following steps.

- For each *Cell*, ascertain if $(Cell \in \Pi^{(1)})$, $(Cell \in \Pi^{(2)})$ or $(Cell \in \Pi^{(3)})$.
- Compute inner and outer approximations of the resulting *Cell* partitions, as follows:

If $Cell \in \Pi^{(1)}$, then

$$\text{Inn}(Cell) = \text{Inn}(fR(\check{u}D, VD)),$$
$$\text{Out}(Cell) = \text{Out}(fR(UD, VDt^*)).$$

If $Cell \in \Pi^{(3)}$, then

$$\text{Inn}(Cell) = \text{Inn}(fR(\check{u}D, VD)),$$
$$\text{Out}(Cell) = \text{Out}(fR(UD, VDt^*)) \vee \text{Out}(fR(UD, \text{Dual}(V)D)).$$

It is important to note that to compute the inner approximation $\text{Inn}(fR(\check{u}D, VD))$, the *D*-transformation over a variable of a U partition (*Strip*) requires the variable to be totally monotonic along the U partition.
- Compute inner and outer approximations of *Strip*, that is,

$$\text{Inn}(Strip) = \bigwedge_{\{Cell^{(1,3)} \text{ in } Strip\}} \text{Inn}(Cell^{(1,3)}),$$

$$\text{Out}(Strip) = \bigwedge_{\{Cell^{(1)} \text{ in } Strip\}} \text{Out}(Cell^{(1)}),$$

where $Cell^{(1,3)}$ is a *Cell* belonging to either of the partitions $\Pi^{(1)}$ or $\Pi^{(3)}$.
- Compute global inner and outer approximations, that is

$$\text{Inn} = \bigvee_{\{Strip^{(1)} \text{ in } StripSet\}} \text{Inn}(Strip^{(1)}),$$

$$\text{Out} = \bigvee_{\{Strip^{(1,3)} \text{ in } StripSet\}} \text{Out}(Strip^{(1,3)}),$$

where $Strip^{(1)}$ is a *Strip* containing at leat one $Cell^{(1)}$, and $Strip^{(3)}$ is a *Strip* not containing any $Cell^{(1)}$ and containing at leat one $Cell^{(3)}$.
- Concerning the bisection strategy, do not bisect a *Cell* or a *Strip* partition only if the objective function is totally monotonic with respect to its components (u, v) and it is consistent with respect to the involved constraints.

Example 10.2.7 Given the constrained minimax optimization problem

$$\min_{x_1 \in [0,6]'} \max_{x_2 \in [2,8]'} x_1^2 + x_2^2 + 2x_1 x_2 - 20x_1 - 20x_2 + 100,$$

subject to the constraints

$$g_1(x_1, x_2) = (x_1 - 5)^2 + (y - 3)^2 - 4 \geq 0,$$
$$g_2(x_1, x_2) = (x_1 - 5)^2 + (y - 3)^2 - 16 \leq 0.$$

For $\varphi = 10^{-5}$ and $\epsilon = 10^{-5}$, the following result is obtained in 0.7 s.
- *Minimax* : $[1.102542, 1.102572]'$.
- *MinimaxList* : $\{([4.142899, 4.142944]', [4.807053, 4.807030]'),$
  $([4.142899, 4.143036]', [6.907135, 6.907089]'),$
  $([4.142944, 4.143036]', [6.907043, 6.907043]')\}$.

Example 10.2.8 Given the constrained continuous minimax optimization problem inspired from [15]

$$\min_{x \in [-3.14, 3.14]'} \max_{y \in [-3.14, 3.14]'} (\cos(y) + \cos(2y + x))^2,$$

subject to the constraints

$$g_1(x, y) = y - x(x + 6.28) \leq 0,$$
$$g_2(x, y) = y - x(x - 6.28) \leq 0.$$

For $\epsilon = 10^{-6}$ and $\varphi = 10^{-6}$, the following result is obtained in 0.8 s.
- *Minimax* : $[8.586377e - 03, 8.586695e - 03]'$.
- *MinimaxList* : $\{([-0.437082, -0.437081]'; [-2.553834; -2.553831]'),$
  $([-0.4370827, -0.4370812]'; [-3.140000, -2.747500]')\}$.

Remark 10.2.4 This algorithm is applicable, with minor and obvious changes, when the variables take values belonging to discrete sets instead of continuous intervals.

10.3 Solution Sets

The second question is the characterization of the solution sets of systems when the unknowns and coefficients belong to certain given domains, with given logical specifications.

Usually quantifiers arise in situations of uncertainty and if the uncertainties can be represented by means of intervals, the domains are intervals. An interval contains a set of real numbers with different meanings: in some cases it means that one or several, but unknown, values belonging to the interval have a property and in other cases it means that all the values have the property. By combining these meanings, many problems can be stated as logical problems. Universal quantification must be used when some parameters are unknown and the predicates have to hold for every possible parameter value, and existential quantification when some parameters can be chosen. So in this case some inputs are uncontrolled disturbances, hence to be universally quantified, but some take values inside prescribed intervals, hence to be existentially quantified. The selection of the coefficients leads to a logical definition such as (10.1), where the universal quantifiers precede the existential ones and the domains are intervals on the real line.

In control systems or in decision making situations described by antagonistic games, uncertain controls or perturbations are often represented by parameters which conflict among themselves because the represented actions can be mutually compensating. For example considering the input–output relationship in a system under interval uncertainty in the form

$$f(a,x) = b,$$

where f is a function whose components are expressions in the variables a and x related by elementary operators, a problem of control can be [85]: for what system states x, for any perturbations given for the values of the variables $a_{k+1},\ldots,a_r$ inside some intervals $A'_{k+1},\ldots,A'_r$ and for any given outputs $b_1,\ldots,b_l$ inside some intervals $B'_1,\ldots,B'_l$, the corresponding controls $a_{k+1},\ldots,a_r$ inside the intervals $A_{k+1},\ldots,A_r$ make that the system output is inside certain intervals $B'_{l+1},\ldots,B'_s$. When all the inputs and outputs are determined the solution of the problem is equivalent to find the following AE-solution set

$$\Xi = \{x \in \mathbb{R}^n \mid \forall (a_1 \in A'_1)\ldots \forall(a_k \in A'_k)\ \forall(b_1 \in B'_1)\ldots \forall(b_l \in B'_l)$$
$$\exists(a_{k+1} \in A'_{k+1})\ldots \exists(a_r \in A'_r)\ \exists(b_{l+1} \in B'_{l+1})\ldots \exists(b_s \in B'_s)$$
$$f(a,x) = b\}$$

An AE-solution set can be consider as a particular case of a more general Σ-solution set obtained combining the quantifiers $\forall$ and $\exists$ in any order,

$$\Sigma = \{x \in \mathbb{R}^n \mid Q(a,A)\ Q(b,B) f(a,x) = b\},$$

where Q$\in \{\forall, \exists\}$. This kind of general Σ-solution sets arise, for example in multi-step decision-making processes under interval uncertainties and they are very related to problems of minimax in operations research where models of decision may be reduced to problems of finding values of, for example [12]

$$\min_{x \in X'} \max_{y \in Y'} \min_{z \in Z'} f(x,y,z).$$

(additional examples about problems where of Σ and AE-solution sets appear can be found in [85]).

Let us consider the problem of characterizing a set of solutions of a AE-*solution set* in the form:

$$\Xi = \{x \in B' \mid (\forall u \in U')(\exists v \in V') f(x,u,v) = 0\}, \qquad (10.12)$$

where f is a real or a vectorial continuous function from $\mathbb{R}^k$ to $\mathbb{R}^m$ and B', U', V' interval vectors of suitable dimensions.

10.3 Solution Sets

10.3.1 Pavings

The first characterization of the solution set is by means of *pavings*, i.e., intervals contained and covering Ξ. This characterization leads to solution sets when the constraints depend on some parameters about which only their belonging to certain intervals is known. A branch and bound algorithm over $\mathbb{R}^n$ will divide it in boxes which can be classified in three classes:

(a) Boxes contained in Ξ. They are the pavings of Ξ.
(b) Boxes contained in the complementary $\overline{\Xi}$. These are boxes to be discarded.
(c) Boxes non-decidable, in other case. These boxes will be subject to subsequent divisions.

As the semantic theorems to apply for prove the involved inclusions are different for real or vectorial functions, it is necessary to split the problem for both kind of functions.

10.3.1.1 f Is a Real Function

In a first step, let us consider that f is a real continuous function f. The algorithm is based on Modal Interval Analysis and branch-and-bound techniques, and referred as *Quantified Set Inversion* (QSI) algorithm [37], because it is inspired by the well-known Set Inversion Via Interval Analysis (SIVIA) algorithm [42, 43].

The QSI algorithm is designed to characterize a AE-solution set

$$\Xi = \{x \in B' \mid (\forall u \in U')(\exists v \in V') \ f(x, u, v) = 0\}, \quad (10.13)$$

and it is based on a branch-and-bound process over the free-variables vector x of the logical formula which define (10.13), together with two bounding rules used to determine if a resulting box X' from the bisection procedure over B' is included in the solution set, $X' \subseteq \Xi$ (*InsideQSI* rule), or if X' does not intersect with the solution set, $X' \cap \Xi = \emptyset$ (*OutsideQSI* rule).

The first bounding rule is

$$InsideQSI : X' \subseteq \Xi \Leftrightarrow (\forall x \in X')(\forall u \in U') \ (\exists v \in V') \ f(x, u, v) = 0.$$

Notice that the first-order logic formula contained in this *InsideQSI* rule can be reduced to a modal interval inclusion by means of the *-Semantic Theorem:

$$Out(f^*(X, U, V)) \subseteq [0, 0] \Rightarrow f^*(X, U, V) \subseteq [0, 0]$$
$$\Leftrightarrow (\forall x \in X')(\forall u \in U')(\exists v \in V') \ f(x, u, v) = 0$$
$$\Leftrightarrow X' \subseteq \Xi, \quad (10.14)$$

Algorithm 3 QSI algorithm

Require: ϵ.
Ensure: Ξ_{Inn} and Ξ_{Out} of the solution set.
1: List=$\{X'\}$; $\Xi_{\text{Inn}} = \{\emptyset\}$; $\Delta\Xi = \{\emptyset\}$;
2: **while** *List not empty* **do**
3:     Dequeue X' from $List$;
4:     **if** *InsideQSI* is true for X' **then**
5:         Enqueue X' to Ξ_{Inn};
6:     **else if** *OutsideQSI* is true for X' **then**
7:         Do nothing;
8:     **else if** $d(X') < \epsilon$ **then**
9:         Enqueue X' to $\Delta\Xi$;
10:    **else**
11:        Bisect X' and enqueue the resulting boxes to $List$;
12:    **end if**
13: **end while**
14: Enqueue Ξ_{Inn} and $\Delta\Xi$ to Ξ_{Out};

with X, U proper intervals, V improper intervals and Out($f^*(X, U, V)$) an outer approximation of the *-semantic extension of f.

The second bounding rule is

$$OutsideQSI : X' \cap \Xi = \emptyset \Leftrightarrow (\forall x \in X')\neg((\forall u \in U')(\exists v \in V')\; f(x, u, v) = 0))$$
$$\Leftrightarrow (\forall x \in X')(\exists u \in U')(\forall v \in V')\; f(x, u, v) \neq 0.$$

Again, the first-order logic formula of the *OutsideQSI* rule can be reduced to a modal interval inclusion:

$$\text{Inn}(f^*(X, U, V)) \not\subseteq [0, 0] \Rightarrow f^*(X, U, V) \not\subseteq [0, 0]$$
$$\Leftrightarrow \neg((\forall u \in U')(\exists v \in V')(\exists x \in X')\; f(x, u, v) = 0)$$
$$\Leftrightarrow (\exists u \in U')(\forall v \in V')(\forall x \in X')\; f(x, u, v) \neq 0$$
$$\Rightarrow (\forall x \in X')(\exists u \in U')(\forall v \in V')\; f(x, u, v) \neq 0$$
$$\Leftrightarrow X' \cap \Xi = \emptyset, \quad (10.15)$$

with U being proper intervals, X, V improper ones, and $\text{Inn}(f^*(X, U, V))$ an inner approximation of the *-semantic extension of f.

Finally, if any of the bounding rules is fulfilled, the box X' is considered as undefined and is bisected.

Algorithm 3 shows the QSI algorithm in pseudo-code form:
where

- $List$: List of boxes;
- Ξ_{Inn}: List of boxes such that $\Xi_{\text{Inn}} \subseteq \Xi$;
- Ξ_{Out}: List of boxes such that $\Xi \subseteq \Xi_{\text{Out}}$;

10.3 Solution Sets

- Enqueue: The result of adding a box to a list;
- Dequeue: The result of extracting a box from a list;
- $d(X')$: Function returning the widest relative width of X' with respect to the original box;
- ϵ: A real value representing the desired precision.

Example 10.3.1 As an example of applying the quantified set inversion algorithm to a real physical system, let us consider the problem of the determination of the mean stream velocity of the waters of a reach of a river [77]. Let us assume that the velocity depends only on the river flow and the bed slope, modelled by means of the following time-invariant black-box model.

$$v_m = a_1 f + a_2 s f,$$

where f (m^3/s) is the river flow rate, s (m/m) is the bed slope (both are the model inputs), v_m (m/s) is the average velocity (it is the model output), and a_1 and a_2 are the model parameters (positive quantities without any physical interpretation). The term in sf represents the synergy acting as an increase of velocity when flow and slope increase.

The problem is to find values for a_1 and a_2 which are consistent with the experimental uncertain values obtained by means of experimental measurements of the mean velocity in different points under different flow conditions, and represented by the intervals of Table 10.1. Figure 10.2 shows the experimental data, where the rectangles represent the intervals of uncertainty associated to the couple of data (flow, velocity).

The solution set of parameters is

$$\Xi = \{(a_1, a_2) \in \mathbb{R}^2 \mid (\exists s^i \in S_{exp}^i)\,(\exists f^i \in F_{exp}^i)\,(\exists v_m^i \in V_{exp}^i)\; v_m^i = a_1 f^i + a_2 s^i f^i\},$$

for all $i = 1, \ldots, 27$, which ensures that each one of the 27 outputs experimental intervals intersects the set of possible trajectories defined by any $(a_1, a_2) \in \Xi_3$ and some input $f^i \in F_{exp}^i$, $s^i \in S_{exp}^i$.

After applying the QSI algorithm, the resulting solution set Ξ is shown in Fig. 10.3.

In the case of the $\geq$ or $\leq$ relations, the *-semantic theorem states the equivalences

$$f^*(X, U, V) \subseteq (-\infty, 0] \;\Leftrightarrow\; (\forall x \in X')(\forall u \in U')(\exists v \in V')\, f(u, v) \leq 0$$

and

$$f^*(X, U, V) \subseteq [0, +\infty) \;\Leftrightarrow\; (\forall x \in X')(\forall u \in U')(\exists v \in V')\, f(u, v) \geq 0.$$

Therefore for the solution set

$$\Xi = \{x \in B' \mid (\forall u \in U')(\exists v \in V')\; f(x, u, v) \geq 0\}, \tag{10.16}$$

Table 10.1 Experimental intervals of slopes, flow rates and mean velocities

S_{exp}^i	F_{exp}^i	V_{exp}^i
[0.003449, 0.003451]	[2.80, 3.20]	[0.133, 0.158]
[0.003449, 0.003451]	[8.60, 9.80]	[0.319, 0.373]
[0.003449, 0.003451]	[10.00, 11.30]	[0.385, 0.452]
[0.003449, 0.003451]	[8.10, 9.20]	[0.312, 0.366]
[0.003449, 0.003451]	[7.30, 8.30]	[0.292, 0.344]
[0.003449, 0.003451]	[10.20, 11.60]	[0.377, 0.444]
[0.002699, 0.002701]	[12.30, 13.90]	[0.534, 0.632]
[0.002699, 0.002701]	[9.70, 11.00]	[0.373, 0.440]
[0.002699, 0.002701]	[20.10, 22.70]	[0.744, 0.873]
[0.002699, 0.002701]	[5.70, 6.50]	[0.271, 0.325]
[0.002699, 0.002701]	[5.10, 5.80]	[0.242, 0.290]
[0.001689, 0.001691]	[7.80, 8.80]	[0.300, 0.353]
[0.001689, 0.001691]	[15.00, 17.00]	[0.576, 0.681]
[0.001689, 0.001691]	[8.70, 9.90]	[0.347, 0.413]
[0.005289, 0.005291]	[13.20, 15.00]	[0.507, 0.600]
[0.005289, 0.005291]	[16.60, 18.80]	[0.638, 0.753]
[0.005289, 0.005291]	[17.30, 19.60]	[0.692, 0.817]
[0.005289, 0.005291]	[10.30, 11.70]	[0.412, 0.488]
[0.005289, 0.005291]	[6.70, 7.60]	[0.291, 0.345]
[0.005289, 0.005291]	[8.10, 9.20]	[0.337, 0.400]
[0.005289, 0.005291]	[18.30, 20.70]	[0.703, 0.828]
[0.005289, 0.005291]	[15.60, 17.70]	[0.600, 0.708]
[0.005289, 0.005291]	[6.30, 7.20]	[0.252, 0.300]
[0.005289, 0.005291]	[8.60, 9.70]	[0.344, 0.405]
[0.005289, 0.005291]	[6.60, 7.50]	[0.275, 0.326]
[0.006779, 0.006781]	[28.60, 32.30]	[1.191, 1.404]
[0.006779, 0.006781]	[27.50, 31.10]	[1.146, 1.352]

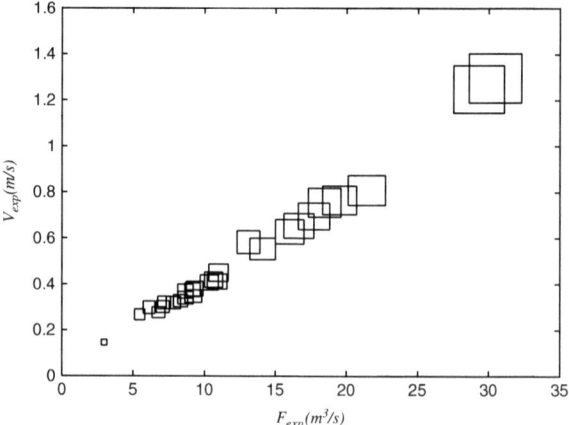

Fig. 10.2 Intervals (F_{exp}, V_{exp})

10.3 Solution Sets

Fig. 10.3 $\varXi$ paving

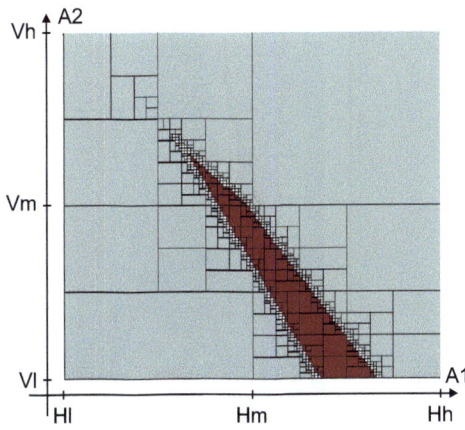

rules (10.14) and (10.15) become

$$\text{Out}(f^*(X, U, V)) \subseteq [0, +\infty) \Rightarrow X' \subseteq \varXi$$
$$\text{Inn}(f^*(X, U, V)) \nsubseteq [0, +\infty) \Rightarrow X' \cap \varXi = \emptyset.$$

Analogously, for the solution set

$$\varXi = \{x \in B' \mid (\forall u \in U')(\exists v \in V') \ f(x, u, v) \leq 0\}, \quad (10.17)$$

rules (10.14) and (10.15) become

$$\text{Out}(f^*(X, U, V)) \subseteq (-\infty, 0] \Rightarrow X' \subseteq \varXi$$
$$\text{Inn}(f^*(X, U, V)) \nsubseteq (-\infty, 0] \Rightarrow X' \cap \varXi = \emptyset.$$

Remark 10.3.1 Nevertheless, inequality predicates could also be expressed as equality predicates by introducing variables, for example

$$f(u, v) \leq 0 \Leftrightarrow f(u, v) - a = 0,$$

where $a \in (-\infty, 0]$ is a slack variable.

Example 10.3.2 For the solution set

$$\varXi = \{(x_1, x_2) \in ([-10, 10]', [-10, 10]') \mid$$
$$(\forall u \in [-1, 1]')(\exists v \in [-2, 2]') \ x_1 u - x_2 v^2 \sin x_1 \geq 0\},$$

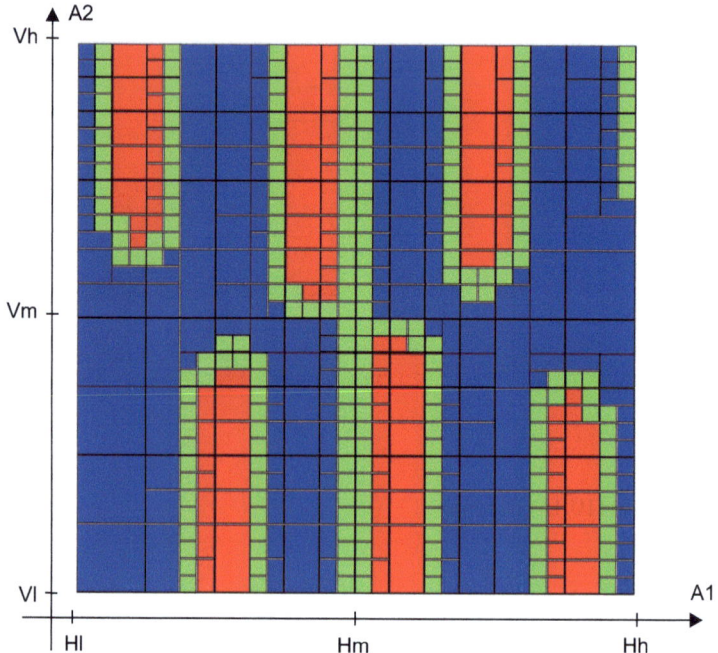

Fig. 10.4 Graphical output for Example 10.3.2

Figure 10.4 shows a graphical result of the QSI algorithm for an $\epsilon = 0.05$ in 40 s on a Pentium IV M 1.5 GHz, where red boxes are included in the solution set, blue boxes are outside of the solution set, and green boxes are undefined.

Remark 10.3.2 The complexity of the algorithm is exponential because of its branch-and-bound nature. It guarantees termination for non ill-posed problems and a finite precision. It is sound because it provides a continuous guaranteed inner approximation to the solution set and it is complete because provides an outer approximation of the solution set.

10.3.1.2 f Is a Vectorial Function

Suppose that, in the Definition (10.12) of the solution set, f is a vectorial continuous function $f = (f_1, \ldots, f_m)$, i.e., the AE-solution set is

$$\varXi = \{x \in \boldsymbol{B'} \mid (\forall \boldsymbol{u} \in \boldsymbol{U'})(\exists \boldsymbol{v} \in \boldsymbol{V'}) f(x, \boldsymbol{u}, \boldsymbol{v}) = 0\}, \qquad (10.18)$$

with f a function from $\mathbb{R}^k$ to $\mathbb{R}^m$. The QSI algorithm can be applied only when the components functions $f_1, \ldots, f_m$ do not share any existentially quantified variable v-component. If they do not, the QSI algorithm must be run m times independently

10.3 Solution Sets

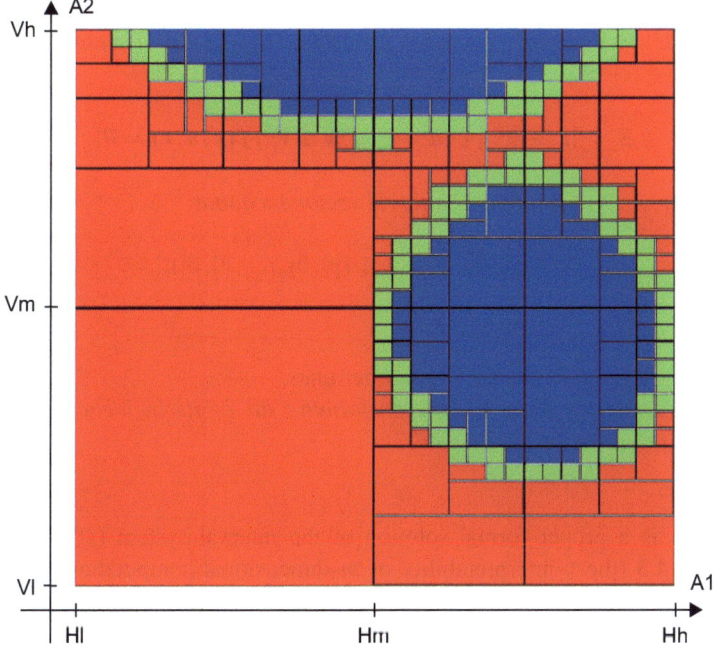

Fig. 10.5 Graphical output for Example 10.3.3

for each of the f-components and the AE-solution set is the intersection of all the obtained partial solution sets. If there exist shared existentially quantified variables, QSI can not be directly applied, except in some special cases.

Example 10.3.3 For the solution set

$$\Xi = \{(x_1, x_2) \in ([-10, 10]', [-10, 10]') \mid$$
$$(\exists v \in [-0.5, 0.5]')(-x_1 + x_2 v + x_1^2 \geq 0 \land -x_2 + (1 + x_1^2)v + v^3 \geq 0)\},$$

the QSI algorithm, in its original form, can not be applied due to the presence of a shared existentially quantified variable v but it can be replaced by the equivalent solution set

$$\Xi = \{(x_1, x_2) \in ([-10, 10]', [-10, 10]') \mid$$
$$(\exists v \in [-0.5, 0.5]') \min(-x_1 + x_2 v + x_1^2, -x_2 + (1 + x_1^2)v + v^3) \geq 0\}.$$

Figure 10.5 shows a graphical representation of the solution provided by the QSI algorithm for an $\epsilon = 0.05$ in 49 s on a Pentium IV M 1.5 GHz.

In the general case, here are two results to decide whether an interval is or not a paving for a AE-solution set.

Theorem 10.3.1 *Let U' and V' be interval vectors and f be a continuous function from $\mathbb{R}^k \times U' \times V'$ in $\mathbb{R}^m$, such that $f = (f_1, \ldots, f_m)$. Let us consider the AE-solution set*

$$\Xi = \{x \in \mathbb{R}^k \mid (\forall u \in U')(\exists v \in V') f(x, u, v) = 0\}.$$

If $X = (X_1, \ldots, X_k)$ is a proper interval vector such that

$$fR(X, U, VT^*) \subseteq ([0,0], \ldots, [0,0]), \tag{10.19}$$

where

1) U *are proper intervals and* V *improper ones,*
2) VT^* *is obtained from* V *by transforming all of its incidences (in all the equations) but one into their duals,*

then $X' \subseteq \Xi$.

Proof If X is a proper formal solution of the interval system (10.19), then by Theorem 4.4.3 (the *-interpretability of m-dimensional computations) applied to the system of inclusions

$$\begin{cases} f_1 R(X_1, \ldots, X_k, U, VT^*) \subseteq [0, 0] \\ f_2 R(X_1, \ldots, X_k, U, VT^*) \subseteq [0, 0] \\ \ldots\ldots\ldots\ldots\ldots\ldots\ldots\ldots\ldots\ldots \\ f_m R(X_1, \ldots, X_k, U, VT^*) \subseteq [0, 0]. \end{cases}$$

As $f_i R(X_1, \ldots, X_k, U, VT^*)$ $(i = 1, \ldots, m)$ are, respectively, interpretable computations of $f_i^*(X_1, \ldots, X_k, U, V)$, the formula

$$(\forall x_1 \in X'_1) \ldots (\forall x_k \in X'_k)(\forall u \in U')(\exists v \in V') f(x_1, \ldots, x_k, v, w) = 0,$$

is valid. So, every $(x_1, \ldots, x_k) \in \Xi$ and $X' \subseteq \Xi$. ∎

Corollary 10.3.1 *Let U' and V' be interval vectors and f be a continuous function from $\mathbb{R}^k \times U' \times V'$ in $\mathbb{R}^m$ such that $f = (f_1, \ldots, f_m)$. Let us consider the AE-solution set*

$$\Xi = \{x \in \mathbb{R}^k \mid (\forall u \in U')(\exists v \in V') f(x, u, v) = 0\}.$$

If $x = (x_1, \ldots, x_k)$ is a point such that

$$fR(x, U, VT^*) \subseteq ([0,0], \ldots, [0,0]), \tag{10.20}$$

where

1) U *are proper intervals and* V *improper ones,*

10.3 Solution Sets

2) VT^* is obtained from V by transforming all of its incidences (in all the equations) but one into their duals,

then $x \in \varXi$.

Proof This is the previous theorem for the particular case of $X = [x, x]$. ∎

Remark 10.3.3 This corollary provides the condition for a point to belong to a solution set and, therefore, can be used to characterize the AE-solution set by means of points.

Theorem 10.3.2 *Let U' and V' be interval vectors and f be a continuous function from $\mathbb{R}^k \times U' \times V'$ in $\mathbb{R}^m$ such that $f = (f_1, \ldots, f_m)$. Let us consider the AE-solution set*

$$\varXi = \{x \in \mathbb{R}^k \mid (\forall u \in U')(\exists v \in V') f(x, u, v) = \mathbf{0}\}.$$

If for some i

$$\operatorname{Inn}(f_i^*(X, U, V)) \not\subseteq [0, 0], \tag{10.21}$$

with U proper intervals and X, V improper ones, then $X' \cap \varXi = \emptyset$.

Proof For such i, the *-semantic theorem implies

$$\operatorname{Inn}(f_i^*(X, U, V)) \not\subseteq [0, 0] \Rightarrow f_i^*(X, U, V) \not\subseteq [0, 0]$$
$$\Leftrightarrow \neg((\forall u \in U')(\exists v \in V')(\exists x \in X')\, f_i(x, u, v) = 0)$$
$$\Leftrightarrow (\exists u \in U')(\forall v \in V')(\forall x \in X')\, f_i(x, u, v) \neq 0$$
$$\Rightarrow (\forall x \in X')\, \neg((\forall u \in U')(\exists v \in V')\, f_i(x, u, v) = 0)$$
$$\Rightarrow X' \cap \varXi = \emptyset. \qquad \blacksquare$$

So the QSI algorithm can be adapted to the vectorial case by means of the branch-and-bound process and applying the rules of selection and rejection. The first rule is

$$\text{Inside QSI}: X' \subseteq \varXi \Leftarrow \begin{cases} f_1 R(X_1, \ldots, X_k, U, VT^*) \subseteq [0, 0] \\ f_2 R(X_1, \ldots, X_k, U, VT^*) \subseteq [0, 0] \\ \ldots\ldots\ldots\ldots\ldots\ldots\ldots\ldots\ldots\ldots \\ f_m R(X_1, \ldots, X_k, U, VT^*) \subseteq [0, 0], \end{cases}$$

The second bounding rule is

$$\text{Outside QSI}: X' \cap \varXi = \emptyset \Leftarrow \operatorname{Inn}(f_i^*(X, U, V)) \not\subseteq [0, 0] \text{ for some } i = 1, \ldots, m$$

Finally, if any of the bounding rules is fulfilled, the box X' is considered as undefined and it is bisected.

10.3.2 Interval Estimations

The characterization of a $\varXi$ solution set by means of pavings, that is, intervals contained in $\varXi$, is not the only possible one, and perhaps it is not always the most convenient. When the result of the parameter estimation is used to simulate with an interval model, the parameters must be intervals and, therefore, it can be more useful to find an interval inner estimate, that is, an interval of parameters contained in $\varXi$, or an interval outer estimate, an interval containing $\varXi$, depending on the solution requirements.

An *inner interval estimate* of $\varXi$ is any interval X' such that $X' \subseteq \varXi$. An *outer interval estimate* of $\varXi$ is any interval X' such that $X' \supseteq \varXi$. The *hull* is the $\subseteq$-minimum interval which contains $\varXi$ and a *weak inner interval estimate* is any interval containing solutions and contained in the hull.

Obviously, a characterization by pavings also provides inner and outer interval estimates of the set $\varXi$. Any box of the paving is an inner interval estimate, and by operating with these pavings, different inner estimates can be obtained.

Nevertheless, different interval inner estimates can be also considered and sometimes it is necessary to find an interval inner estimate fulfilling some criteria, such as maximum volume, maximum diagonal, or others.

Consider the particular case when $m = n$ and the equations $f_i(x, u, v) = 0$ are the linear equations $a_{i1}x_1 + a_{i2}x_2 + \ldots + a_{in}x_n = b_i$, and consider any set of n proper intervals $Y_1, Y_2, \ldots, Y_n$ satisfying the inclusions

$$\begin{cases} A_{11} * Y_1 + A_{12} * Y_2 + \ldots + A_{1n} * Y_n \subseteq B_1 \\ A_{21} * Y_1 + A_{22} * Y_2 + \ldots + A_{2n} * Y_n \subseteq B_2 \\ \ldots\ldots\ldots\ldots\ldots\ldots\ldots\ldots\ldots\ldots\ldots\ldots\ldots\ldots\ldots \\ A_{n1} * Y_1 + A_{n2} * Y_2 + \ldots + A_{nn} * Y_n \subseteq B_n, \end{cases}$$

where $A_{i_1 j_1}, \ldots, A_{i_p j_p}, B_{k_{q+1}}, \ldots, B_{k_n}$ are proper intervals and $A_{i_{p+1} j_{p+1}}, \ldots, A_{i_n j_n}, B_{k_1}, \ldots, B_{k_q}$ are uni-incident improper. Then the n-dimensional interval $(Y_1', Y_2' \ldots, Y_n')$ satisfies $(Y_1', Y_2' \ldots, Y_n') \subseteq \varXi$ and turns out to be an interval inner estimate of the solution set $\varXi$. Therefore, starting from an initial solution $\mathbf{Y}^{(0)}$ of the system $\mathbf{A} * \mathbf{X} \supseteq \mathbf{B}$, the Jacobi algorithm yields, provided it converges, a sequence of interval vectors $\mathbf{Y}^{(0)} \supseteq \mathbf{Y}^{(1)} = \mathfrak{J}(\mathbf{Y}^{(0)}) \subseteq \mathbf{Y}^{(2)} = \mathfrak{J}(\mathbf{Y}^{(1)}) \supseteq \ldots$ which, in line with the properties of the Jacobi interval operator, arranges that $\mathbf{Y}^{(t)}$ is a formal solution of $\mathbf{A} * \mathbf{X} \supseteq \mathbf{B}$ if t is even, and a formal solution of $\mathbf{A} * \mathbf{X} \subseteq \mathbf{B}$ if t is odd. So the interval vectors of the sequence $(\mathbf{Y}^{(1)}, \mathbf{Y}^{(3)}, \mathbf{Y}^{(5)}, \ldots)$, and every other one obtained starting from a different initial solution, are pavings if they are proper intervals.

Example 10.3.4 Let us suppose given a system of two linear equations and two unknowns

$$\begin{cases} a_{11}x_1 + a_{12}x_2 = b_1 \\ a_{21}x_1 + a_{22}x_2 = b_2. \end{cases}$$

10.3 Solution Sets

The set-intervals of variation of each variable are $A'_{11} = [2,4]'$, $A'_{12} = [1,2]'$, $A'_{21} = [-1,2]'$, $A'_{22} = [2,4]'$, $B'_1 = [0,2]'$ and $B'_2 = [2,3]'$. The problem is to find an inner and an outer estimate for the AE-solution set, for example

$$\Xi = \{x \in \mathbb{R}^2 \mid (\forall a_{12} \in [1,2]')(\forall b_2 \in [2,3]')(\exists a_{11} \in [2,4]')(\exists a_{21} \in [-1,2]')$$
$$(\exists a_{22} \in [2,4]')(\exists b_1 \in [0,2]') \ (a_{11}x_1 + a_{12}x_2 = b_1, a_{21}x_1 + a_{22}x_2 = b_2)\}.$$

An inner estimate can be obtained by finding the proper interval formal solutions, if they exist, of the linear system

$$\begin{cases} [4,2] * X_1 + [1,2] * X_2 \subseteq [0,2] \\ [2,-1] * X_1 + [4,2] * X_2 \subseteq [3,2]. \end{cases}$$

The Jacobi algorithm applied to the system

$$\begin{cases} [4,2] * X_1 + [1,2] * X_2 = [0,2] \\ [2,-1] * X_1 + [4,2] * X_2 = [3,2] \end{cases}$$

yields the sequence

$X^{(0)} = ([-0.0909091, -0.0909091], [0.848485, 0.848485])$,

$Y^{(0)} = X^{(1)} = ([-0.848485, 0.575758], [0.477273, 1.59091])$,

$Y^{(1)} = ([-0.238636, -0.295455], [0.75, 1])$,

$Y^{(2)} = ([-0.375, 0], [0.676136, 1.29545])$,

$Y^{(3)} = ([-0.338068, -0.147727], [0.75, 1])$,

$Y^{(4)} = ([-0.375, 0], [0.713068, 1.14773])$,

$Y^{(5)} = ([-0.356534, -0.0738636], [0.75, 1])$,

$Y^{(11)} = ([-0.372692, -0.0092233], [0.75, 1])$,

$Y^{(21)} = ([-0.374928, -0.000289], [0.75, 1])$,

$Y^{(31)} = ([-0.374998, -9.0003e-6], [0.75, 1])$,

$Y^{(47)} = ([-0.375, 0], [0.75, 1])$,

which provides the inner estimate $([-0.375, 0]', [0.75, 1]') \subseteq \Xi_{\alpha\beta}$. All Y^i for i an odd integer greater than 1 are also inner estimates and the limit of the sequence $(Y^{(1)}, Y^{(3)}, Y^{(5)}, \ldots)$ is a $\subseteq$-maximal inner estimate.

When the Jacobi algorithm does not converge to a proper solution, the mixed integer or non-linear optimization techniques described in Chap. 6 can be used and non-linear optimization will be necessary when f is not a linear function.

Modal Interval Analysis allows of transforming the logical problem into an algebraic one in a very direct way. The transformed algebraic problem is set-theoretical and it can be solved with interval techniques and/or linear programming techniques, if the functions are linear, or with optimization tools. The restrictions are inequalities obtained from the interval inclusions of each equation, which are the algebraic translation provided by the modal interval analysis of the semantic statement of the problem.

For example let us consider the parameter identification problem defined by the solution set

$$\Xi = \{p \mid (\forall i \in \{1,\ldots,n\}) \, (\exists u^i \in U^i_{exp}) \, (\exists y^i \in Y^i_{exp}) \, y^i = f(p,u^i)\}.$$

An interval inner estimate of Ξ is an interval P such that the logical formulas

$$(\forall p \in P) \, (\exists u^i \in U^i_{exp}) \, (\exists y^i \in Y^i_{exp}) \, y^i = f(p,u^i), \quad \text{for all } i = 1,\ldots,n$$

are true, which is equivalent to the following n modal interval inclusions

$$f^*(P, U^i_{exp}) \subseteq Y^i_{exp} \qquad i = 1,\ldots,n$$

with P and Y^i_{exp} proper intervals and U^i_{exp} improper ones.

To find an inner interval estimate of Ξ of maximum volume, it is necessary to solve a non-linear optimization problem. If $P = (P_1,\ldots, P_k)$ with $P_j = [\underline{p}_j, \overline{p}_j]$, the problem is:

maximize: $\prod_{j=1}^{k}(\overline{p}_j - \underline{p}_j)$

subject to

1. The value $p \in P$ must have physical sense.
2. The intervals P_j must be proper intervals.
3. $f^*(P, U^i_{exp}) \subseteq Y^i_{exp}$ for all $i = 1,\ldots,n$, with P and Y^i_{exp} proper intervals and U^i_{exp} improper ones.

As the function to be maximized is a real function, any optimization procedure of numerical real analysis can be used to solve the problem.

Example 10.3.5 Coming back to the previous Example 10.3.1 for the mean velocity in the reach of the river Ter, where the model is

$$v_m = a_1 f + a_2 s f$$

and the same semantic statement for the parameter identification. The problem is to obtain an interval $A = (A_1, A_2)$ with maximum volume, which, when presented as a non-linear optimization problem, is:

maximize: $(\overline{a}_1 - \underline{a}_1) * (\overline{a}_2 - \underline{a}_2)$

10.3 Solution Sets

Fig. 10.6 Simulation with the obtained parameters

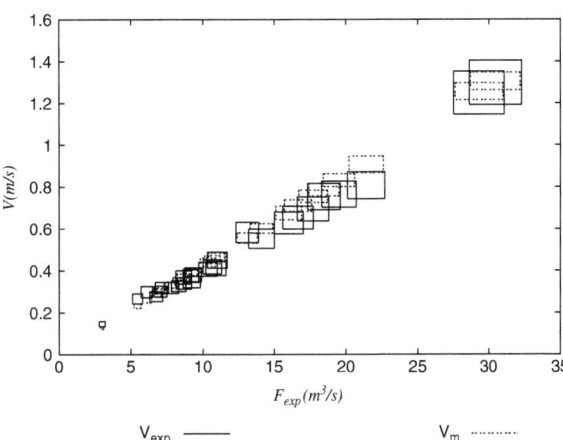

subject to

1. $A_1 \geq [0,0]$, $A_2 \geq [0,0]$.
2. A_1 and A_2 must be proper intervals.
3. $f^*(A_1, A_2, S^i_{exp}, F^i_{exp}) \subseteq V^i_{exp}$ for $i = 1, \ldots, 27$, with A_1, A_2, V^i_{exp} proper intervals and S^i_{exp}, F^i_{exp} improper ones.

Using the optimization library GRG2©[58] the result is the interval $A = (A_1, A_2)$, where

$$A_1 = [0.041724, 0.042548] \quad A_2 = [0, 0.238903].$$

With this interval for the parameters, the results of a simulation with the set of data of Table 10.1 compared with the experimental intervals for the velocity are shown in Fig. 10.6.

The semantics associated to this interval of parameters A is

$$(\forall (a_1, a_2) \in A) \, (\exists s^i \in S^i_{exp}) \, (\exists f^i \in F^i_{exp}) \, (\exists v^i_m \in V^i_{exp}) \, v^i_m = a_1 f^i + a_2 s^i f^i,$$

for all $i = 1, \ldots, 27$.

The join of all the non-decidable intervals, and any other interval containing this join, is an interval outer estimate. This join is a good approximation (up to the fixed precision for the QSI algorithm in the paving process) to the hull of the solution set. Obviously the exact hull is not attainable due to the unavoidable rounding, but with optimization techniques it is possible to approximate the hull with an error less than the accuracy of the algorithm used.

Optimization techniques can be used to get weak inner estimates. By Corollary 10.3.1, a point belongs to Ξ if

$$x \in \Xi \Leftarrow \begin{cases} f_1 R(x_1, \ldots, x_k, U, VT^*) \subseteq [0, 0] \\ f_2 R(x_1, \ldots, x_k, U, VT^*) \subseteq [0, 0] \\ \ldots \ldots \ldots \ldots \ldots \ldots \ldots \ldots \ldots \ldots \\ f_m R(x_1, \ldots, x_k, U, VT^*) \subseteq [0, 0], \end{cases}$$

So, finding a weak inner estimate of Ξ can be put in the form of solving the $2k$ mathematical programming problems:

$$\min x_1, \max x_1, \min x_2, \max x_2, \ldots, \min x_k, \max x_k$$

subjecting all of them to the same restrictions:

$$\text{Inf}(f_1 R(x_1, \ldots, x_i, \ldots, x_k, U, VT^*)) \geq 0,$$
$$\text{Sup}(f_1 R(x_1, \ldots, x_i, \ldots, x_k, U, VT^*)) \leq 0,$$
$$\text{Inf}(f_2 R(x_1, \ldots, x_i, \ldots, x_k, U, VT^*)) \geq 0,$$
$$\text{Sup}(f_2 R(x_1, \ldots, x_i, \ldots, x_k, U, VT^*)) \leq 0,$$
$$\ldots \ldots \ldots \ldots \ldots \ldots \ldots \ldots \ldots \ldots \ldots \ldots \ldots$$
$$\text{Inf}(f_m R(x_1, \ldots, x_i, \ldots, x_k, U, VT^*)) \geq 0,$$
$$\text{Sup}(f_m R(x_1, \ldots, x_i, \ldots, x_k, U, VT^*)) \leq 0.$$

The solution of these $2k$ problems will provide a weak inner estimate of Ξ. If for all the syntax trees of $f_1, \ldots, f_m$ the variables v are uni-incident, all the rational computations $f_1 R, \ldots, f_m R$ are optimal, and the corresponding optimization algorithm gives global optima for all of them, the transformation from V to VT^* is not necessary and the solutions define the hull of Ξ (up to roundings).

In the particular case when $m = n$ and the equations $f_i(x, u, v) = 0$ are the linear equations $a_{i1}x_1 + a_{i2}x_2 + \ldots + a_{in}x_n = b_i$, it would be necessary to solve the $2n$ corresponding programming problems:

$$\min x_1, \max x_1, \min x_2, \max x_2, \ldots, \min x_k, \max x_k$$

subjecting all of them to the same restrictions:

$$\text{Inf}(A_{11}x_1 + A_{12}x_2 + \ldots + A_{1i}x_i + \ldots + A_{1n}x_n) \geq \text{Inf}(B_1),$$
$$\text{Sup}(A_{11}x_1 + A_{12}x_2 + \ldots + A_{1i}x_i + \ldots + A_{1n}x_n) \leq \text{Sup}(B_1),$$
$$\text{Inf}(A_{21}x_1 + A_{22}x_2 + \ldots + A_{2i}x_i + \ldots + A_{2n}x_n) \geq \text{Inf}(B_2),$$
$$\text{Sup}(A_{21}x_1 + A_{22}x_2 + \ldots + A_{2i}x_i + \ldots + A_{2n}x_n) \leq \text{Sup}(B_2),$$
$$\ldots \ldots \ldots \ldots \ldots \ldots \ldots \ldots \ldots \ldots \ldots \ldots \ldots$$
$$\text{Inf}(A_{n1}x_1 + A_{n2}x_2 + \ldots + A_{ni}x_i + \ldots + A_{nn}x_n) \geq \text{Inf}(B_n),$$
$$\text{Sup}(A_{n1}x_1 + A_{n2}x_2 + \ldots + A_{ni}x_k + \ldots + A_{nn}x_n) \leq \text{Sup}(B_n).$$

10.3 Solution Sets

Due to the definition of multiplication of a real number by an interval, none of these problems is a linear programming problem since the restrictions are not totally linear functions but piece-wise linear functions. So it will be necessary either to solve them as non-linear problems, following the previous approach, or to introduce 0–1 variables, one for each real variable x_i ($i = 1, \ldots, n$), to convert them to mixed integer problems. In this case, as all the coefficients and right-hand sides are uni-incident and all the rational computations are optimal, the solution of all these $2n$ problems will provide the hull of Ξ.

In the particular case when $m = n$ and the equations $f_i(x, u, v) = 0$ are the linear equations $a_{i1}x_1 + a_{i2}x_2 + \ldots + a_{in}x_n = b_i$, it would be necessary to solve the $2n$ corresponding programming problems:

$$\min x_1, \ \max x_1, \ \min x_2, \ \max x_2, \ \ldots, \ \min x_k, \ \max x_k$$

subjecting all of them to the same restrictions:

$$\text{Inf}(A_{11}x_1 + A_{12}x_2 + \ldots + A_{1i}x_i + \ldots + A_{1n}x_n) \geq \text{Inf}(B_1),$$

$$\text{Sup}(A_{11}x_1 + A_{12}x_2 + \ldots + A_{1i}x_i + \ldots + A_{1n}x_n) \leq \text{Sup}(B_1),$$

$$\text{Inf}(A_{21}x_1 + A_{22}x_2 + \ldots + A_{2i}x_i + \ldots + A_{2n}x_n) \geq \text{Inf}(B_2),$$

$$\text{Sup}(A_{21}x_1 + A_{22}x_2 + \ldots + A_{2i}x_i + \ldots + A_{2n}x_n) \leq \text{Sup}(B_2),$$

$$\ldots\ldots\ldots\ldots\ldots\ldots\ldots\ldots\ldots\ldots\ldots\ldots\ldots\ldots\ldots\ldots\ldots\ldots\ldots\ldots\ldots\ldots\ldots$$

$$\text{Inf}(A_{n1}x_1 + A_{n2}x_2 + \ldots + A_{ni}x_i + \ldots + A_{nn}x_n) \geq \text{Inf}(B_n),$$

$$\text{Sup}(A_{n1}x_1 + A_{n2}x_2 + \ldots + A_{ni}x_k + \ldots + A_{nn}x_n) \leq \text{Sup}(B_n).$$

Due to the definition of multiplication of a real number by an interval, none of these problems is a linear programming problem since the restrictions are not totally linear functions but piece-wise linear functions. So it will be necessary either to solve them as non-linear problems, following the previous approach, or to introduce 0–1 variables, one for each real variable x_i ($i = 1, \ldots, n$), to convert them to mixed integer problems. In this case, as all the coefficients and right-hand sides are uni-incident and all the rational computations are optimal, the solution of all these $2n$ problems will provide the hull of Ξ.

Example 10.3.6 Let us consider a linear $n \times n$ system with coefficients defined by the intervalized matrix

$$\mathbf{A} = \begin{pmatrix} 1 & 2 & 3 & \cdots & n-1 & n \\ 2 & 2 & 3 & \cdots & n-1 & n \\ 3 & 3 & 3 & \cdots & n-1 & n \\ \vdots & \vdots & \vdots & \ddots & \vdots & \vdots \\ n-1 & n-1 & n-1 & \cdots & n-1 & n \\ n & n & n & \cdots & n & n \end{pmatrix}$$

uniformly widened by ± 0.00001 and all the n right-hand sides are intervals bounded by 0.999 and 1.001. Let us consider, for example the united solution set

$$\Xi_{uni} := \{x \in \mathbb{R}^n \mid (\exists a_{11} \in A'_{11}) \ldots (\exists a_{nn} \in A'_{nn}) \, (\exists b_1 \in B'_1) \ldots (\exists b_n \in B'_n) \, \mathbf{Ax} = \mathbf{b}\}$$

A weak inner estimate can be obtained from the interval system in which the coefficients are improper intervals $[a_{ij} + 0.00001, a_{ij} - 0.00001]$ and the right-hand sides are equal to the proper interval $[0.999, 1.001]$. By considering it as a non-linear programming problem, in the case of $n = 10$ the solution is (using the GRG2 © optimization software, www.optimalmethods.com),

$$\begin{aligned}X = &([-.0020027157, .0020023119], [-.0040052708, .0040047840], \\ &[-.0040054313, .0040049443], [-.0040052709, .0040049443], \\ &[-.0040054312, .0040049443], [-.0040052804, .0040049443], \\ &[-.0040054309, .0040046444], [-.0040052779, .0040046266], \\ &[-.0040054281, .0040043241], [.0980975654, .1019020469]),\end{aligned}$$

with a CPU time of 2 s in a workstation HP 9000 with a processor PA-RISC 50 MHz. The solution as mixed integer programming problems (using the LINDO © linear and integer programming software, www.lindo.com) is

$$\begin{aligned}X = &([-.002001204, .002001645], [-.004003203, .004003042], \\ &[-.004003363, .004003042], [-.004003290, .004003042], \\ &[-.004003363, .004003042], [-.004003290, .004003042], \\ &[-.004003363, .004003042], [-.004003108, .004003024], \\ &[-.004002894, .004002687], [.09809904, .1019010]),\end{aligned}$$

with a CPU time of 135 s in an Intel Celeron 500 MHz, which is the hull (up to rounding) of the solution set Ξ. In the case of $n = 30$, the CPU times are 37 s and close to 60×220 s, respectively. In the case of $n = 50$, the CPU times are 310 s and close to $100 \times 1{,}200$ s, respectively.

Now let us consider, for example the tolerable solution set

$$\Xi_{tol} = \{x \in \mathbb{R}^n \mid (\forall a_{11} \in A'_{11}) \ldots (\forall a_{nn} \in A'_{nn}) \, (\exists b_1 \in B'_1) \ldots (\exists b_n \in B'_n) \, \mathbf{Ax} = \mathbf{b}\}$$

A weak inner estimate can be obtained from the interval system in which the coefficients are proper intervals $[a_{ij} - 0.00001, a_{ij} + 0.00001]$ and the right-hand sides are equal to the proper interval $[0.999, 1.001]$. By considering it as a non-linear programming problem (using the GRG2 © optimization software), in the case of $n = 10$ the solution is

10.3 Solution Sets

$$X = ([-.0019980803, .0019980822], [-.0039960054, .0039960422],$$
$$[-.0039958000, .0039959707], [-.0039959964, .0039960273],$$
$$[-.0039958683, .0039959563], [-.0039960071, .0039959236],$$
$$[-.0039960325, .0039959914], [-.0039960230, .0039960112],$$
$$[-.0039959931, .0039961398], [.0981017657, .1018981786]),$$

with a CPU time of 2 s in a workstation HP 9000 with a processor PA-RISC 100 MHz. The solution as mixed integer programming problems (using the LINDO © linear and integer programming software) is

$$X = ([-.001999001, .001998999], [-.003997842, .003997838],$$
$$[-.003997842, .003997838], [-.003997842, .003997838],$$
$$[-.003997842, .003997838], [-.003997842, .003997838],$$
$$[-.003997846, .003997842], [-.003997853, .003997849],$$
$$[-.003997933, .003998005], [.09810092, .1018991]),$$

with a CPU time of 150 s in a Intel Celeron 500 MHz, which is the hull (up to rounding) of the solution set $\varXi$. In the case of $n = 30$, the CPU times are 65 s and close to 60×450 s, respectively. In the case of $n = 50$, the CPU times are 360 s and close to 100×780 s, respectively.

Example 10.3.7 (From [65]) Let us consider, for example the solution set

$$\varXi = \{x \in \mathbb{R}^n \mid Q(a_{ij}, A_{ij})(\exists b_i \in B'_i) \, \mathbf{A}\mathbf{x} = \mathbf{b}\}.$$

with $B'_i = [-1, 1]'$, $A'_{ij} = [0, 2]'$ if $i \neq j$, $A'_{ij} = [t, t]'$ if $i = j$ $(i, j = 1, \ldots, n)$, Q is an existential quantifier when $i = j$ or $i = j - 1$ or $i = j + 1$ and a universal quantifier in other case and each universal quantifier precedes to all the existential quantifiers.

A weak inner estimate of $\varXi$ can be obtained from the interval system **A*X=B** with

$$\mathbf{A} = \begin{pmatrix} [t,t] & [2,0] & [0,2] & \ldots & [0,2] & [0,2] \\ [2,0] & [t,t] & [2,0] & \ldots & [0,2] & [0,2] \\ [0,2] & [2,0] & [t,t] & \ldots & [0,2] & [0,2] \\ \ldots & \ldots & \ldots & \ldots & \ldots & \ldots \\ [0,2] & [0,2] & [0,2] & \ldots & [t,t] & [2,0] \\ [0,2] & [0,2] & [0,2] & \ldots & [2,0] & [t,t] \end{pmatrix} \text{ and } \mathbf{B} = \begin{pmatrix} [-1,1] \\ [-1,1] \\ [-1,1] \\ \ldots \\ [-1,1] \\ [-1,1] \end{pmatrix}.$$

By considering it as a non-linear programming problem, in the case of $n = 10$ and $t = 12$ the solution is (using the GRG2 © optimization software),

$$X = ([-.0493460794, .0508658012], [-.0616031482, .0590925454],$$
$$[-.0762287990, .0705556843], [-.0563179030, .0544155900],$$
$$[-.0563179030, .0544155900], [-.0563179030, .0544155900],$$
$$[-.0563179030, .0544155900], [-.0611468829, .0705556843],$$
$$[-.0647274327, .0592073076], [-.0493460794, .0457792387]),$$

with a CPU time of 5 s in a workstation HP 9000 with a processor PA-RISC 50 MHz. The solution as mixed integer programming problems (using the LINDO © linear and integer programming software) is

$$X = ([-.1, .1], [-.1125, .1125],$$
$$[-.1125, .1125], [-.1125, .1125],$$
$$[-.1125, .1125], [-.1125, .1125],$$
$$[-.1125, .1125], [-.1125, .1125],$$
$$[-.1125, .1125], [-.1, .1]),$$

with a CPU time of 160 s in a Intel Celeron 500 MHz, which is the hull (up to rounding) of the solution set $\varXi$.

10.4 A Semantic View of Control

Control problems, together with an extensive class of technical problems involving uncertainties, can be formalized by means of the predicate logic and, therefore, they are suitable for being solved by modal interval techniques, thanks to the semantic extensions f^* and f^{**} and their corresponding semantic theorems. Both semantic theorems state an equivalence between a logical formula involving intervals and functional predicates (where we assume the universal quantifiers precede the existential ones) with an interval inclusion. The solution to these inclusions can be obtained, in some easy cases, by means of interval arithmetic computations, but in other more general cases the f^*-algorithm will be needed. Often the logical definition of the problem is stated as the characterization of a solution set which can be made by means of the QSI algorithm, based on the f^*-algorithm as well.

Models usually are inaccurate, i.e., they are approximate representations of the systems. This is a consequence of the modelling procedure which, usually, involves hypotheses, assumptions, simplifications, linearizations, etc. A modern view of control sees feedback as a tool for uncertainty management [63]. A complementary way to manage uncertainty is including it in the models by making them accurate but imprecise. This is what interval models do.

For instance, a model such as the one of Eqs. (10.5),

10.4 A Semantic View of Control

$$y(k+1) = ay(k) + bu(k)$$

with $k+i = (n+i)\Delta t$, if, for example $a \in A = [0.2, 0.4]'$ and $b \in B = [0.4, 0.6]'$, it becomes an interval model. It can be seen as a set of models. It is obvious that the necessary control variable for different systems belonging to the set of systems represented by the interval model is different. Therefore, a single value of the control variable can not control all the systems, but the set of necessary control variables is bounded and it can be represented by an interval. The semantics at the core of the problem of searching an interval for the control variable $U(k)$ is

$$(\forall a \in [0.2, 0.4]')(\forall b \in [0.4, 0.6]')(\exists u(k) \in U(k)')$$
$$y_{sp}(k+1) = ay(k) + bu(k). \tag{10.22}$$

If $y_{sp}(k+1) = 7$ and $y(k) = 2$, and whichever are the values of a and b (in their respective given intervals), the solution for the interval $U(k)'$ such that the necessary, but perhaps unknown, control variable is in $U(k)'$ is the range of the real function

$$u(k) = \frac{y_{sp}(k+1) - ay(k)}{b}. \tag{10.23}$$

which is easily obtainable by the interval computation

$$U(k)' = \frac{7 - 2\,[0.2, 0.4]'}{[0.4, 0.6]'} = \frac{7 - [0.4, 0.8]'}{[0.4, 0.6]'} = \frac{[6.2, 6.6]'}{[0.4, 0.6]'} = [10.33, 16.5]'$$
$$\tag{10.24}$$

made with outer rounding to satisfy the logical condition (10.22). So, as was seen in the Introduction, for $a = 0.3$ and $b = 0.5$, it is $u(k) = 12.8 \in [10.33, 16.5]'$.

Different quantified logical formulas can be used to express different control problems in a formal way. For instance, it can be interesting to know the set of necessary control variables $U(k)$ for achieving all the setpoints $y_{sp}(k+1)$ belonging to an interval, such as $[6, 8]'$,

$$(\forall a \in [0.2, 0.4]')\,(\forall b \in [0.4, 0.6]')\,(\forall y_{sp}(k+1) \in [6, 8]')\,(\exists u(k) \in U(k)')$$
$$y_{sp}(k+1) = a_1 y(k) + b_1 u(k). \tag{10.25}$$

The solution of this problem is the range of the real function defined in (10.23), also easily obtainable by the interval computation

$$U(k)' = \frac{[6, 8]' - 2\,[0.2, 0.4]'}{[0.4, 0.6]'} = \frac{[6, 8]' - [0.4, 0.8]'}{[0.4, 0.6]'} = \frac{[5.2, 7.6]'}{[0.4, 0.6]'} \subseteq [8.66, 19]'$$

computed with outer rounding. Notice that this result obviously includes the one of (10.24), as now the setpoint is a set and it is desired to have the possibility of achieving all its values.

But it can be also interesting to know the set of values for the necessary control variable $U(k)$ to achieve only some setpoint $y_{sp}(k+1)$ belonging to the interval $[6, 8]'$

$$(\forall a \in [0.2, 0.4]') \, (\forall b \in [0.4, 0.6]') \, Q(u(k), U(k)) \, (\exists y_{sp}(k+1) \in [6, 8]')$$
$$y_{sp}(k+1) = a_1 y(k) + b_1 u(k). \tag{10.26}$$

The solution of this problem is out of reach of classical interval arithmetic. Modal Interval Analysis provides logical tools to solve this kind of problems. According to the *-semantic theorem, this logical condition is equivalent to finding an outer estimate of the u^*-extension of the function (10.23) to the intervals $A = [0.2, 0.4]$, $B = [0.4, 0.6]$ and $Y_{sp}(k+1) = [8, 6]$. As its rational extension $uR(A, B, Y_{sp}(k+1))$ is optimal in the domains involved, the solution is

$$U(k) = \frac{[8, 6] - 2\,[0.2, 0.4]}{[0.4, 0.6]} = \frac{[8, 6] - [0.4, 0.8]}{[0.4, 0.6]} = \frac{[7.2, 5.6]}{[0.4, 0.6]} \subseteq [12, 14], \tag{10.27}$$

which is included in the result of (10.24). This is obvious because now the setpoint constraint has been relaxed. The semantics (10.26) becomes

$$(\forall a \in [0.2, 0.4]') \, (\forall b \in [0.4, 0.6]') \, (\exists y_{sp}(k+1) \in [6, 8]') \, (\exists u(k) \in [12, 14]')$$
$$y_{sp}(k+1) = a_1 y(k) + b_1 u(k). \tag{10.28}$$

An interesting result is obtained if the setpoint is even more relaxed, for instance if the setpoint is any value in $[5, 8]'$:

$$U(k) = \frac{[8, 5] - 2\,[0.2, 0.4]}{[0.4, 0.6]} = \frac{[8, 5] - [0.4, 0.8]}{[0.4, 0.6]} = \frac{[7.2, 4.6]}{[0.4, 0.6]} \subseteq [12, 11.5] \tag{10.29}$$

The modality of the result has changed so the semantics have changed and 10.26 becomes

$$(\forall a \in [0.2, 0.4]') \, (\forall b \in [0.4, 0.6]') \, (\forall u(k) \in [11.5, 12]') \, (\exists y_{sp}(k+1) \in [6, 8]')$$
$$y_{sp}(k+1) = a_1 y(k) + b_1 u(k)$$

which means that any value of the control variable in $[11.5, 12]'$ produces the desired results. This is an important difference with regard to the previous examples in which the necessary control variable was unknown but bounded.

10.4.1 Measurements and Uncertainty

A new step is to allow the true value of the output of the system at time k, i.e., $y(k)$, to be not accurately known. This is due to the uncertainty associated with the measuring procedure, which usually involves noise, analog to digital conversion errors, etc. The proposed approach is to consider the uncertainty in the measurements converting the real-valued measurements into interval measurements. Therefore, if the inaccuracy of the measurements can be bounded by the interval $P' = [\underline{p}, \overline{p}]'$, the measurement $y_m(k)$ is converted into the interval measurement

$$Y_m(k)' = y_m(k) + P' \qquad (10.30)$$

For instance, assume as in the previous example the true value of $y(k)$ is 2, but the measured value is $y_m(k) = 2.03$ and the inaccuracy of the measurement can be bounded by $P = [-0.1, 0.1]'$. Then the interval measurement is

$$Y_m(k)' = y_m(k) + P' = 2.03 + [-0.1, 0.1]' = [1.93, 2.13]' \qquad (10.31)$$

which, obviously, includes the true value of the output.

The control variables of the previous cases must be computed taking into account this inaccuracy. For instance,

$$(\forall a \in [0.2, 0.4]') \, (\forall b \in [0.4, 0.6]') \, (\forall y(k) \in [1.93, 2.13]') \, Q(u(k), U(k))$$
$$(\exists y_{sp}(k+1) \in [5, 8]') \, y_{sp}(k+1) = a \, y(k) + b \, u(k). \qquad (10.32)$$

This logical proposition is equivalent to finding an outer estimate of the u^*-extension of the function (10.23) to the intervals $A = [0.2, 0.4]$, $B = [0.4, 0.6]$, $Y_{sp}(k+1) = [8, 6]$ and $Y_m(k) = [1.93, 2.13]$. As its rational extension $uR(A, B, Y_{sp}(k+1), Y_m(k))$ is optimal in the domains involved, the solution is

$$U(k) = \frac{[8, 5] - [0.2, 0.4] * [1.93, 2.13]}{[0.4, 0.6]} = \frac{[7.148, 4.614]}{[0.4, 0.6]} \subseteq [11.91, 11.54],$$

computed with outer rounding, with the semantics meaning

$$(\forall a \in [0.2, 0.4]') \, (\forall b \in [0.4, 0.6]') (\forall y(k) \in [1.93, 2.13]')(\forall u(k) \in [11.54, 11.91]')$$
$$(\exists y_{sp}(k+1) \in [5, 8]') \, y_{sp}(k+1) = a \, y(k) + b \, u(k).$$

Comparing this result with the one obtained in (10.29), it can be observed that the inaccuracy of the measurements obviously gives a narrower $U(k)$ when it is a universal interval. If $U(k)$ were an existential interval, this inaccuracy would make it wider.

10.4.2 An Application to Temperature Control

The presented control method is applied now to the control of the temperature of a house with central heating/cooling. This method allows, in addition, designing the actuator, i.e., to determine the necessary power.

Building the model, the main component of the system is the house. Assuming to begin with that the temperature in the house is greater than the temperature outside the house, there is an energy (heat) flow from inside to outside. The heat is transmitted by conduction, convection and radiation. Assuming that all forms of heat transmission are considered in the thermal resistance of the house, the heat losses are:

$$p_{io} = \frac{t_{in} - t_{out}}{r_{th}}, \tag{10.33}$$

where

- p_{io} is power: the amount of heat transmitted from inside to outside per time unit.
- t_{in} is the inside temperature.
- t_{out} is the outside temperature.
- r_{th} is the thermal resistance of the house: windows, doors, walls, roof, etc.

Central heating tries to compensate for these power losses and to maintain the inside temperature at the desired value (setpoint). This temperature varies depending on the difference between the heat power produced by the central heating and the lost power:

$$\frac{dt_{in}}{dt} = \frac{p_{h/c} - p_{io}}{\sum_i m_i c_i}, \tag{10.34}$$

where

- m_i is the mass of each element which is inside the house: air, furniture, etc.
- c_i is its specific heat.
- $p_{h/c}$ is the power of the heating/cooling system.

The discrete model that allows computing the necessary heating power is

$$p_{h/c}(k) = \frac{t_{in}(k) - t_{out}(k)}{r_{th}} + (t_{in}(k+1) - t_{in}(k)) \frac{\sum_i m_i c_i}{\Delta t}. \tag{10.35}$$

Let us suppose that r_{th} and $\sum_i m_i c_i$ are uncertain parameters taking values within some given intervals R_{th} and MC and that $t_{in}(k)$ and $t_{out}(k)$ are measured variables with values inside certain intervals $T_{in}(k)$ and $T_{out}(k)$.

Let us consider the problem of finding an interval of control $P_{h/c}(k)$ able to give enough power to reach, in one step, from any temperature $t_{in}(k)$ inside an interval $T_{in}(k)$ to any temperature $t_{in}(k+1)$ inside an interval $T_{in}(k+1)$, when

10.4 A Semantic View of Control

the outside temperature is any value of the interval $T_{out}(k)$ and the values of the model parameters are in their respective intervals R_{th} and MC. Semantically stated, this problem is

$$(\forall r_{th} \in R'_{th}) \, (\forall m \in MC') \, (\forall t_{in}(k) \in T_{in}(k)') \, (\forall t_{out}(k) \in T_{out}(k)')$$
$$(\forall t_{in}(k+1) \in T_{in}(k+1)') \, (\exists p_{h/c}(k) \in P_{h/c}(k)')$$
$$p_{h/c}(k) = \frac{t_{in}(k) - t_{out}(k)}{r_{th}} + (t_{in}(k+1) - t_{in}(k)) \frac{\sum_i m_i c_i}{\Delta t}, \qquad (10.36)$$

which is similar to problem (10.25). The solution $P_{h/c}(k)'$ is the range of the function (10.35) but, in this case, this range is not reachable by means of simple interval arithmetic computations due to the multi-incidences of some of the variables in (10.35). Nevertheless in the context of classical intervals, there exist branch-and-bound algorithms to compute the range of the function with a given accuracy.

But it is also interesting to know the set of values for the necessary control variable $p_{h/c}(k)$ to achieve any of the acceptable temperatures in an set $T_{out}(k)'$. Semantically stated this problem is

$$(\forall r_{th} \in R'_{th}) \, (\forall m \in MC') \, (\forall t_{in}(k) \in T_{in}(k)') \, (\forall t_{out}(k) \in T_{out}(k)')$$
$$Q(p_{h/c}(k), P_{h/c}(k)) \, (\exists t_{in}(k+1) \in T_{in}(k+1)')$$
$$p_{h/c}(k) = \frac{t_{in}(k) - t_{out}(k)}{r_{th}} + (t_{in}(k+1) - t_{in}(k)) \frac{\sum_i m_i c_i}{\Delta t}, \qquad (10.37)$$

where Q is a universal or existential quantifier, depending on the modality of the computed interval control $P_{h/c}(k)$. This problem is similar to problem (10.26), which was out of the reach of classic intervals and its solution is an outer estimate of the $p^*_{h/c}$ extension of the function (10.35) to the proper intervals R_{th}, MC, $T_{out}(k)$, $T_{in}(k)$ and the improper interval $T_{in}(k+1)$. As its rational extension

$$p_{h/c} R(R_{th}, MC, T_{out}(k), T_{in}(k), T_{in}(k+1))$$

is not, in general, optimal in the domains involved due to the multi-incidences of some variables in (10.35), the solution could not be obtained by means of simple modal interval arithmetic, but it was necessary to use the f^*-algorithm to compute the involved interval extensions.

For example let us consider typical values for the parameters of the model:

- $r_{th} \in R_{th} = [0.001, 0.01] \, \frac{K}{W}$,
- $\sum_i m_i c_i \in MC = [5 \cdot 10^6, 6 \cdot 10^6] \, \frac{J}{K}$,
- $\Delta t = 120s$,

Let us also suppose that the outside temperature is $T_{out}(k) = [4, 5]$, the rooms's initial temperature is $T_{in}(k) = [18, 18.1]$ and the rooms's desired final temperature is $T_{in}(k+1) = [18.2, 18.7]$. The solution to the one-step first problem (10.36), i.e., the *-extension of the function (10.35), turns out to be

$$p_{h/c}(k) = [5{,}476.67, 49{,}000] \text{ W}$$

This result means

$$(\forall r_{th} \in [0.001, 0.01]') \, (\forall \sum_i m_i c_i \in [5 \cdot 10^6, 6 \cdot 10^6]') \, (\forall t_{in}(k) \in [18, 18.1]')$$

$$(\forall t_{out}(k) \in [4, 5]') \, (\forall t_{in}(k+1) \in [18.2, 18.7]') \, (\exists p_{h/c}(k) \in [5{,}476.67, 49{,}000]')$$

$$p_{h/c}(k) = \frac{t_{in}(k) - t_{out}(k)}{r_{th}} + (t_{in}(k+1) - t_{in}(k)) \frac{\sum_i m_i c_i}{\Delta t},$$

i.e., a power between 5,476.67 and 49,000 W is necessary to reach a determined room temperature in 120 s. The precise value depends on the desired temperature and on the precise values of the uncertain parameters. This result can be used either to choose a value for the control action, that is, the heating power, or to design the actuator, that is, the heater. If the available power is already fixed, this result can be used to assess its usefulness. In this case, if the available power is less than 5,476.67 W, none of the temperatures between $18.2 + 273.15$ and $18.7 + 273.15$ K can be reached, at least in 120 s. If the available power is between 5,476.67 and 49,000 W, some temperatures can be reached, depending on the true values of the uncertain parameters. Finally, if the available power is greater than 49 kW, any of these temperatures can be reached, so the heater is useful even in the worst case.

The solution to the second problem (10.37), i.e., the *-extension of the function (10.35) to the same proper intervals R_{th}, MC, Δt, $T_{in}(k)$, $T_{out}(k)$ and the improper interval $T_{in}(k+1) = [18.3, 18.2]$, turns out to be

$$p_{h/c}(k) = [9{,}643.33, 24{,}000] \text{ W}$$

This means that

$$(\forall r_{th} \in [0.001, 0.01]') \, (\forall m \in [5 \cdot 10^6, 6 \cdot 10^6]') \, (\forall t_{in}(k) \in [18, 18.1]')$$

$$(\forall t_{out}(k) \in [4, 5]') \, (\exists p_{h/c}(k) \in [9{,}643.33, 24{,}000]') \, (\exists t_{in}(k+1) \in [18.2, 18.3]')$$

$$p_{h/c}(k) = \frac{t_{in}(k) - t_{out}(k)}{r_{th}} + (t_{in}(k+1) - t_{in}(k)) \frac{\sum_i m_i c_i}{\Delta t},$$

i.e., an unknown power between 9,643.33 and 24,000 W will bring the room's temperature to an unknown value between $18.2 + 273.15$ and $18.7 + 273.15$ K in 120 s, whatever are the true values of the uncertain parameters in this period of time. This result is obviously included in the previous one, as now any of the final temperatures belonging to the interval is accepted. This is the computation that must be performed when any temperature considered as comfortable is acceptable.

The solution to the second problem (10.37), i.e., the *-extension of the function (10.35) to the same proper intervals R_{th}, MC, Δt, $T_{in}(k)$, $T_{out}(k)$ and the improper interval $T_{in}(k+1) = [18.7, 18.2]$, turns out to be

$$p_{h/c}(k) = [26{,}310, 24{,}000]\, W.$$

This means that

$(\forall r_{th} \in [0.001, 0.01]')\ (\forall m \in [5 \cdot 10^6, 6 \cdot 10^6]')\ (\forall t_{in}(k) \in [18, 18.1]')$
$(\forall t_{out}(k) \in [4, 5]')\ (\forall p_{h/c}(k) \in [26{,}310, 24{,}000]')\ (\exists t_{in}(k+1) \in [18.2, 18.7]')$

$$p_{h/c}(k) = \frac{t_{in}(k) - t_{out}(k)}{r_{th}} + (t_{in}(k+1) - t_{in}(k))\frac{\sum_i m_i c_i}{\Delta t},$$

i.e., any power between 24,000 and 26,310 W will bring the room's temperature to an unknown value between $18.2 + 273.15$ and $18.7 + 273.15$ K in 120 s, whatever are the true values of the uncertain parameters in this period of time. This result is very interesting because there is a guarantee on the final temperature and this guarantee is maintained whatever are the true values of the uncertain parameters and the heating power that is being applied.

This methodology is easily extended to a multi-step case by repeating the same computations for the successive steps.

Remark 10.4.1 The obtained results are useful in several ways. One of them is the control of an uncertain system. The control action is not precisely determined but it is bounded, so it is known if the set-point is reachable or not with the available actuator. Another way is the design of the actuator. In any case these tools help the engineers to assess the possibilities of already designed controlled systems or to design new ones.

10.5 Concluding Remarks

The applications presented in this chapter can be seen as simple examples meant to explain the applicability of modal intervals in different fields and cannot be considered as real-world problems, which are out of the essentially theoretical scope of this monograph. Real-world problems are usually difficult to introduce and they deserve a detailed study in a book focused on the application of models intervals. Applications in control [69, 91], fault detection [3, 4, 78], fuzzy design [9], fault diagnosis [78], mechanical engineering [45, 50–52, 93–95], biomedicine [10, 19, 38, 74], computer graphics [35, 36], constraint programming [31], fuzzy logic [7], etc are examples of different fields where modal intervals have been successfully used. Finally, modal intervals have also aroused interest in the theoretical domain and have been recently revisited and reformulated [29, 30].

References

1. G. Alefeld, J. Herzberger, *Introduction to Interval Computations* (Academic, New York, 1983)
2. M.A. Amouzegar, A global optimization method for nonlinear bilevel programming problems. IEEE Trans. Syst. Man Cybern. Part B **29**, 771–777 (1999)
3. J. Armengol, J. Vehí, L. Travé-Massuyès, M.Á. Sainz, Application of modal intervals to the generation of error-bounded envelopes. Reliable Comput. **7**(2), 171–185 (2001)
4. J. Armengol, J. Vehí, L. Travé-Massuyès, M.Á. Sainz, Application of multiple sliding time windows to fault detection based on interval models, in *12th International Workshop on Principles of Diagnosis DX 2001. San Sicario, Italy*, ed. by Sh. McIlraith, D. Theseider Dupré (2001), pp. 9–16
5. P. Basso, Optimal search for the global maximum of functions with bounded seminorm. SIAM J. Numer. Anal. **9**, 888–903 (1985)
6. E. Baumann, Optimal centered form. BIT **28**, 80–87 (1987)
7. R.J. Bhiwani, B.M. Patre, Solving first order fuzzy equations: A modal interval approach, in *2009 2nd International Conference on Emerging Trends in Engineering and Technology (ICETET)* (2009), pp. 953–956
8. F. Bierbaum, K.P. Schwierz, Moore methods and applications of interval analysis A *Bibliography on Interval Mathematics* (SIAM, Philadelphia, 1986)
9. J. Bondia, A. Sala, A. Picó, M.A. Sainz, Controller design under fuzzy pole-placement specifications: An interval arithmetic approach. IEEE Trans. Fuzzy Syst. **14**(6), 822–836 (2006)
10. R. Calm, M. García-Jaramillo, J. Bondia, M.A. Sainz, J. Vehí, Comparison of interval and monte carlo simulation for the prediction of postprandial glucose under uncertainty in tipe 1 diabetes mellitus. Comput. Methods Progr. Biomed. **104**, 325–332 (2011)
11. F. Cellier, *Continuous System Modeling* (Springer, Berlin, 1991)
12. J.M. Danskin, *The Theory of Max-Min and Its Applications to Weapons Allocation Problems* (Springer, Berlin, 1967)
13. H. Dawood, *Theories of Interval Arithmetic* (Lambert Academic, Saarbrucken, 2011)
14. L.H. de Figueiredo, J. Stolfi, Affine arithmetic: concepts and applications. Numer. Algorithms **37**(1), 147–158 (2004)
15. V.F. Demyanov, V.N. Malozemov, *Introduction to Minimax* (Dover, New York, 1990)
16. A.V. Demyanov, V.F. Demyanov, V.N. Malozemov, Minmaxmin problems revisited. Optim. Methods Softw. (OMS) **17**(5), 783–804 (2002)
17. G.D. Erdmann, A new minimax algorithm and its applications to optics problems. Ph.D. thesis, University of Minnesota, USA, 2003
18. GAO/IMTEC-92-26, Patriot missile defense: Software problem led to system failure at Dhahran, Saudi Arabia (1992). http://www.fas.org/spp/starwars/gao/im92026.htm

19. M. García-Jaramillo, R. Calm, J. Bondía, J. Vehí, Prediction of postprandial blood glucose under uncertainty and intra-patient variability in type 1 diabetes: a comparative study of three interval models. Comput. Methods Programs Biomed. **108**, 325–332 (2012)
20. E. Gardenyes, Numerical information and modal intervals, in *26th Sciencie Week*, November 1986
21. E. Gardenyes, H. Mielgo, Modal intervals: functions, in *Polish Symposium on Interval and Fuzzy Mathematics. Poznan, Poland* (1986)
22. E. Gardenyes, A. Trepat, The interval computing system SIGLA-PL/I(0). Freibg. Intervall Ber. **79**(8) (1979)
23. E. Gardenyes, A. Trepat, Fundamentals of Sigla, an interval computing system over the completed set of intervals. Computing **24**, 161–179 (1980)
24. E. Gardenyes, A. Trepat, J.M. Janer, Sigla-PL/1: Development and applications, in *Interval Mathematics*, ed. by K.L.E. Nickel (Academic, New York, 1980), pp. 301–315
25. E. Gardenyes, A. Trepat, H. Mielgo, Present perspective of the SIGLA interval system. Freibg. Intervall Ber. **82**(9), 1–65 (1982)
26. E. Gardenyes, H. Mielgo, A. Trepat, Modal intervals: Reasons and ground semantics, in *Interval Mathematics*. Lecture Notes in Computer Sciences, vol. 212 (Springer, Heidelberg, 1986), pp. 27–35
27. E. Gardenyes, M.A. Sainz, L. Jorba, R. Calm, R. Estela, H. Mielgo, A. Trepat, Modal intervals. Reliable Comput. **7**(2), 77–111 (2001)
28. J. Gleick, A bug and a crash. The New York Times Magazine **1** (1996)
29. A. Goldstein, Modal intervals revisited. Part 1: A generalized interval natural extension. Reliable Comput. **16**, 130–183 (2012)
30. A. Goldstein, Modal intervals revisited. Part 2: A generalized interval mean-value extension. Reliable Comput. **16**, 184–209 (2012)
31. C. Grandón, G. Chabert, B. Neveu, Generalized interval projection: a new technique for consistent domain extension, in *Proceedings of the 20th International Joint Conference on Artifical Intelligence* (IJCAI'07), San Francisco, CA (Morgan Kaufmann, Los Altos, 2007), pp. 94–99
32. R.T. Gregory, D.L. Karney, *A Collection of Matrices for Testing Computational Algorithms* (Wiley Interscience/Wiley, New York, 1969)
33. E. Hansen, *Global Optimization Using Interval Analysis* (Marcel Dekker, New York, 1992)
34. E. Hansen, W. Walster, *Global Optimization Using Interval Analysis*, 2nd edn, revised and expanded (Marcel Dekker, New York, 2004)
35. N. Hayes, System and method to compute narrow bounds on a modal interval spherical projection (Patent Number PCT/US2006/038871), 2007
36. N. Hayes, System and method to compute narrow bounds on a modal interval polynomial function (Patent Number US2008/0256155A1), 2009
37. P. Herrero, Quantified Real Constraint Solving Using Modal Intervals with Applications to Control. Ph.D. thesis, Ph.D. dissertation 1423, University of Girona, Girona (Spain), 2006
38. P. Herrero, R. Calm, J. Vehí, J. Armengol, P. Georgiou, N. Oliver, C. Tomazou, Robust fault detection system for insulin pump therapy using continuous glucose monitoring. J. Diabetes Sci. Technol. **6** (2012)
39. N.J. Higham, *Accuracy and Stability of Numerical Algorithms* (SIAM, Philadelphia, 1996)
40. B. Jakobsen, F. Rosendahl, The Sleipner Platform accident. Struct. Eng. Int. **4**(3), 190–193 (1994)
41. L. Jaulin, Reliable minimax parameter estimation. Reliable Comput. **7**(3), 231–246 (2001)
42. L. Jaulin, E. Walter, Guaranted bound-error parameter estimation for nonlinear models with uncertain experimental factors. Automatica **35**, 849–856 (1993)
43. L. Jaulin, E. Walter, Set inversion via interval analysis for nonlinear bounded-error estimation. Automatica **29**(4), 1053–1064 (1993)

44. L. Jaulin, M. Kieffer, O. Didrit, E. Walter, Applied Interval Analysis: With Examples in Parameter and State Estimation, Robust Control and Robotics (Springer, London, 2001)
45. S. JongSok, Q. Zhiping, W. Xiaojun, Modal analysis of structures with uncertain-but-bounded parameters via interval analysis. J. Sound Vib. **303**, 29–45 (2007)
46. L. Jorba, Intervals de Marques (in catalan). Ph.D. thesis, Facultad de Matemáticas, Universidad de Barcelona, Spain, 2003
47. W. Kahan, How futile are mindless assessments of roundoff in floating-point computation 95 (2006 in progress). Available in http://www.cs.berkeley.edu/wkahan/Mindless.pdf
48. E. Kaucher, Algebraische orweiterungen der intervallrechnung unter erhaltung der ordnungs und verbandsstrukturen. Comput. Suppl. **1**, 65–79 (1977)
49. E. Kaucher, Interval analysis in the extended interval space IR. Comput. Suppl. **2**, 33–49 (1979)
50. S. Khodaygan, M.R. Movahhedy, Tolerance analysis of assemblies with asymmetric tolerances by unified uncertaintyŰaccumulation model based on fuzzy logic. Int. J. Adv. Manuf. Technol. **53**, 777–788 (2011)
51. S. Khodaygan, M.R. Movahhedy, M. Saadat Fomani, Tolerance analysis of mechanical assemblies based on modal interval and small degrees of freedom (mi-sdof) concepts. Int. J. Adv. Manuf. Technol. **50**, 1041–1061 (2010)
52. S. Khodaygan, M.R. Movahhedy, M. Saadat Foumani, Fuzzy-small degrees of freedom representation of linear and angular variations in mechanical assemblies for tolerance analysis and allocation. Mech. Mach. Theory, **46**(4), 558–573 (2011)
53. C. Kirjer-Neto, E. Polak, On the conversion of optimization problems with max-min constraints to standard optimization problems. SIAM J. Optim. **8**(4), 887–915 (1998)
54. V. Kreinovich, Why intervals? A simple limit theorem that is similar to limit theorems from statistics. Reliable Comput. **1**(1), 33–40 (1995)
55. V.M. Kreinovich, V. Nesterov, N.A. Zheludeva, Interval methods that are guaranteed to underestimate (and the resulting new justification of kaucher arithmetic). Reliable Comput. **2**(2), 119–124 (1996)
56. L. Kupriyanova, Inner estimation of the united solution set of interval algebraic system. Reliable Comput. **1**(1), 15–41 (1995)
57. L. Kupriyanova, Finding inner estimates of the solution sets to equations with interval coefficients. Ph.D. thesis, Saratov State University, Saratov, Russia, 2000
58. L.S. Lawsdon, A.D. Waren, *GRG2 User's Guide* (Prentice hall, Englewood Cliffs, 1982)
59. S. Markov, E. Popova, C. Ullrich, On the solution of linear algebraic equations involving interval coefficients. Iterative Methods Linear Algebra IMACS Ser. Comput. Appl. Math. **3**, 216–225 (1996)
60. R.E. Moore, Interval arithmetic and automatic error analysis in digital computing. Ph.D. thesis, Stanford University, USA, 1962
61. R.E. Moore, *Interval analysis* (Prentice-Hall, Englewood Cliffs, 1966)
62. R.E. Moore, *Methods and Applications of Interval Analysis* (Studies in Applied Mathematics (SIAM), Philadelphia, 1979)
63. R.M. Murray, K.J. Åström, S.P. Boyd, R.W. Brockett, G. Stein, Future directions in control in an information-rich world. IEEE Control Syst. Mag. (2003)
64. V. Nesterov, Interval and twin arithmetics. Reliable Comput. **3**(4), 369–380 (1997)
65. A. Neumaier, *Interval Methods for Systems of Equations* (Cambridge University Press, Cambridge, 1990)
66. K. Nickel, Verbandtheoretische grundlagen der intervallmathematik. Lect. Notes Comput. Sci. **29**, 251–262 (1975)
67. K. Nickel, Optimization using interval mathematics. Freibg. Intervall Ber. **1**, 25–47 (1986)
68. M. Nogueira, A. Nandigam, Why intervals? because if we allow other sets, tractable problems become intractable. Reliable Comput. **1**(4), 389–394 (1998)
69. P. Herrero, L. Jaulin, J. Vehí, M.A. Sainz Guaranteed set-point computation with application to the control of a sailboat. Int. J. Control Autom. Syst. **8**, 1–7 (2010)

70. E. Polak, in *Optimization. Algorithms and Consistent Approximations*. Applied Mathematical Sciences, vol. 124 (Springer, New York, 1997)
71. S. Ratschan, Applications of quantified constraint solving (2002). http://www.mpi-sb.mpg.de/~ratschan/appqcs.html
72. H. Ratschek, *Nichtnumerische Aspekte der Interval-Mathematik*, vol. 29 (Springer, Heidelberg, 1975)
73. H. Ratschek, J. Rokne, *Computer Methods for the Range of Functions* (Ellis Horwood, Chichester, 1984)
74. A. Revert, R. Calm, J. Vehí, J. Bondia, Calculation of the best basal-bolus combination for postprandial glucose control in insulin pump therapy. IEEE Trans. Biomed. Eng. **58**, 274–281 (2011)
75. S.M. Rump, R.E. Moore, Algorithm for verified inclusions - theory and practice, in *Reliability in Computing*, (Academic, San Diego, 1988), pp. 109–126
76. B. Rustem, M. Howe, *Algorithms for Worst-Case Design and Applications to Risk Management* (Princeton University Press, Princeton, 2002)
77. M.Á. Sainz, J.M. Baldasano, Modelo matemático de autodepuración para el bajo Ter (in spanish). Technical report, Junta de Sanejament, Generalitat de Catalunya, 1988
78. M.Á. Sainz, J. Armengol, J. Vehí, Fault diagnosis of the three tanks system using the modal interval analysis. J. Process Control **12**(2), 325–338 (2002)
79. S.P. Shary, Solving the linear interval tolerance problem. Math. Comput. Simul. **39**(2), 145–149 (1995)
80. S.P. Shary, Algebraic approach to the interval linear static identification, tolerance and control problems, or one more application of kaucher arithmetic. Reliable Comput. **2**(1), 3–33 (1996)
81. S.P. Shary, *Algebraic Solutions to Interval Linear Equations and Their Applications* (Akademie, Berlin, 1996), pp. 224–233
82. S.P. Shary, Algebraic approach in the "outer problem" for interval linear equations. Reliable Comput. **3**, 103–135 (1997)
83. S.P. Shary, Modal intervals, in *Interval Gauss-Seidel Method for Generalized Solution Sets to Interval Linear Systems. Applications of Interval Analysis to Systems and Control* (Universitat de Girona, Spain, 1999), pp. 67–81
84. S.P. Shary, Outer estimation of generalized solution sets to interval linear systems. Reliable Comput. **5**, 323–335 (1999)
85. S.P. Shary, A new technique in systems analysis under interval uncertainty and ambiguity. Reliable Comput. **8**, 321–418 (2002)
86. SIGLA/X group, Modal intervals. basic tutorial, in *Proceedings of MISC 99. Applications of Interval Analysis to Systems and Control* (Universitat de Girona, Spain, 1999), pp. 157–227
87. T. Sunaga, Theory of an interval algebra and its applications to numerical analysis. RAAG Mem. **2**, 547–564 (1958)
88. P. Thieler, Technical calculations by means of interval mathematics (1985), pp. 197–208
89. A. Trepat, Completación reticular del espacio de intervalos. Ph.D. thesis, Facultad de Matemáticas, Universidad de Barcelona, Spain, 1982
90. A. Trepat, E. Gardenyes, J.M. Janer, Approaches to simulation and to the linear problem in the sigla system. Freibg. Intervall Ber. **81**(8) (1980)
91. J. Vehí, J. Rodellar, M.Á. Sainz, J. Armengol, Analysis of the robustness of predictive controllers via modal intervals. Reliable Comput. **6**(3), 281–301 (2000)
92. Y. Wang, Semantic tolerance modeling based on modal interval, in *NSF Workshop on Reliable Engineering Computing*, Savannah, Georgia, 2006
93. Y. Wang, Closed-loop analysis in semantic tolerance modeling. J. Mech. Des. **130**, 061701–061711 (2008)
94. Y. Wang, Interpretable interval constraint solvers in semantic tolerance analysis. Comput. Aided Des. Appl. **5**, 654–666 (2008)
95. Y. Wang, Semantic tolerance modeling with generalized intervals. J. Mech. Des. **130**, 081701–081708 (2008)
96. M. Warmus, Calculus of approximations. Bull. Acad. Pol. Sci. **Cl.III,IV**, 253–259 (1956)

97. M. Warmus, Approximations and inequalities in the calculus of approximations. classification of approximate numbers. Bull. Acad. Pol. Sci. **9**, 241–245 (1961)
98. S. Zuche, A. Neumaier, M.C. Eiermann, Solving minimax problems by interval methods. BIT **30**, 742–751 (1990)

Index

Symbols

AE-solution set 280
$\chi(A)$ 124
$\leq$-proper twin 160
$\subseteq$-proper twin 163
c-tree optimal 104
f^* algorithm 178
k-Uniform monotonicity 69
**-semantic extension 43, 245
**-semantic theorem 57, 253
*-semantic extension 43, 244
*-semantic theorem 53, 251

A

Accepted interval predicates 240
Accepted predicates 20
Admissible operator 199
Amplification of dependence 10
Anti-symmetry 196
Arithmetic operators 122
Associated interval 193
Associated mark 193

B

Bounding criteria 178
Breaking point 10

C

Canonical coordinates 22
Canonical notation for an interval of marks 233
Cell 173, 273
Centre 191
Centred operator 199
Chebyshev approximation 265
Classic interval 1
CLIP operator 66
Co-predicates of intervals of marks 232
Co-predicates, interval, set of 240
Comparable marks 195
Compatible granularity 193
Completeness 183
Composed symmetries 170
Computed operator 257
Condition of significance 192
Constrained minimax problem 276
Control interval 18
Control of temperature 302
Control variable 267
Convergence condition 151
Crossed twin 169

D

Degenerate interval 18
Degenerate intervals of marks 231
Difference of intervals of marks 259
Difference of marks 209
Digital granularity 191
Digital immersion 233
Digital numbers 7
Digits 190
Distributive law 6
Domain 19

Dual   26
Dual operator   233

E

Effective tolerance   194
Electrical circuit   95
Elementary symmetries   170
Equality, material   234
Equality, weak   235
Equivalent optimality   96
External shadow   193

F

Feedback   267
Floating point   191
Formal solution   143
Function of marks   221

G

Granularity   191
GRG2   154, 293, 296, 297

H

Hull   290

I

Imprecision index   193
Improper interval   17
Improper intervals of marks   231
Improper operator   34, 240
Improper twin   169
Inclusion twin relation   167
Inclusion, material   234
Inclusion, weak   235
Indiscernibility margin   193
Inequality, weak   235
Infimum   22, 161, 167, 239
Inmersion   233
Inner approximation for f^* 175

Inner interval estimate   290
Inner rounding   15, 70
Internal shadow   193
Interval accepted predicate   32
Interval co-predicates   31, 242
Interval models   298
Interval predicates   31, 242
Interval projection   233
Interval property   106
Interval rejected co-predicate   32
Interval-semantics of marks   220
Intervals of marks   230

J

Jacobi interval operator   145
JM-commutativity   45, 50, 246, 249
Join   29, 238, 242

L

Lateral optimality   103
Left monotonic associativity   78
Less or equal twin relation   167
Less than or equal to   25
LID (limited identity) operator   65
LINDO   153, 296–298
Linear interval system   144
List, Strip's   273

M

Mantissa   190
Mark   191
Mark operator over $\mathbb{M}(t, n)$ 199
Mark operator over $I^*(\mathbb{R})$ 198
Mark predicates   230
Mark-semantics   220
Material equality   234
Material greater than or equal to   235
Material inclusion   234
Material inequality   235
Material less than or equal to   235
Materially equal   195
Materially greater than or equal to   196
Materially less than   197
Materially less than or equal to   196
Max   29, 239

Index

Maximal approach 201
Maximum 22, 201
Meet 29, 238, 242
Meet-join operator 30, 239
Metric approach 201
Min 29, 239
Min–max theorem 45
Minimal approach 200
Minimax 268
Minimax algorithm 272
Minimum 22, 203
Modal coordinates 19
Modal equality 23
Modal inclusion 23
Modal interval 19
Modal interval extension 43
Modal interval of marks 231
Modal quantifier 20, 232
Modal rounding 27
Modal syntactic *-extension 59
Modal syntactic extension 258
Modal syntactic function 60, 258
Modal syntactic operator 60, 257
Modality 19
Multi-incidence 74

N

Non-convergence condition 152
Number of digits 191

O

One-variable operators 62
Optimality 74, 182
Outer approximation for f^* 175
Outer interval estimate 290
Outer rounding 14, 70

P

Parameter α 193
Partial monotonicity 62
Partition 173, 273
Pavings 281
Point-wise interval 18
Poor computational extension 41, 244
Predicates of intervals of marks 232

Predicates, interval, set of 240
Product of intervals of marks 259
Product of marks 204
Proper interval 17
Proper intervals of marks 231
Proper operator 34, 240
Proper twin 166, 169
Proper-transposed twin 169
Punctual twin 169

Q

QSI 281
Quantified set inversion 281
Quantifiers 12
Quotient of intervals of marks 260
Quotient of marks 205

R

Ratschek's function 124
Rejected co-predicates 26
Rejected interval co-predicates 240
Right monotonic associativity 79

S

Saddle points set 48, 249
Saddle value 49, 249
SDP 48
SDV 49
Semantic approach 200
Semantic for a function of marks 223
Semantics 5
Set of interval co-predicates 240
Set of interval predicates 240
Set of modal intervals 19
Set of twins 169
Set-theoretical extension 256
Set-theoretical inclusion twin 167
Set-theoretical operator 255
SIVIA 281
Solution set 143, 158
Soundness 183
Spike function 214
Split modality 101
Split optimality 102
Stopping criteria 179

Strip   173, 273
Sub-distributive law   6, 10
Sum of intervals of marks   259
Sum of marks   207, 209
Supremum   22, 161, 167, 239
Syntactic tree   3
System $I(\mathbb{R})$ 1

T

Tolerable solution set   158, 296
Tolerance   177, 191
Tolerance interval   17
Total monotonicity   88
Transitivity   196
Tree-optimality   82, 93
Twin inclusion for f^* 174
Twin join   171
Twin lower bound   171
Twin meet   171
Twin symmetries   170
Twin upper bound   171
Twin-interval   169
Twins   160
Two tanks   107
Two-variable operators   62
Type of a mark   192

U

Unconstrained minimax problem   268
Uni-incidence   74
Uni-incident list of vectors   98
Uniform monotonicity   62, 64
United extension   4, 243
United solution set   158

V

Validity index   193

W

Weak equality   235
Weak inclusion   235
Weak inequality   235
Weak inner interval estimate   290
Weakly equal   195
Weakly greater than or equal to   196, 236
Weakly less than   197
Weakly less than or equal to   196, 235
Width   11

LECTURE NOTES IN MATHEMATICS   Springer

Edited by J.-M. Morel, B. Teissier; P.K. Maini

Editorial Policy (for the publication of monographs)

1. Lecture Notes aim to report new developments in all areas of mathematics and their applications - quickly, informally and at a high level. Mathematical texts analysing new developments in modelling and numerical simulation are welcome.

   Monograph manuscripts should be reasonably self-contained and rounded off. Thus they may, and often will, present not only results of the author but also related work by other people. They may be based on specialised lecture courses. Furthermore, the manuscripts should provide sufficient motivation, examples and applications. This clearly distinguishes Lecture Notes from journal articles or technical reports which normally are very concise. Articles intended for a journal but too long to be accepted by most journals, usually do not have this "lecture notes" character. For similar reasons it is unusual for doctoral theses to be accepted for the Lecture Notes series, though habilitation theses may be appropriate.

2. Manuscripts should be submitted either online at www.editorialmanager.com/lnm to Springer's mathematics editorial in Heidelberg, or to one of the series editors. In general, manuscripts will be sent out to 2 external referees for evaluation. If a decision cannot yet be reached on the basis of the first 2 reports, further referees may be contacted: The author will be informed of this. A final decision to publish can be made only on the basis of the complete manuscript, however a refereeing process leading to a preliminary decision can be based on a pre-final or incomplete manuscript. The strict minimum amount of material that will be considered should include a detailed outline describing the planned contents of each chapter, a bibliography and several sample chapters.

   Authors should be aware that incomplete or insufficiently close to final manuscripts almost always result in longer refereeing times and nevertheless unclear referees' recommendations, making further refereeing of a final draft necessary.

   Authors should also be aware that parallel submission of their manuscript to another publisher while under consideration for LNM will in general lead to immediate rejection.

3. Manuscripts should in general be submitted in English. Final manuscripts should contain at least 100 pages of mathematical text and should always include

   - a table of contents;
   - an informative introduction, with adequate motivation and perhaps some historical remarks: it should be accessible to a reader not intimately familiar with the topic treated;
   - a subject index: as a rule this is genuinely helpful for the reader.

   For evaluation purposes, manuscripts may be submitted in print or electronic form (print form is still preferred by most referees), in the latter case preferably as pdf- or zipped ps-files. Lecture Notes volumes are, as a rule, printed digitally from the authors' files. To ensure best results, authors are asked to use the LaTeX2e style files available from Springer's web-server at:

   ftp://ftp.springer.de/pub/tex/latex/svmonot1/ (for monographs) and
   ftp://ftp.springer.de/pub/tex/latex/svmultt1/ (for summer schools/tutorials).

Additional technical instructions, if necessary, are available on request from lnm@springer.com.

4. Careful preparation of the manuscripts will help keep production time short besides ensuring satisfactory appearance of the finished book in print and online. After acceptance of the manuscript authors will be asked to prepare the final LaTeX source files and also the corresponding dvi-, pdf- or zipped ps-file. The LaTeX source files are essential for producing the full-text online version of the book (see http://www.springerlink.com/openurl.asp?genre=journal&issn=0075-8434 for the existing online volumes of LNM). The actual production of a Lecture Notes volume takes approximately 12 weeks.

5. Authors receive a total of 50 free copies of their volume, but no royalties. They are entitled to a discount of 33.3 % on the price of Springer books purchased for their personal use, if ordering directly from Springer.

6. Commitment to publish is made by letter of intent rather than by signing a formal contract. Springer-Verlag secures the copyright for each volume. Authors are free to reuse material contained in their LNM volumes in later publications: a brief written (or e-mail) request for formal permission is sufficient.

Addresses:
Professor J.-M. Morel, CMLA,
École Normale Supérieure de Cachan,
61 Avenue du Président Wilson, 94235 Cachan Cedex, France
E-mail: morel@cmla.ens-cachan.fr

Professor B. Teissier, Institut Mathématique de Jussieu,
UMR 7586 du CNRS, Équipe "Géométrie et Dynamique",
175 rue du Chevaleret
75013 Paris, France
E-mail: teissier@math.jussieu.fr

For the "Mathematical Biosciences Subseries" of LNM:

Professor P. K. Maini, Center for Mathematical Biology,
Mathematical Institute, 24-29 St Giles,
Oxford OX1 3LP, UK
E-mail: maini@maths.ox.ac.uk

Springer, Mathematics Editorial, Tiergartenstr. 17,
69121 Heidelberg, Germany,
Tel.: +49 (6221) 4876-8259

Fax: +49 (6221) 4876-8259
E-mail: lnm@springer.com

MIX
Papier aus verantwortungsvollen Quellen
Paper from responsible sources
FSC® C105338

If you have any concerns about our products,
you can contact us on
ProductSafety@springernature.com

In case Publisher is established outside the EU,
the EU authorized representative is:
**Springer Nature Customer Service Center GmbH
Europaplatz 3, 69115 Heidelberg, Germany**

Printed by Libri Plureos GmbH
in Hamburg, Germany